SQL for DB2 and SQL/DS Application Developers

SQL for DB2 AND SQL/DS Application Developers

Jonathan S. Sayles

QED Information Sciences, Inc.
Wellesley, Massachusetts

P.O. Box 82-181
Wellesley, MA 02181

Library of Congress Catalog Number: 89-10322
International Standard Book Number: 0-89435-303-9

Printed in the United States of America
90 91 92 10 9 8 7 6 5 4 3 2 1

Library of Congress Cataloging-in-Publication Data

Sayles, Jonathan S.
SQL for DB2 and SQL/DS application developers / Jonathan S. Sayles.
492 p. 17.92 x 25.6 cm.
ISBN 0-89435-303-9
1. SQL (Computer program language) 2. Relational databases.
3. IBM Database 2 (Computer system)
I. Title.
QA76.73.S67S29 1989
005.75'6 -- dc 20

For information about our audio products, write us at:
Newbridge Book Clubs, 3000 Cindel Drive, Delran, NJ 08370

Table of Contents

Preface

Let's see now, you're (select any from the following list which apply):

( ) too busy
( ) too old
( ) too young
( ) overextended
( ) already way over budget
( ) not convinced of the need to overcome the inertia to learn, or get some training in, SQL and relational database technology.

I sympathize. After all, the "law of diminishing amusement" takes a sizeable amount of octane out of our professional enthusiasm over the years. And it isn't long before the prospect of learning a new computer language rates right up there with doing income taxes.

But you see, learning SQL will be different. It's:

1. *Easy to learn.* Little did you know that you've already begun your first class in SQL! By reading the opening sentence of this preface, you were exposed to four SQL reserved words (SELECT, ANY, FROM, WHERE). What's more, they mean the same thing to SQL as they do to you.

2. *Easy to use.* Whether you're running a query written in SQL or writing a program that contains embedded SQL statements, the language is consistent, logical, and straightforward.

3. *Necessary.* SQL is IBM's strategic direction in relational database languages and the ANSI standard for relational languages. SQL-based DBMS and 4GL application development tools are proliferating like rabbits on the market, in micro, mini, and mainframe products such as Gupta's *SQLBase®*, *ORACLE*, *INFORMIX*, *INGRES*, *SQL/DS*, *dBASE IV*, and (lest we forget) *DB2*. In short, SQL is the first language since COBOL that virtually everyone in the industry "agrees on."

This book will take the reader along the smoothest, easiest route to the pinnacle of SQL mastery. In it I have embedded examples that provide living testimony to SQL "gotchas," rules of thumb for safe SQL coding and testing, and lots of information on SQL performance techniques.

SQL for DB2 and SQL/DS presents an SQL computer-assisted course that is tailored for DP personnel and business end users. It covers the heart of SQL instructional material. Along the way, you'll learn what can be done with SQL, how it is used, and how to use it to accomplish production-level SQL query processing. All of the exercises may be typed in and executed on a real SQL system.

With this book you'll learn faster because you will use SQL to learn SQL. In each chapter, you'll get an explanation of a command or function from the book (which is also your personal SQL reference manual). Then you'll conduct a dry run by writing the statement in the book. Finally, you may practice by using SQL to access the data tables. An answer key is provided so that you can check your answers.

You learn by doing. You may use either a PC-based SQL system on your personal computer if you purchased the diskette, or DB2 or SQL/DS on a mainframe. It doesn't matter which is used, because the book uses the standard IBM format of the native SQL language. If you do not have the diskette, you may purchase it from QED. This will run on an IBM PC or PS/2 and contains an SQL system. To order, please send $160.00 to:

QED Information Sciences, Inc.
170 Linden Street
Wellesley, MA 02181

I have designed this course for application programmers, database administrators, technical support personnel, systems designers and administrators, information center personnel and business end users. Basically, anyone involved with relational database systems can benefit from the material in this course. A note to programmers—all examples and exercises will be given using native or interactive SQL, as opposed to embedded SQL. If you are going to be writing programs using SQL you should read my book *Embedded SQL for DB2.* It is available from QED. If you are using DB2 or SQL/DS, you may need help from your training department or the appropriate technical person to get you started. You will follow the instructions in Chapter 3 to load the data.

If you are using a mainframe, you should be aware that some SQL statements in this book require special authority to execute. If you execute an exercise and receive an error message telling you that you don't have the authority to run this type of statement, you will need to contact either your training coordinator, information center personnel or database administration personnel to get the authority. Further, the authority may not be available to you, depending on your company's Security and Authorization Policy.

I wish to dedicate this book to one of my best friends and instructional inspirations, Dr. Melvin Goldstein of the University of Hartford. Mel is the kind of teacher who comes along once or twice in a lifetime, and dazzles you with his insight, intelligence, wit and understanding.

I would like to thank my DB2 and SQL/DS students of the past three years. "From my students I've been taught" couldn't be truer, and many improvements to the previous version of this book can be directly attributed to one or more DP'ers asking a new question or putting a different spin on some old dusty observation I was making.

Finally, I would also like to thank my wife Maryellen, and my sons Christopher and Jamison for their patience with me as I labored hour by hour creating, revising, correcting and consolidating the updates to this book.

SQL COMMAND INDEX

Chapter 1
RELATIONAL TERMS AND CONCEPTS

Here we present SQL, its capabilities, relational terms, and concepts. Specifically, relational data objects—what they are, what they do, and how they interact with the system—are discussed from a practical, as well as a theoretical point of view.

The following topics are covered:

1.1 What is SQL?
1.2 SQL Processing Capabilities
1.3 Tables
1.4 Databases
1.5 Indexes
1.6 Views
1.7 Synonyms
1.8 Physical Data Objects — Tablespaces, STOGROUPS, DBSPACES and STORPOOLS
1.9 The Big Picture
1.10 The DB2 Big Picture

1.1 WHAT IS SQL?

The name SQL stands for Structured Query Language. Quite simply, SQL (pronounced "sequel") is a data access language (see Figure 1.1). Like any language, it is used for communication. Who is it you are communicating with when you use SQL? You are talking to a database manager. This could be DB2, SQL/DS, SQLBase, ORACLE, INFORMIX, INGRES, or any number of relational database management systems (RDBMS) that support the SQL language. These database systems understand SQL, and as you learn to converse in SQL, you will find yourself doing some fairly sophisticated data processing.

SQL is easy to learn. Despite the fact that SQL is a computer programming language, it is much simpler than traditional programming languages like COBOL, BASIC, FORTRAN, or APL. You will probably find learning SQL to be a straightforward process. Why?

First, SQL uses a semi free-form, English-like structure for its commands. You SELECT your data, FROM some table, WHERE certain conditions are met. Or you INSERT your VALUES INTO some table, or DELETE rows FROM some table, WHERE your conditions are met. It is very logical and easy to follow.

- Simple, flexible, and powerful data access language
- Standard relational language
- English-like query language
- SQL is supported in the following relational products:

  DB2
  SQL/DS
  SQLBase
  QMF
  ORACLE
  INFORMIX
  VAX Rdb
  SYBASE
  SAS
  FOCUS
  REXX (on VM and TSO)
  XDB

Figure 1.1. What is SQL?

Second, SQL decides how to get your data to and from the database. All you have to specify is what you want, and SQL does the rest. This is called being nonnavigational and it promises large productivity gains for the data processing industry. Sometimes entire programs in traditional data processing systems can be replaced with a single SQL query.

Third, most relational DBMSs support both interactive and static SQL processing. That is, SQL statements can be executed in an interactive fashion — where you talk directly to the database manager— or SQL statements can be embedded in traditional computer programming languages like COBOL. This is necessary because SQL was originally intended to be used with other programming languages. By itself, SQL has no commands for screen dialog, or for more than crude report formatting, so this dual-mode feature is very important in any kind of formal data processing application. Through embedded SQL statements, DB2 and SQL/DS data can be accessed by interactive and batch programs. Although embedded SQL statements themselves are very similar to their interactive counterparts, we will not be covering them in this book.*

Finally, SQL processes data at the "set level," meaning your updates will change a set of records (rows), and your query output will come back as a set of records (a results table). This is as opposed to the traditional record1 . . . record2 . . . record3 . . . mode of processing data files. While generally adding to the power and ease of relational systems, this "set level" processing concept poses a few considerations for anyone used to the record1 . . . record2 . . . record3 . . . processing frame of mind. That is why these will be discussed fully as the book progresses. Now let's see what kinds of things SQL allows you to do.

1.2 SQL PROCESSING CAPABILITIES

SQL is composed of a data definition language, a data manipulation language, and a data control language (see Figure 1.2). These three languages support the complete spectrum of relational data processing activity. In fact, in most SQL-based products, all access to the data is through SQL and only through SQL.

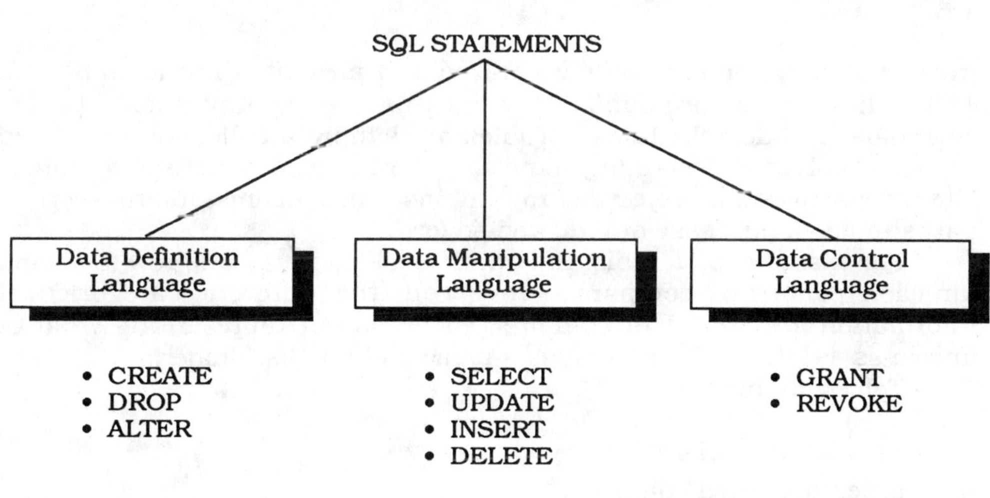

Figure 1.2. SQL processing capabilities.

---

* See *Embedded SQL for DB2 Application Design and Programming* published by QED.

Data definition allows creation, deletion, and modification of the data structures needed for our system. These structures include tables, databases, indexes, views—the "structural" parts of the database system.

Data manipulation is divided into three categories: retrieving data, manipulating data, and updating data. *Retrieving* data means querying the database. A selection of data stored on tables is presented on the screen. *Manipulating* data refers to the SQL features that allow us to perform statistical functions on data such as averaging and summing columns, and other arithmetic functions like multiplying columns. *Updating* data refers to inserting and deleting rows on tables and changing values in columns; in other words, database maintenance.

Data control allows us definition of a security mechanism or scheme for protecting the data in a system from unauthorized access. It consists of *Granting* and *Revoking* processing privileges on our data objects to other system users. Data access statements allow for as simple or elaborate a set of controls as is needed for the system.

An important feature of SQL statements is that they can be executed interactively; that is, they may be executed any time the system is operating. To delete an existing table and define a totally new structure does not require the herculean effort it currently takes using hierarchical or network database systems. Relational tables are simple to define, and once defined, simple to modify and use.

1.3 TABLES

In relational systems, data is organized and presented in tables (Figure 1.3). These tables are similar to tables used every day (phone books, calendars, airline schedules). Tables are simply a collection of related rows and columns describing some singular business processing "entity" like a department, a project, a group of insurance claims, information on part suppliers, purchase orders, and so forth.

For example, as a table, the phone book contains one row (or more) for each person or company. In a *row* the different categories of information are stored in *columns*, each *column* representing what is known as a *data attribute* (such as a name and phone number).

Tables are made up of:

- unordered, unnamed rows
- ordered, named columns

1.3.1 Rows

Rows are the horizontal components of a table (Figure 1.4). They are unnamed and unordered. This means that there is no way to specify a particular table row by its position in a table (i.e., 1st row, last row, etc.). Rows are accessed by data value only.

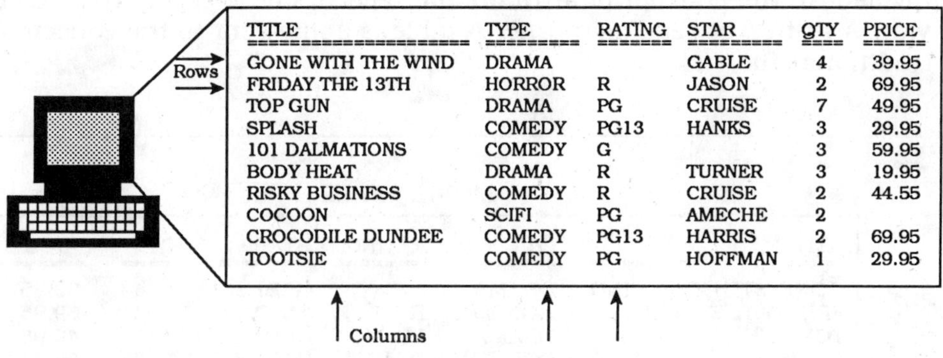

TITLE	TYPE	RATING	STAR	QTY	PRICE
GONE WITH THE WIND	DRAMA		GABLE	4	39.95
FRIDAY THE 13TH	HORROR	R	JASON	2	69.95
TOP GUN	DRAMA	PG	CRUISE	7	49.95
SPLASH	COMEDY	PG13	HANKS	3	29.95
101 DALMATIONS	COMEDY	G		3	59.95
BODY HEAT	DRAMA	R	TURNER	3	19.95
RISKY BUSINESS	COMEDY	R	CRUISE	2	44.55
COCOON	SCIFI	PG	AMECHE	2	
CROCODILE DUNDEE	COMEDY	PG13	HARRIS	2	69.95
TOOTSIE	COMEDY	PG	HOFFMAN	1	29.95

Relational Table describing MOVIES

- Similar to phone book

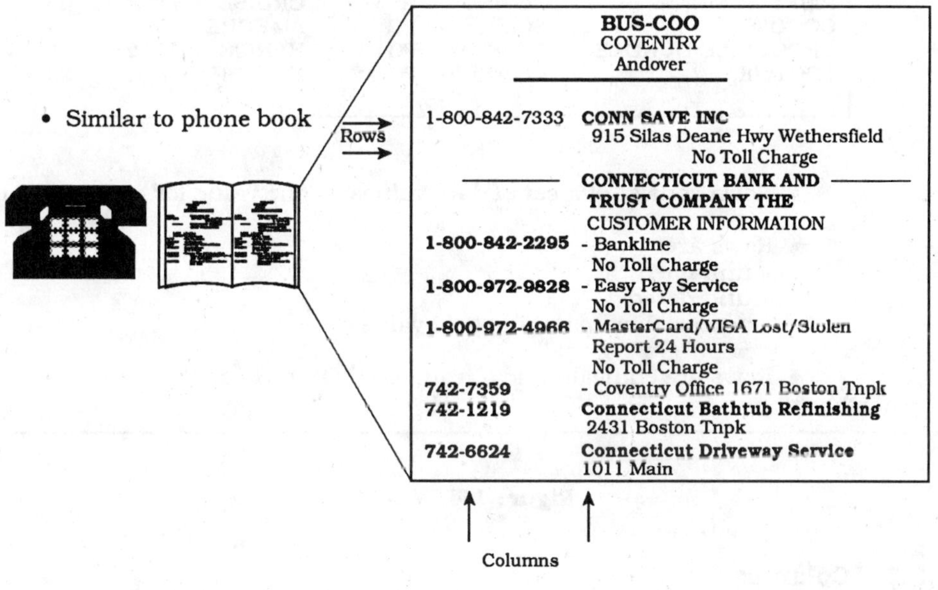

BUS-COO
COVENTRY
Andover

1-800-842-7333 **CONN SAVE INC**
915 Silas Deane Hwy Wethersfield
No Toll Charge

CONNECTICUT BANK AND TRUST COMPANY THE
CUSTOMER INFORMATION
1-800-842-2295 - Bankline
No Toll Charge
1-800-972-9828 - Easy Pay Service
No Toll Charge
1-800-972-4966 - MasterCard/VISA Lost/Stolen
Report 24 Hours
No Toll Charge
742-7359 - Coventry Office 1671 Boston Tnpk
742-1219 **Connecticut Bathtub Refinishing**
2431 Boston Tnpk
742-6624 **Connecticut Driveway Service**
1011 Main

Figure 1.3. Tables.

Data processing personnel can note that rows are roughly equivalent to the concept of a traditional record, one logically related set of values written or read as a unit. A table is equivalent to the concept of a traditional file.

MOVIE TABLE

	TITLE	TYPE	RATING	STAR	QTY	PRICE
R➤	GONE WITH THE WIND	DRAMA		GABLE	4	39.95
O➤	FRIDAY THE 13TH	HORROR	R	JASON	2	69.95
W➤	TOP GUN	DRAMA	PG	CRUISE	7	49.95
S➤	SPLASH	COMEDY	PG13	HANKS	3	29.95
	101 DALMATIONS	COMEDY	G		3	59.95
	BODY HEAT	DRAMA	R	TURNER	3	19.95
	RISKY BUSINESS	COMEDY	R	CRUISE	2	44.55
	COCOON	SCIFI	PG	AMECHE	2	
	CROCODILE DUNDEE	COMEDY	PG13	HARRIS	2	69.95
	TOOTSIE	COMEDY	PG	HOFFMAN	1	29.95

- Rows consist of a set of data values — one value for each column
- Rows are:
  - unnamed
  - unordered
  - accessed by matching field values
- Rows are equivalent to a traditional OS record

Figure 1.4. Rows.

1.3.2 Columns

Columns are the vertical components of a table (Figure 1.5). Unlike rows, columns are named and given a default order by the SQL statement that defines a table (CREATE). This means that there can be a "positional" (left to right) aspect for columns in our query statements. We will be discussing the significance of this as we go along.

MOVIE TABLE

TITLE	TYPE	RATING	STAR	QTY	PRICE
GONE WITH THE WIND	DRAMA		GABLE	4	39.95
FRIDAY THE 13TH	HORROR	R	JASON	2	69.95
TOP GUN	DRAMA	PG	CRUISE	7	49.95
SPLASH	COMEDY	PG13	HANKS	3	29.95
101 DALMATIONS	COMEDY	G		3	59.95
BODY HEAT	DRAMA	R	TURNER	3	19.95
RISKY BUSINESS	COMEDY	R	CRUISE	2	44.55
COCOON	SCIFI	PG	AMECHE	2	
CROCODILE DUNDEE	COMEDY	PG13	HARRIS	2	69.95
TOOTSIE	COMEDY	PG	HOFFMAN	1	29.95

Columns

- Columns are named collections of data fields
- Columns are vertical components of a table
- Columns are created by:
  - defining a name
  - defining a data attribute

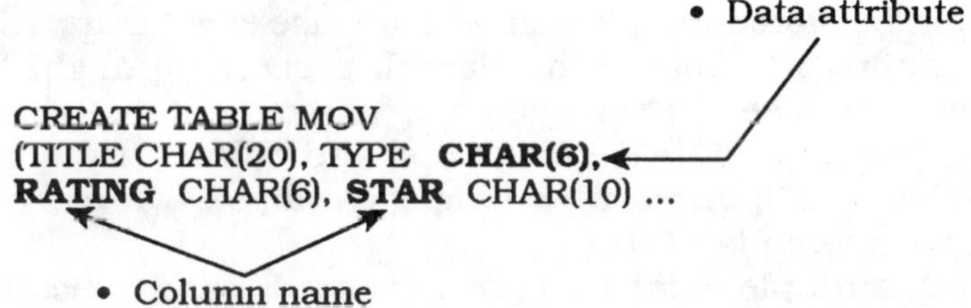

Figure 1.5. Columns.

Every table column has a unique name within the table and a data attribute defined for it as well. Data attributes describe three things about a column:

1. the maximum length or size of the data that the column can hold;
2. the type of data to be stored in the column (both alphabetic and numeric data, or only numeric data);
3. whether the column can hold unknown data values. These special fields are called *null values.* They represent an absence of any data value and are treated in a unique way by SQL. We will talk more about null values later on.

1.3.3 Fields

A *field* refers to the specific data value stored in a table for a particular intersection of a row and a column (see Figure 1.6). Fields are the smallest pieces of data stored in tables and manipulated by SQL.

1.4 DATABASES

The difference between a database, a database management system, and a data access language can be a little confusing at first. Perhaps this description will help.

- A *database* is simply an integrated collection of related files or, in DB2, related tables;
- A *database management system* (DBMS) refers to the entire computerized file-keeping system (in our case DB2, SQL/DS);
- A *data access language* (SQL for instance) is the language system users employ to communicate with DB2 or SQL/DS to accomplish their work.

In relational systems databases function primarily as repositories for groups of related tables. For instance, all the tables in the Personnel system might be stored in a Personnel database (see Figure 1.7). All of the tables containing information on Payroll processing might be stored in the Payroll database. This makes it convenient to bring an entire application system and its tables online and offline. By housing all the related tables and indexes in a single database, we can use a single command to accomplish this.

In DB2, multiple databases are permitted. Each typically contains a single applications systems table and related objects. In SQL/DS there is one database for each copy of the product. In DB2, databases are logical objects that require no space for storage and are used merely to expedite system administration. In SQL/DS the database represents all of the relational objects users have access to. In SQL/DS you connect to and use a single database at a time. In DB2 you may access tables stored across different databases.

MOVIE TABLE

TITLE	TYPE	RATING	STAR	QTY	PRICE
GONE WITH THE WIND	DRAMA		GABLE	4	39.95
FRIDAY THE 13TH	HORROR	R	JASON	2	69.95
TOP GUN	DRAMA	PG	CRUISE	7	49.95
SPLASH	COMEDY	PG13	HANKS	3	29.95
101 DALMATIONS	COMEDY	G		3	59.95
BODY HEAT	DRAMA	R	TURNER	3	19.95
RISKY BUSINESS	COMEDY	R	CRUISE	2	44.55
COCOON	SCIFI	PG	AMECHE	2	
CROCODILE DUNDEE	COMEDY	PG13	HARRIS	2	69.95
TOOTSIE	COMEDY	PG	HOFFMAN	1	29.95

Fields

- A field is the actual content — or data value — at the intersection of a particular row/column
- A field is:
  - the smallest unit of data represented by DB2 and SQL/DS
  - the smallest unit of data manipulated in SQL
- A NULL field denotes absence of data value — for that particular field

Figure 1.6. Fields.

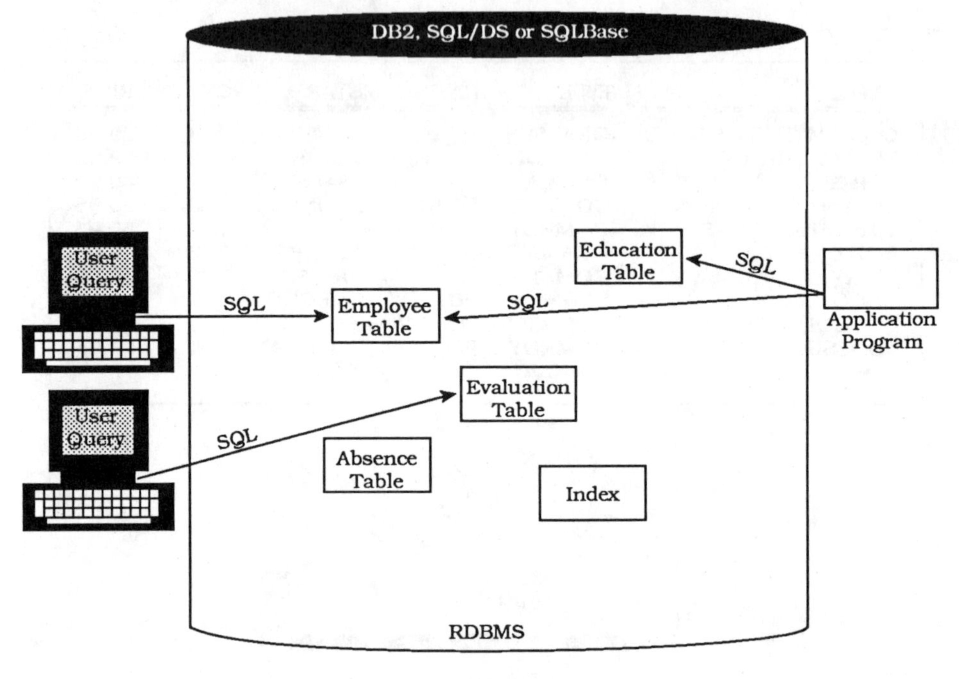

Figure 1.7. Personnel database.

Let's look at a few more relational data objects.

1.5 INDEXES

Indexes are pointers to specific rows (data records) in your table (Figure 1.8). They are logically ordered by the values of a key. For instance, an index created on the key of LAST_NAME would store all the last names in the table along with a pointer to their individual rows. This process is roughly the same as the process that occurs when you use an index in a book. If you wanted to find all the pages where the SQL clause ORDER BY is referenced in the book, you would turn to the index which would list ORDER BY, and all the pages that contain a reference to it. Then you could go directly to the page containing the information on ORDER BY. (This would presumably be much faster than thumbing through this magnum opus page by page.)

Indexes are

- pointers to specific data records in a table — logically ordered by key
- created by DBA's, applications developers, and sometimes users
- used to select access path
- maintained by DB2 and SQL/DS

Indexes:

- speed retrieval of information
- can slow down update processing
- can be used to enforce unique data values on a column
- are stored separately from table
- can be created as:
  UNIQUE/NONUNIQUE CLUSTERED
- can be created at any time, except for clustering index
- are never referenced in SQL statements

DB2 and SQL/DS:

- support multiple table indexes, multiple column indexes

Figure 1.8. Indexes.

Unfortunately, indexes require storage space and can significantly slow down update processing. Requirement trade-offs must be weighed carefully before the choice to create indexes is made.

In spite of this, it is probably a good idea to index one or two key* columns of every table, simply because the retrieval benefits are so great.

Once indexes are created, they are never referenced again in SQL statements. This is different from some of the popular PC DBMS products like dBASE III. DB2, not the system user, decides whether or not to use an index in locating a particular row.

* The word "key" has many connotations in data processing circles. We will be using it, in a limited sense, to describe the key or most important column(s) in a table. Often, key columns store unique identifying values such as employee IDs, social security numbers, policy numbers, etc.

Another feature of indexes is that they can be used to enforce unique values on a column. In other words, if a column (like SOCIAL_SECURITY_NBR) is defined with an index that has the UNIQUE option, duplicate social security numbers will not be allowed in that column.

1.6 VIEWS

A *view* is a data object definition that represents some logical subset of rows and columns from one or more tables (Figure 1.9). To a user, a view looks just like a table. But views are only images or descriptions of table information and do not duplicate or store any physical data. Because of this, views are sometimes referred to as "virtual" tables. Views can contain some or all of the columns and rows from one or more tables.

A view:

- is an imaginary table
- does not represent a physically stored data object
- is a logical subset of one or more rows and columns from one or more real tables
- can simplify query processing
- has limited update capability
- extends or filters user's perception of data
- can be used as part of security procedures
- is often used to promote data/program independence

View definitions are stored in system catalog

Figure 1.9. Views.

Views have limited update capability, and are mainly used:

1. To simplify end-user query processing by representing the results of precoded queries;
2. As part of systems security measures by restricting access to particular row/column subsets.
3. To maintain application/data independence.

The term *view* can be understood from its function; that is, to provide a user a particular look at the information stored on the database. For instance, if user A needs to see only columns one and four from the customer table, we could define a view of that table containing only columns one and four. This would become user A's view of the table, and for all intents and purposes user A might not ever find out that other data even existed on the table.

1.7 SYNONYMS

DB2 and SQL/DS allow for shorthand aliases or nicknames for tables called synonyms. Each user can define a synonym of any table or view they have access authority for. Synonyms can be an aid in managing system data objects. Also by substituting more rational names like Payroll and Project for typical production tables names like:

A8972ND.93JD_9IUDHB123

you can expedite your work, and save typographical errors.

1.8 PHYSICAL DATA OBJECTS — TABLESPACES, STOGROUPS, DBSPACES AND STORPOOLS

It was the intent of the designers of relational database systems that end users (i.e., humans) should be insulated from the physical files that data is stored on. To that end, they devised the strategy similar to what is shown in Figure 1.10:

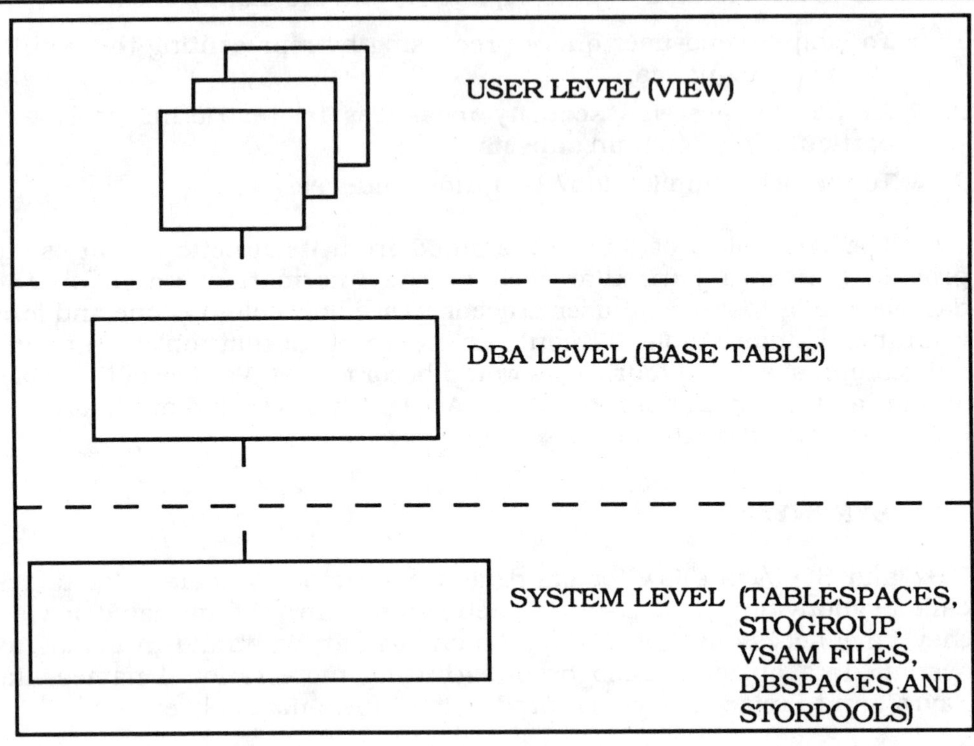

Figure 1.10. Conceptual levels of data object management and usage.

The idea was that system users would be able to access only the data elements they needed using view definitions of base tables. DBAs would define and manage the base tables the views were built upon, and the system would handle all physical file management. It was the creators' intentions that below the DBA level, human intervention would not be necessary.

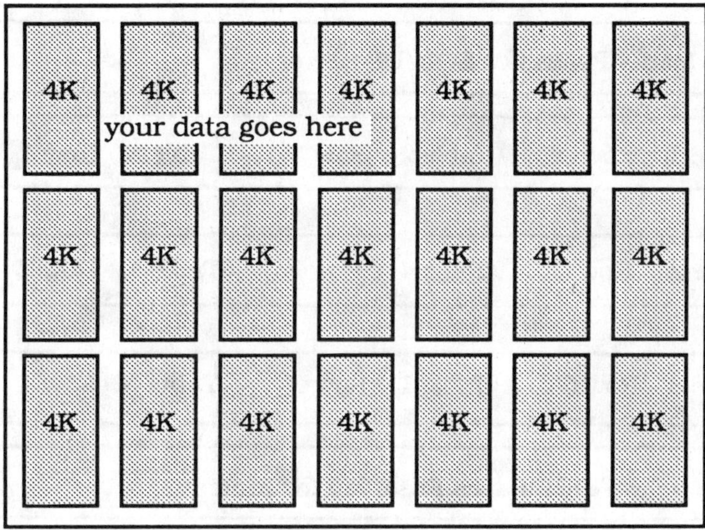

Figure 1.11. DB2 Tablespace.

A Storage Group or STOGROUP is a defined collection of disk pacts from which DB2 can obtain tracks and cylinders for your tables. With STOGROUPS system users do not have to:

- Define files for tables;
- Extend files for tables (if the tables fill up);
- Delete files once the tables are dropped.

DB2 takes care of all of this for you.

A Tablespace is the file space your data resides on. You can almost think of it as a file in the traditional sense. DB2 files are stored on VSAM datasets.**

Each tablespace can hold one or more tables, and you do not have to explicitly define a tablespace for your tables. DB2 will provide one for you (remember the conceptual levels above — no human intervention).

** ESDS (ENTRY SEQUENCED DATA SET) or LDS (LINEAR DATA SET).

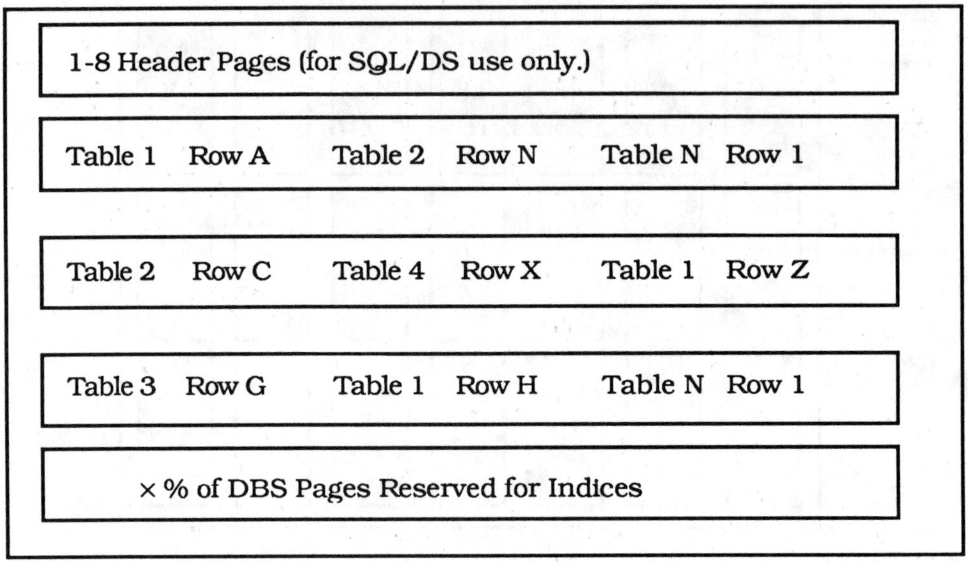

1-8 Header Pages (for SQL/DS use only.)

Table 1	Row A	Table 2	Row N	Table N	Row 1
Table 2	Row C	Table 4	Row X	Table 1	Row Z
Table 3	Row G	Table 1	Row H	Table N	Row 1

x % of DBS Pages Reserved for Indices

Figure 1.12. DBSPACE.

Each DBSPACE can hold one or more tables (up to 255). Figure 1.12 describes an SQL/DS DBSPACE — the rough equivalent of a DB2 tablespace. From an applications perspective the differences between DBSPACES and tablespaces are not too significant. They both store table rows and are created by DBA's. One difference which may not be apparent from the figures, is that SQL/DS DBSPACES store both tables and indexes together. DB2 tablespaces store only tables (indexes are stored as separate VSAM files called indexspaces).

1.9 THE BIG PICTURE

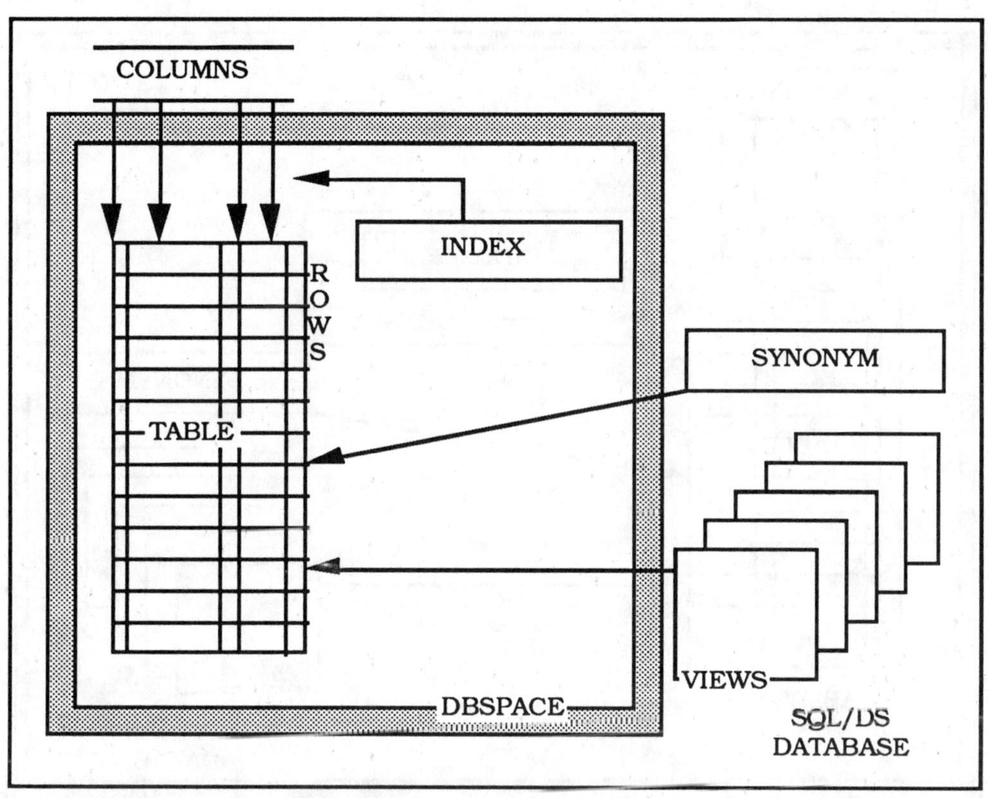

Figure 1.13. The SQL/DS Data Object Big Picture.

As you can see from the above diagram:

- Tables are composed of rows and columns
- VIEWs, Indexes and Synonyms are built on tables
- Tables and Indexes exist in DBSPACES
- Spaces exist in the Databases

1.10 THE DB2 BIG PICTURE

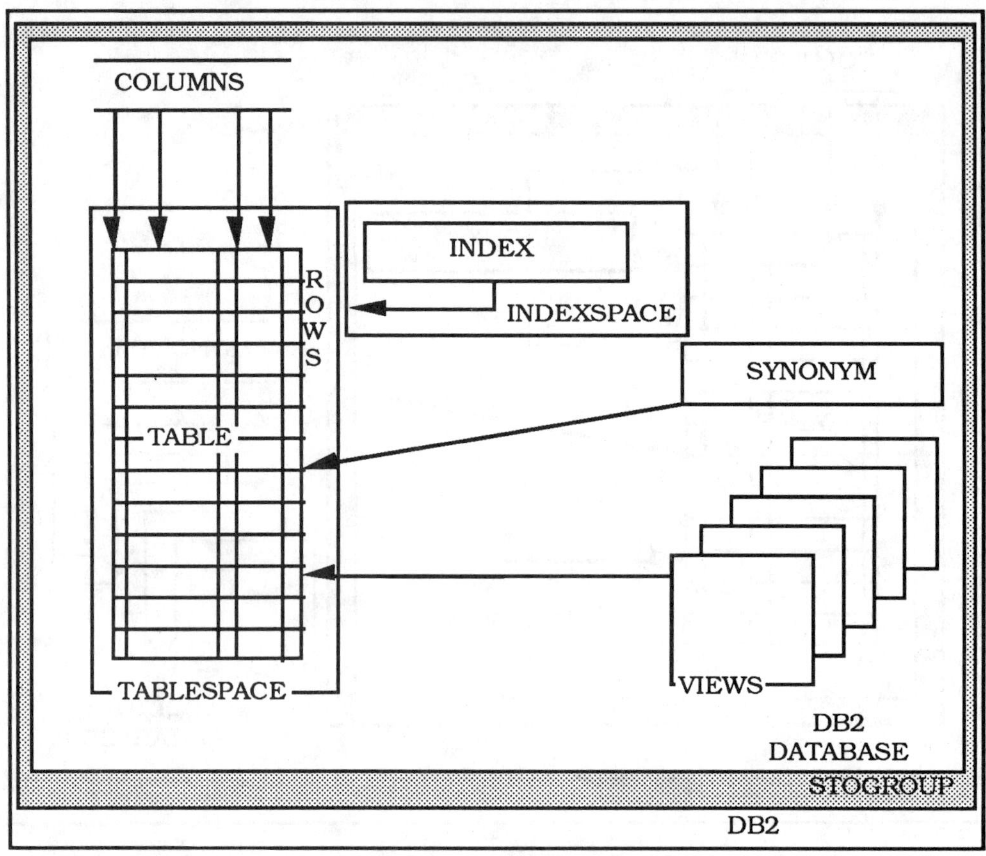

Figure 1.14. The DB2 Big Picture.

As you can see from the above diagram:

- Tables are composed of rows and columns
- Views, indexes and synonyms are built on tables
- Tables exist in tablespaces
- Tablespaces exist in databases
- Databases exist in stogroups

Other configurations are possible, such as multiple tablespaces from the same database existing in different stogroups, etc.

Chapter 2
OVERVIEW AND SYSTEM FEATURES OF DB2 and SQL/DS

Here you will learn about relational database management systems (RDBMS). We will provide enough information for you to understand the concepts and the potential of the relational approach in storing and retrieving data.

The following topics are covered:

2.1 What is a DBMS?
2.2 What is a Relational DBMS?
2.3 RDBMS Components
2.4 DB2 and SQL/DS
2.5 The SQL/DS Catalog
2.6 SQL/DS's Related Products
2.7 The DB2 Environment
2.8 DB2 Catalog
2.9 DB2's Related Products
2.10 DB2I

2.1 WHAT IS A DBMS?

A database management system (DBMS) is a generic term for computer software that acts as an electronic filing or record-keeping system. See Figure 2.1.

Modern DBMSs come in many different classifications, and with many different processing capabilities, but in general they try to accomplish three things:

1. Data consolidation
2. Data sharing
3. Data protection

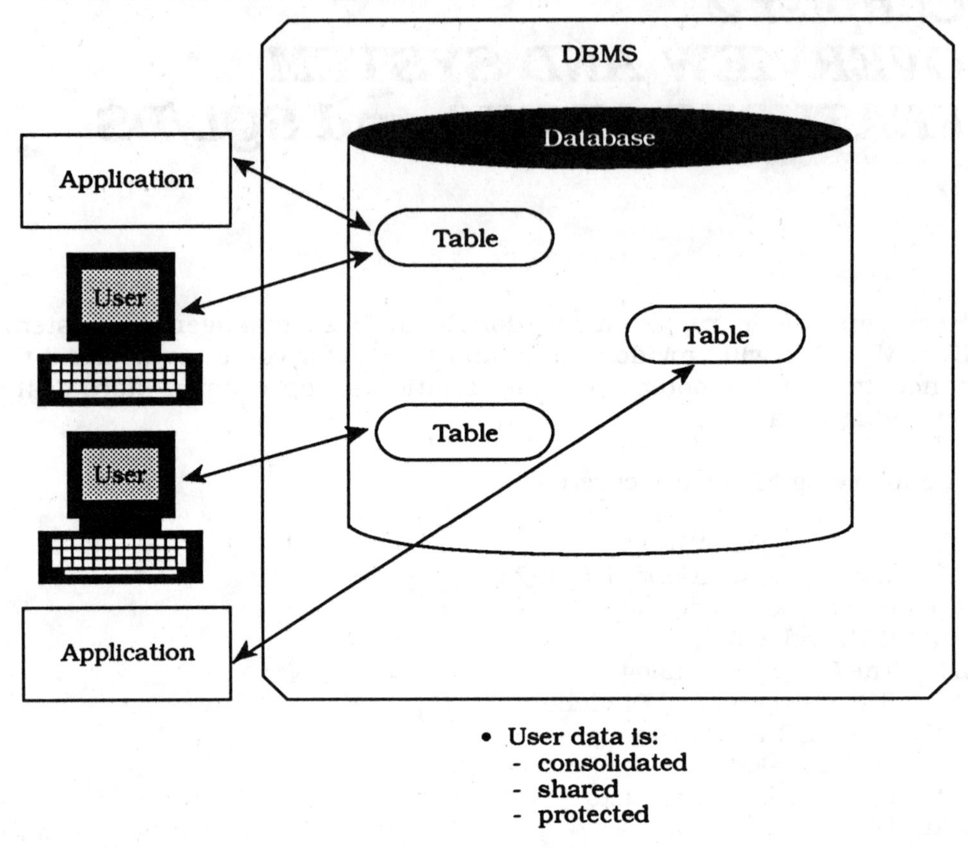

Figure 2.1. What is a database management system?

Data consolidation refers to the combining or unifying of separate data files into a centralized structure, and storing the data in a nonredundant format. A redundant format is a structure that stores the same data items in two or more locations. For instance, if within a company an employee address is stored on not only the payroll file, but a department file, project file, employee history file, etc., you have noncentralized structures carrying redundant information. An integrated (nonredundant) system carries the employee address in one location.

Data sharing refers to the ability of the system to allow multiple users concurrent access to individual pieces of data in the database. (You can think of the database as a "pool" of shareable information.)

Data protection refers to the ability of a DBMS to maintain the integrity of its data in the face of certain types of processing adversity (system crashes, program failures, etc.). If these types of events occur, the DBMS must have the ability to back out (or undo) incomplete or erroneous changes to data stored on the database.

DBMS's organize data in what is known as a data model. You can think of a data model as the infrastructure of the data organization. In other words, how the data is presented to the user. There are three basic data models:

- Hierarchical
- Network
- Relational

Figure 2.2 is a graphic representation of the three models. What are the differences? Hierarchical and network systems use links or pointers to connect records in the system. These systems are called static or

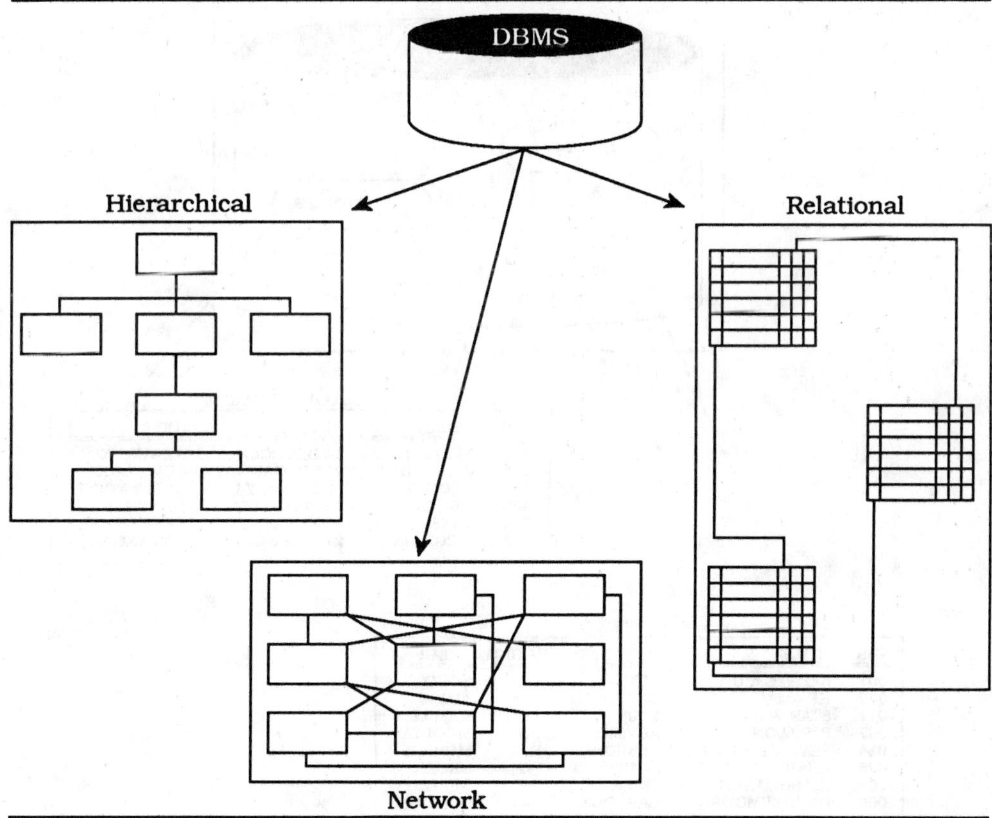

Figure 2.2. Data model comparison.

monolithic because the records are "physically" linked through their definition. The structures are typically complicated to work with and difficult to modify. This disadvantage is tempered by a greater speed of access.

Records (Rows) in relational systems are not physically linked by pointers. Further, relationships between separate tables are based on real field names such as IDs, names, employee numbers, etc. The records in the system are content-addressable in that you access them by matching stored data values.

2.2 WHAT IS A RELATIONAL DBMS?

Now let's take a look at what a relational database management system is.

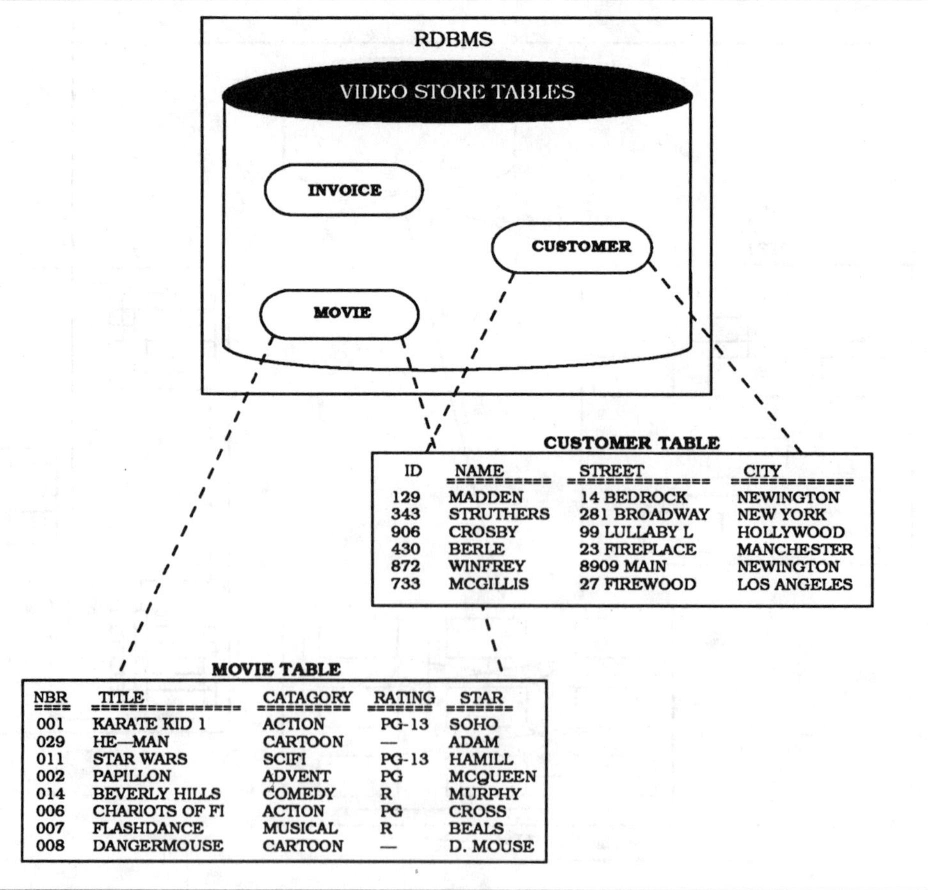

CUSTOMER TABLE

ID	NAME	STREET	CITY
129	MADDEN	14 BEDROCK	NEWINGTON
343	STRUTHERS	281 BROADWAY	NEW YORK
906	CROSBY	99 LULLABY L	HOLLYWOOD
430	BERLE	23 FIREPLACE	MANCHESTER
872	WINFREY	8909 MAIN	NEWINGTON
733	MCGILLIS	27 FIREWOOD	LOS ANGELES

MOVIE TABLE

NBR	TITLE	CATAGORY	RATING	STAR
001	KARATE KID 1	ACTION	PG-13	SOHO
029	HE—MAN	CARTOON	—	ADAM
011	STAR WARS	SCIFI	PG-13	HAMILL
002	PAPILLON	ADVENT	PG	MCQUEEN
014	BEVERLY HILLS	COMEDY	R	MURPHY
006	CHARIOTS OF FI	ACTION	PG	CROSS
007	FLASHDANCE	MUSICAL	R	BEALS
008	DANGERMOUSE	CARTOON	—	D. MOUSE

Figure 2.3. All data is presented to system users as two dimensional tables.

As of this writing, there are over 200 "relational" DBMS products on the market. Of these about 100 products use SQL on a mainframe, mini, or personal computer. True relational products can be distinguished from non- or pseudo-relational systems, by these three criteria:

1. The data is presented to system users as a collection of tables. These tables are a simple, two-dimensional representation of rows and columns. (See Figure 2.3)
2. The system allows users to generate new tables using existing tables. These new, generated tables can be temporary structures (typically when they are the outcome of queries used for reports), or permanent structures. (See Figure 2.4)
3. Relational systems allow three distinct relational operations on tables: (See Figure 2.5)
   - SELECT—accessing specific table rows
   - PROJECT—accessing specific table columns
   - JOIN—combining data stored in two or more separate tables

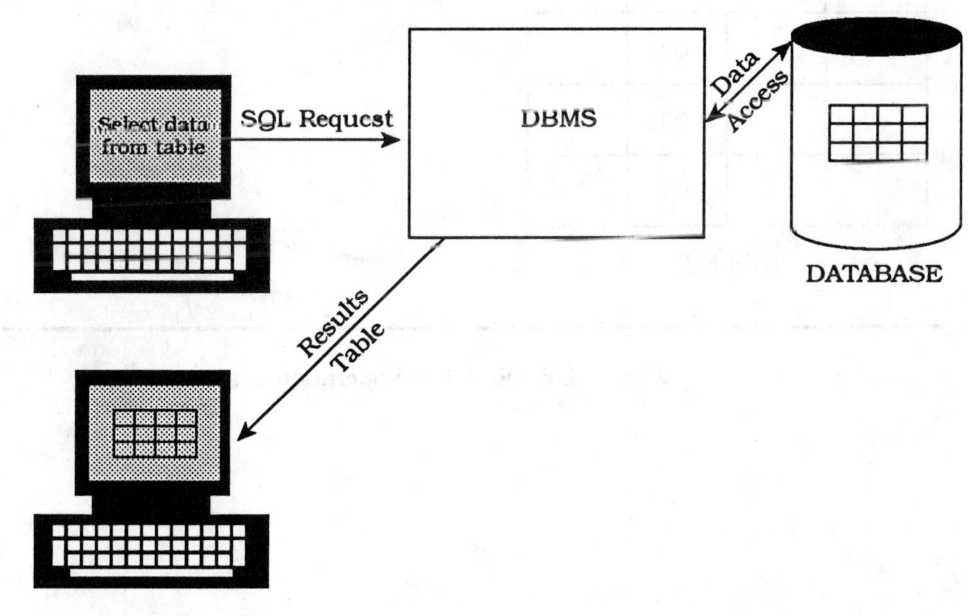

Figure 2.4. Generating new tables from existing tables.

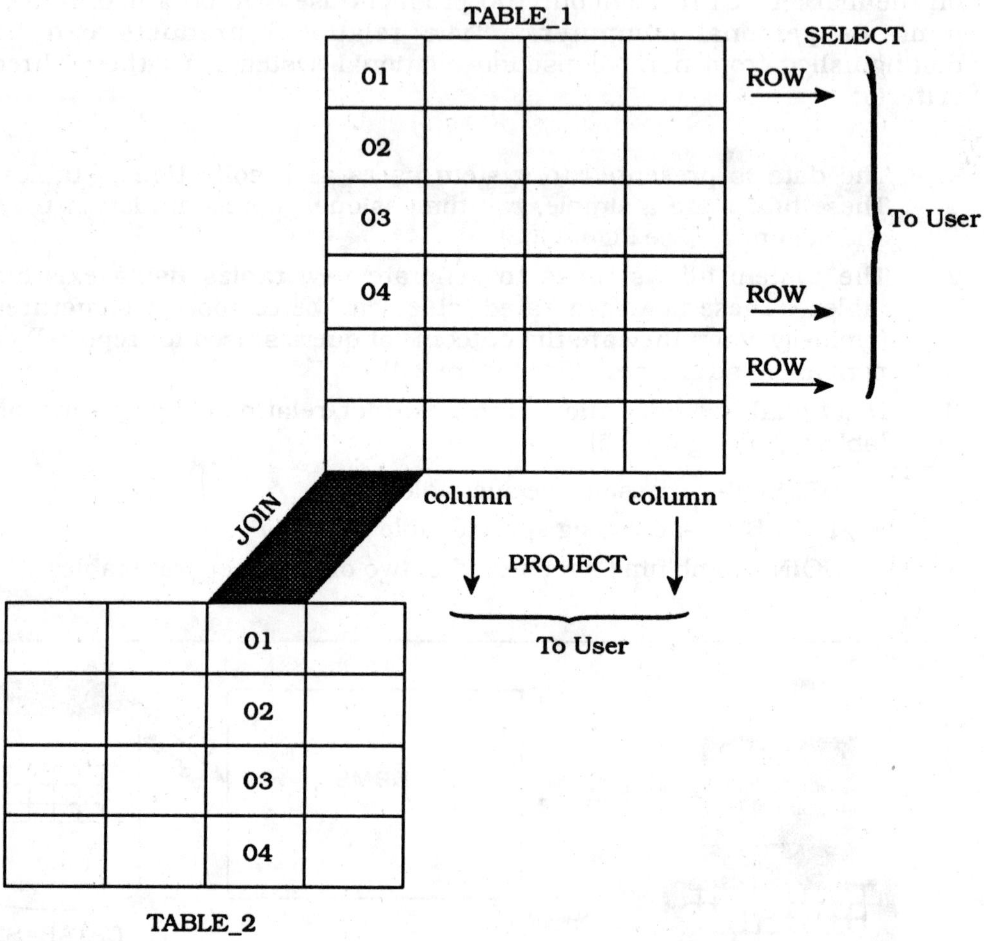

Figure 2.5. Relational operations.

2.3 RDBMS COMPONENTS

There are three basic components of a modern RDBMS (see Figure 2.6):

Front-end user interface products are the software packages that interface with system users such as interactive SQL language interfaces, screen generators, application generators, report generators, forms-based SQL products, etc. These types of packages offer enhanced productivity in the form of application development tools, or end-user decision support tools.

Back-end data management services supply the engine that powers the system, providing data access, data integrity, and data security and control mechanisms.

Data access language is the means of communicating system user requests to the data management services. More and more the language of choice is SQL.

Some RDBMS products such as INGRES and ORACLE have concentrated on DBMS functionality, providing a veritable cornucopia of front-end products. The strength of the IBM offerings (DB2 and SQL/DS) lies in their sophisticated data control and integrity mechanisms.

2.4 DB2 AND SQL/DS

DB2 (Database 2) is IBM's relational database management system for large mainframe computers running under the MVS/SP, MVS/XA and MVS/ESA operating systems. The first commercial release of DB2 was in 1983 and DB2 remains the single most talked about mainframe database product in a decade. Its strengths lie in its data management capabilities and (as of DB2 version 2.1) its online transaction processing speed — DB2 was benchmarked at over 430 transactions per second in a typical debit/credit application.

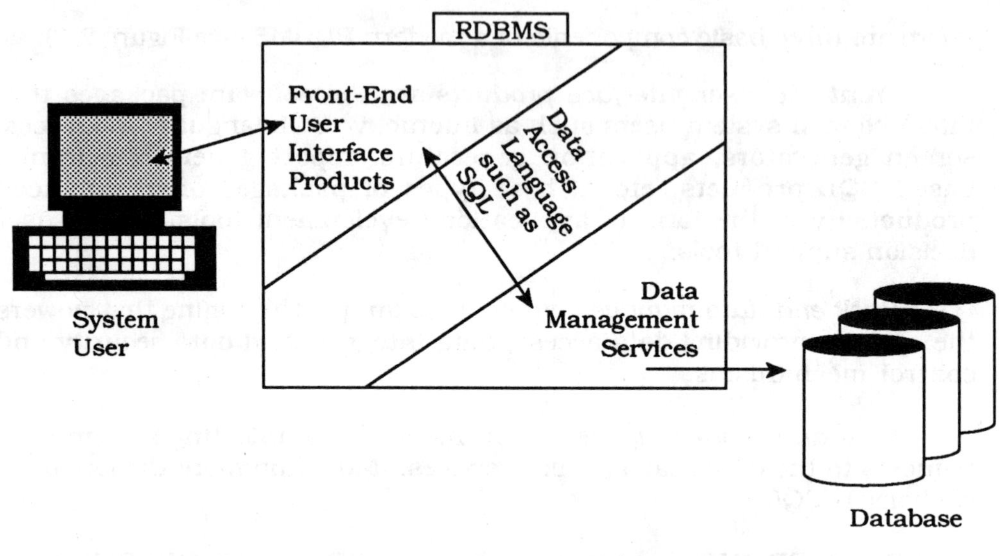

- *User Interface Software:* ←SQL→ • *Data Management Services:*

User Interface Software	Data Management Services
Forms-based Products	Data Integrity/Control Mechanisms
Utility Programs	Physical Data Management processes
Application Development Tools	Query Optimizer
Performance Monitors	Distributed Processing Support Products
Decision Support Products	
4GL Application Generators	
Report Writers	

Figure 2.6. RDBMS components.

SQL/DS — Structured Query Language/Data System — was IBM's first commercially released, full-function Relational Database Management System. The product was announced in January 1981, and as of this writing, the current production release offered by IBM is Version 2 release 2 (V2.2).

The original function of SQL/DS was to provide access — using the Structured Query Language (SQL) — to data stored in a database which supported the relational data model (see section 2.2). To that end, IBM provided access to SQL/DS databases via the following standard processing environments:

VM/SP
DOS/VSE
CICS/DOS/VS

Integrity controls are also built into the product to allow a program running under CICS/DOS/VS to access DL/I databases concurrently with SQL/DS tables. Full read/write integrity against both DBMSs is guaranteed by the COMMIT* processing built into the product.

A typical SQL/DS set-up on the VM operating system is shown in Figure 2.7.

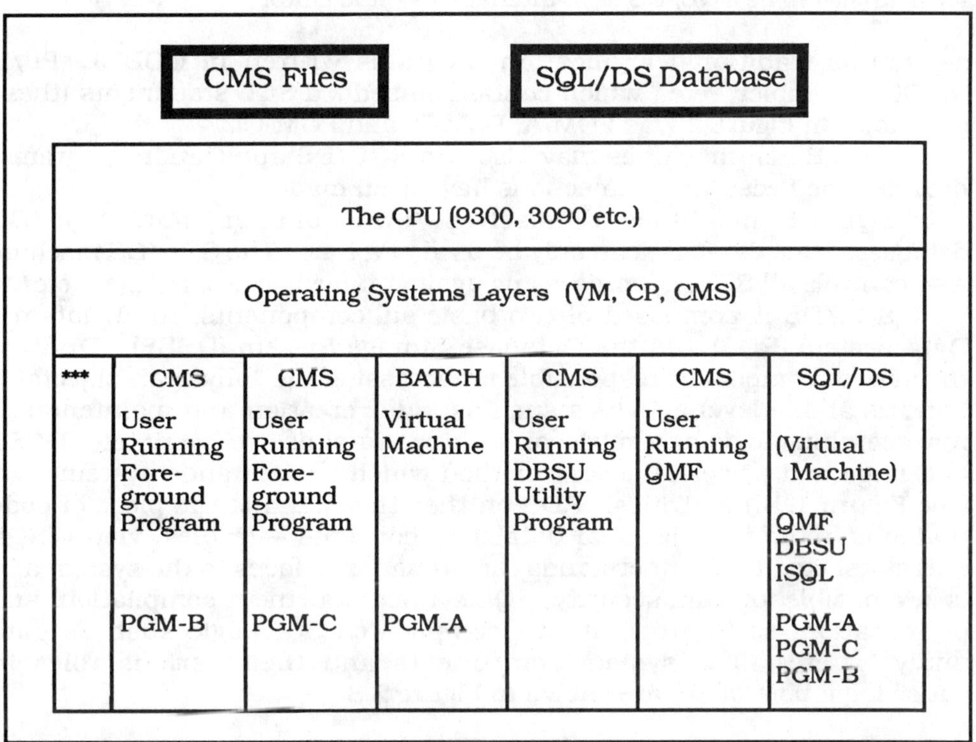

Figure 2.7. Typical SQL/DS set-up on VM/SP.

* COMMIT — the act of making permanent, updates against data in a database.

As you can see from Figure 2.7, SQL/DS on VM/SP runs as its own Virtual Machine. CMS users "connect" to the SQL/DS machine with standard CP links (Typically, linking to virtual address 195, and accessing it as your "Q" disk — but consult with your DBA first). The CMS users can then if authorities permit access products installed with SQL/DS:

- QMF — IBM's query-editor/report writer for relational products
- DBSU — The SQL/DS utility program
- ISQL — The SQL/DS SQL interface* — line editor

as well as traditional application programs written in COBOL, PL/I, REXX, Assembler, etc., which contain embedded SQL statements (these are shown in Figure 2.7 as PGM-A, PGM-B, and PGM-C).

CMS Batch machines may also run SQL/DS application programs, provided the necessary connections have been made.

SQL/DS machines themselves "own" one or more SQL/DS databases. One database can only be used at a time. The SQL/DS machine also controls all SQL/DS processing against tables, views, indexes, etc.

SQL/DS is composed of two basic subcomponents, the Relational Data System (RDS) and the Database Storage System (DBSS). DBSS is the system component responsible for accessing the "physical" objects. It handles all file level activities such as index creation and maintenance, row searching logic and retrieval, locking, database recovery, etc. DBSS is a large and powerful access method which understands system-level (see Figure 1.10) activities. RDS on the other hand is the piece of code that understands "relational database" concepts — tables, views, SQL operators, etc. RDS understands the human interfaces to the system and is responsible for data security, SQL syntax, statement compilation, and optimization, sorting operations, complicated SQL logic such as data analysis, etc. The system components and their various roles in satisfying a user query are shown in Figure 2.8.

---

* A word here is in order regarding any confusion that may exist over the terms SQL/DS and SQL. SQL/DS is the DBMS product licensed by your installation. SQL stands for Structured Query Language, and is the "access language" (see Figure 2.4) you write your queries in. It is regrettable that IBM chose two such similar acronyms for such disparate "things"!

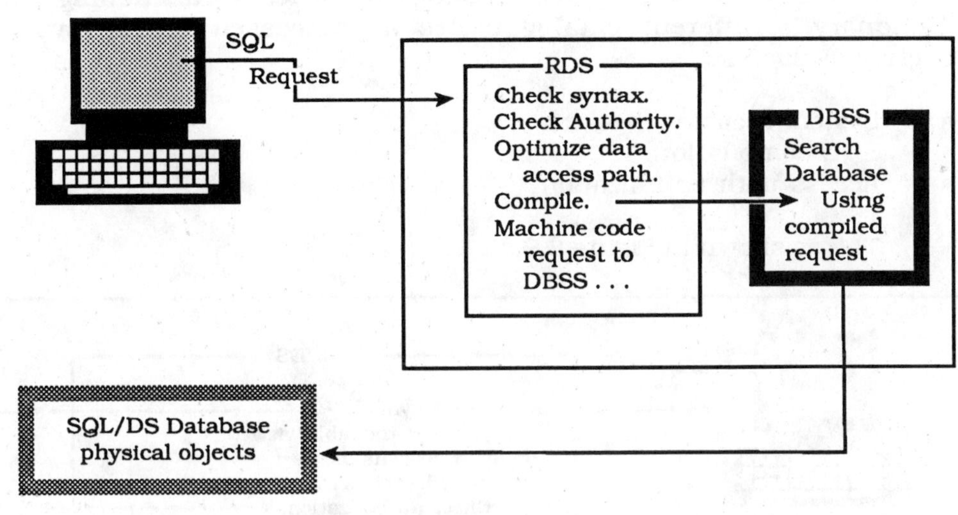

Figure 2.8. SQL/DS system components and responsibilities.

2.5 THE SQL/DS CATALOG

SQL/DS automatically maintains a built-in data dictionary that consists of a set of tables called the system catalog. The system catalog contains "metadata," that is, data about user-data, and is used by SQL/DS to direct and control the operation of the system. In the system catalog tables is information about all the data and data objects that SQL/DS manages.

The system catalog tables themselves can be queried (but not updated) by system users to:

- Assess the status of user and SQL/DS data objects:
  - Reaching out-of-space conditions?
  - Time for reorganization?
  - Condition of indexes?

- Verify application design:
  - Objects and relationships correct?
  - Naming conventions applied?
  - Drop/Analysis impact — (What if we drop this table . . .)?

SQL/DS uses the system catalog as an active functioning data dictionary. Different catalog tables are accessed during system operations such as:

- System security checking
- SQL compilation
- Access path optimization

This is shown in Figure 2.9.

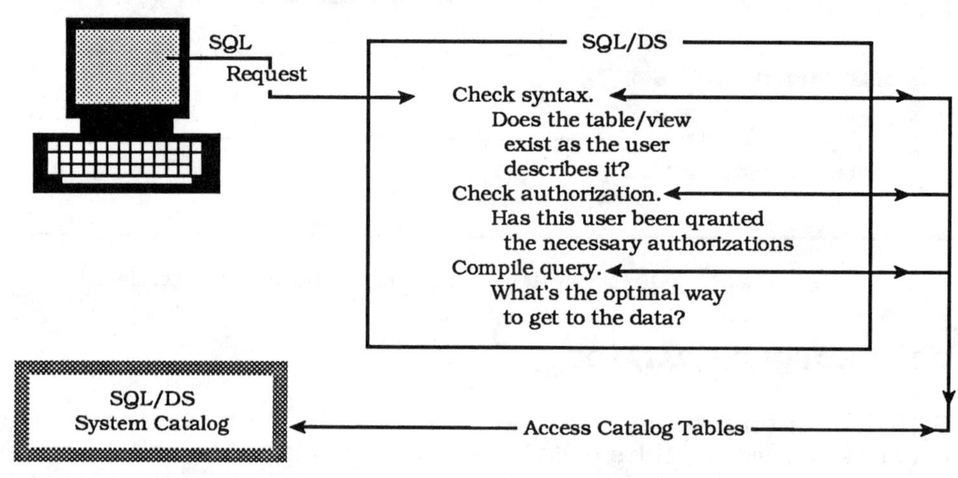

Figure 2.9. System Catalog — used by SQL/DS.

2.6 SQL/DS's RELATED PRODUCTS

Figure 2.10 describes some of the typical add-ons and peripheral products installed with SQL/DS. As you can see, the architecture is conceptually layered with the SQL/DS RDBMS "engine" at the kernel. All access to SQL/DS data is through the SQL language, so all interface products and users requesting services must communicate through SQL. Here are some short descriptions of the various packages and interfaces with SQL/DS.

ISQL: An SQL query interface, consisting of a line editor which allows for SQL query testing, and a few report formatting commands. Not a generally business-user friendly interface, but programmers and DBAs use the ISQL for quick (at least quicker than using QMF) access to SQL/DS for one-shot queries and testing.

DBSU: Data Base Services Utility — A general-purpose utility program that can:
- Execute SQL statements interactively
- Execute SQL statements contained in a file
- Execute data management utilities:
  - Image copy
  - Reorganization
  - Data export/import

QMF: Query Management Facility — A powerful report-writer/query-editor capable of producing production quality reports in an interactive, fill-in-the-blank, PF-key driven fashion. Report programs can be executed interactively or in batch mode. QMF provides true business-user friendly access to SQL/DS data.

4GL Products: Such as SAS and Focus allow users knowledgeable in those products to express queries in their own familiar language. The queries are translated into SQL by the products. Using these packages can allow a company to preserve/extend its investment in existing applications.

REXX: A common IBM command language, REXX programs containing SQL statements can be built quickly and easily. Excellent for application prototyping and system development work suited for compact yet powerful command languages.

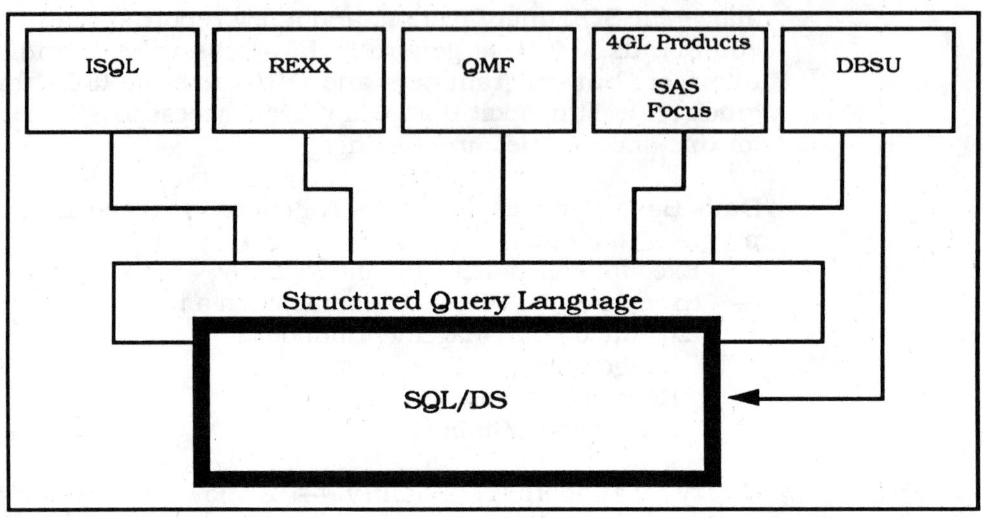

Figure 2.10. SQL/DS related products.

2.7 THE DB2 ENVIRONMENT

Programs that access DB2 data run under the supervision of one of the transaction managers shown in Figure 2.11. A transaction is any complete unit of business processing; a transaction manager is the software responsible for the successful completion of the unit of work involved in a transaction.

- TSO — Time Sharing Option
- IMS DC — Information Management Systems Data Communication
- CICS — Customer Information Control System

The transaction manager provides essential services to DB2 including:

Data Communications — the ability to communicate with users at remote terminals.

Data Integrity — the ability to maintain DB2 or other data in an uncompromised and consistent state.

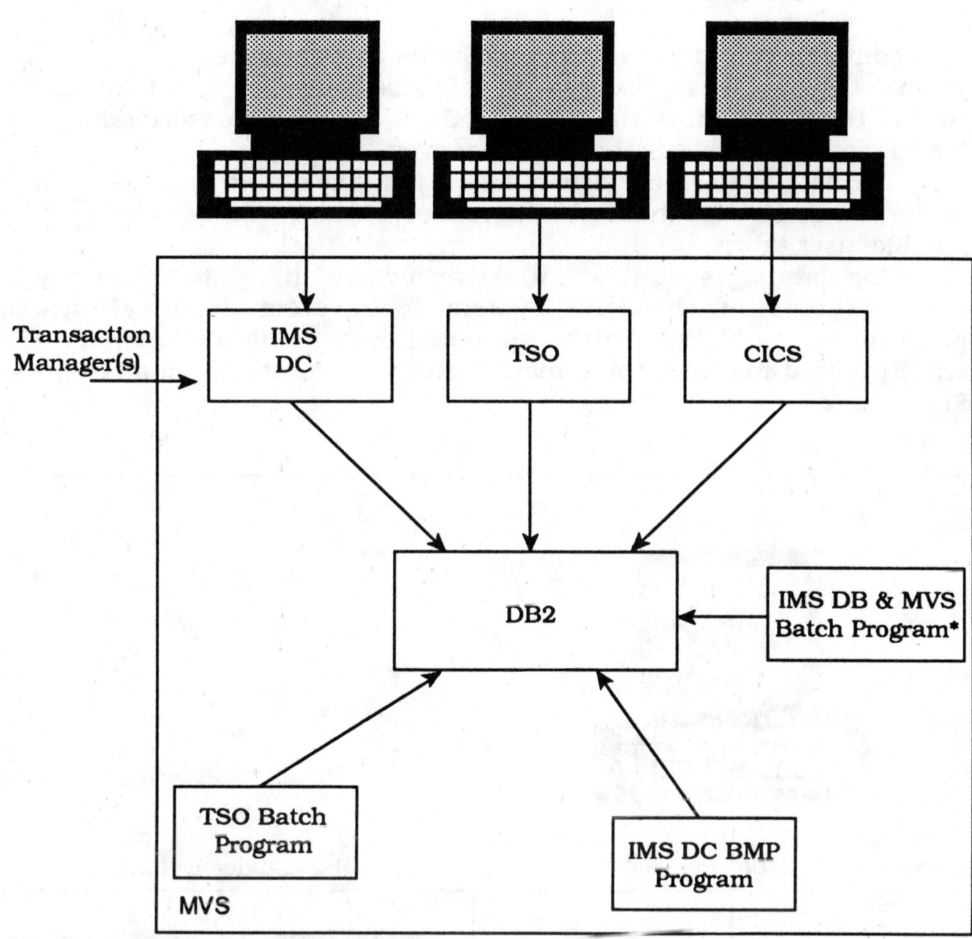

* As of this writing, a special type of connection needs to be made from MVS Batch programs to DB2.

Figure 2.11. DB2 environments. DB2 relationships to TSO, CICS, IMS/DC.

2.8 DB2 CATALOG

DB2 contains basically two different kinds of databases, user databases and a catalog database. The catalog database is used by DB2 to direct and control the operation of the system. On it are stored descriptions of all the data and data objects that DB2 manages.

The catalog itself consists of tables just like a user database. These tables can be queried (but not updated* or defined) with SQL statements, just like user tables.

The catalog is used as an active, functioning data dictionary by various parts of the DB2 system for system security, system performance, and SQL query optimization. (A data dictionary is a structure that maintains information on the content of a database.) See Figure 2.12.

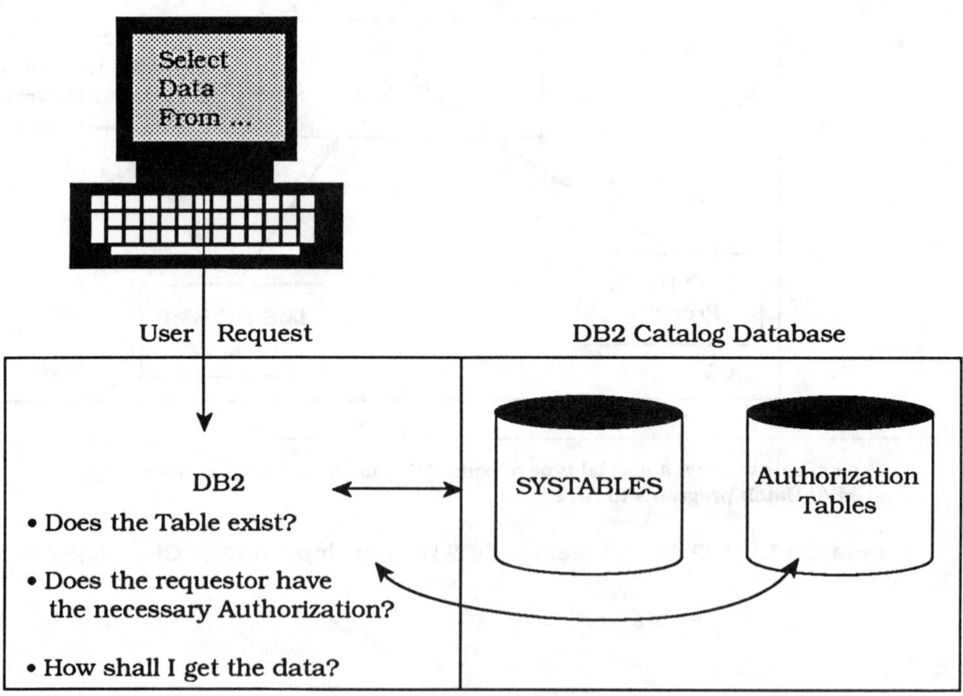

Figure 2.12. Catalog use by DB2.

---

* With DB2 Version 2.1 persons with SYSADM (System Administrator) authorization may update the DB2 catalog for purposes of performance and simulation.

The DB2 catalog consists of around thirty tables such as:

- SYSTABLES — which contain a row of information on each table in the system;
- SYSCOLUMNS — which contain a row of information for each column in each table in the system;
- SYSINDEXES — which contain a row of information for each index on each table in the system.

Most, if not all, of the popular RDBMS products feature a similar catalog of some sort.

2.9 DB2'S RELATED PRODUCTS

As you can see from Figure 2.13, the system is conceptually layered with the DB2 RDBMS at the kernel. This kernel basically functions as a data manager, accepting or rejecting requests for data, accessing the database, and passing the data back and forth between the database and the other interfaces.

The only language the DB2 kernel understands is SQL, so anyone requesting access to data stored on DB2 databases must communicate with DB2 using SQL statements. The following products are:

- DB2I (DB2 Interactive Interface): A user-friendly interface consisting of a main menu and seven options that allow programmers and database administrators to accomplish work interactively and productively in the DB2 environment.
- QMF (Query Management Facility): A front-end/report-writing program that is meant for the true business end-user. QMF consists of an easy to follow collection of PF-key-driven menus and options. It provides user-friendly access to SQL and DB2.
- DXT (Data Extract): A data facsimile (data copy) program that allows us to transfer information from other file organizations (like sequential, indexed, and hierarchical file structures) into a format that is ready for loading into DB2 tables.
- DBEDIT: A custom forms front-end program.
- VSAM TRANSPARENCY: A utility program designed to make DB2 data accessible to application programs that expect to find the data stored as VSAM (virtual sequential access method) files.

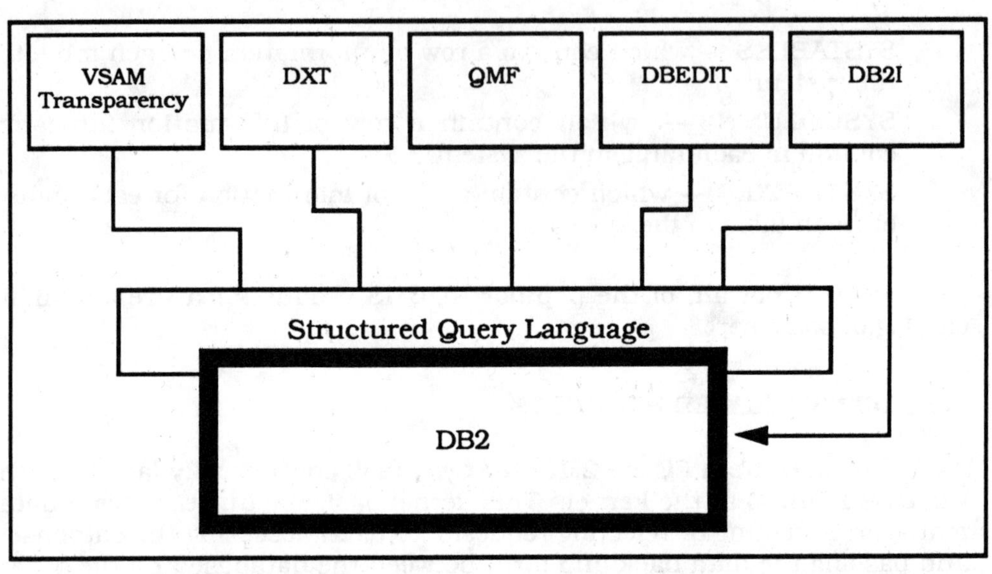

Figure 2.13. DB2 related products.

2.10 DB2I

DB2I is an easy-to-use, menu-driven interface for DB2. It runs under the supervision of TSO. DB2I provides for interactive processing and execution of seven important DB2 facilities (see Figure 2.14):

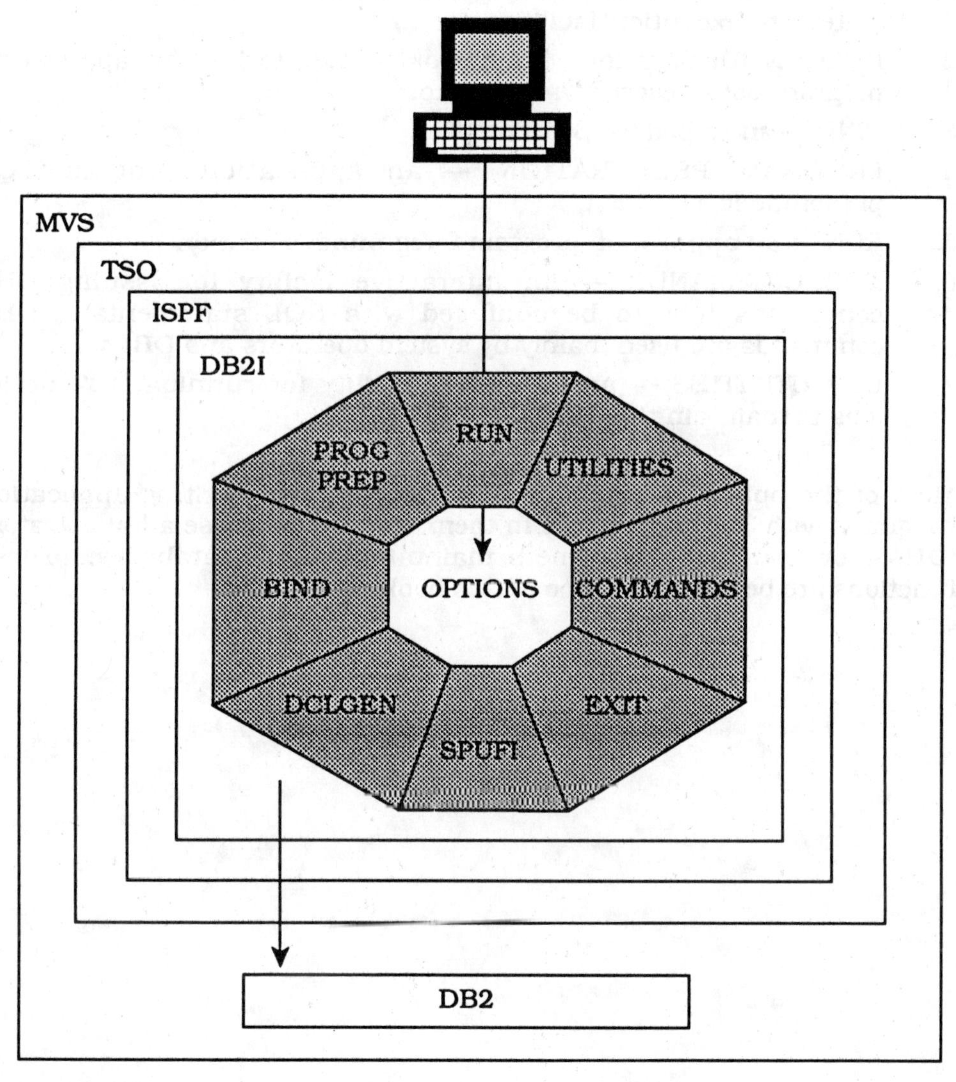

Figure 2.14. DB2I environment.

1. SPUFI (SQL processor using File Input) — an interactive SQL statement execution facility.
2. DCLGEN (Declarations [Copybook] Generator) — an application program data description generator.
3. BIND — an embedded SQL compiler.
4. PROGRAM PREPARATION — an application program/SQL precompiler.
5. RUN — an application program foreground execution.
6. DB2 COMMANDS — an interactive facility for issuing DB2 commands (not to be confused with SQL statements). DB2 commands are used mainly by system operators and DBAs.
7. DB2 UTILITIES — an interactive facility for running DB2 utility jobs to maintain the database.

Most of the options are intended for programmers writing application programs with SQL statements in them, and for database administrators (DBAs) or system programmers maintaining DB2 databases. These functions are beyond the scope of this book.

Chapter 3
GETTING STARTED

This work has been designed to allow you to use it with a SQL system on either a PC or a terminal linked to a mainframe computer. In order to practice the exercises, it will be necessary for you to load the tables for the following case. We will show you how to do this. It is time-consuming and detailed work but you will gain a great deal of experience with SQL if you elect to do it. **If you are using a computer, there will be minor differences in the input statements from those shown here. Adjust your answer to account for these differences.**

Precision is critical in communicating with your computer. A comma in the wrong place can produce erroneous results or no results at all. For this reason, we give you precise instructions on what to type. Take your time and check your answers carefully. Use spaces between your keywords. Try to establish a pattern. As you progress, the work becomes more difficult, and it is important that you type the statements accurately. Table names, column names, and keywords all must be spelled correctly. Punctuation must be correct. This will become easier as you work with SQL. Do not get frustrated if you don't get the correct answer. Sometimes it is very difficult to find your typing error.

If you are using a terminal linked to your mainframe computer, you will have to learn how to access your facility's SQL system. You will need help from your training coordinator, technical support staff member, or information center staff. Installations have site-specific requirements that must be taught locally by your company's support personnel.

You do not have to use the computer. If you wish, you may just write your answers in the appropriate space. If you don't use a computer, you may bypass Sections 3.7 to 3.8.

The following topics are covered:

3.1 The Video Store Case
3.2 The Video Store Tables
3.3 The Invoice (INV) Table
3.4 The Customer (CUST) Table
3.5 The Movie (MOV) Table
3.6 The Invoice-Number Movie-Number (I_M) Table
3.7 Loading the Tables
3.8 Creating the Tables

3.1 THE VIDEO STORE CASE

The case you will use is about a new video store. To keep track of invoices, movies, and customers there is a relational database management system and a video store database that has four tables.

1. **Invoices** (INV): contains data on each individual invoice for movie rentals.
2. **Customers** (CUST): contains data for each customer who has rented a movie.
3. **Movies** (MOV): contains data on each movie that we own.
4. **Invoices/Movies**(I_M): contains an entry (row) for each movie rented, along with its invoice number.

Take a minute to review the database and its tables in Figure 3.1.

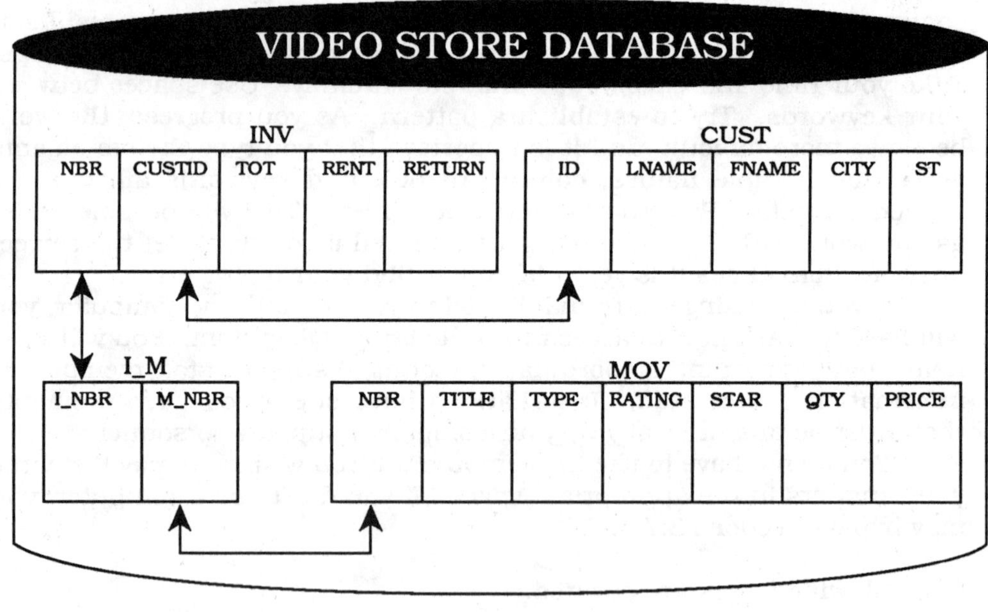

Figure 3.1. The video store database.

3.2 THE VIDEO STORE TABLES

You will have to load the four tables. Take a moment to scan over each individual table in Figure 3.2, noticing:

1. what types of information are stored in each column;
2. the values or content of each row;

3. **any intertable relationships (for instance, notice that customer identification data is carried on the CUST table in the ID column, and on the INV table in the CUSTID column).**

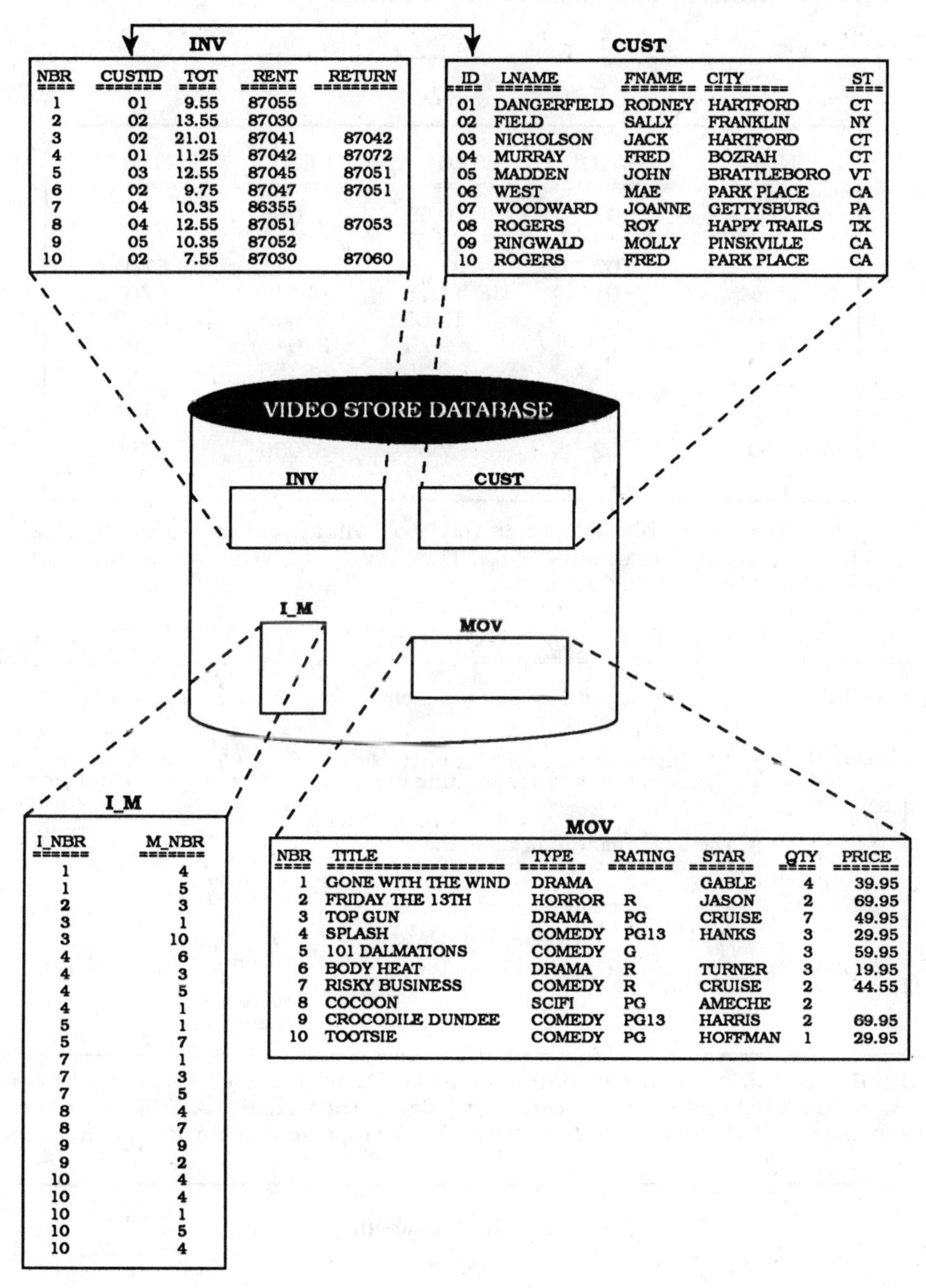

INV

NBR	CUSTID	TOT	RENT	RETURN
1	01	9.55	87055	
2	02	13.55	87030	
3	02	21.01	87041	87042
4	01	11.25	87042	87072
5	03	12.55	87045	87051
6	02	9.75	87047	87051
7	04	10.35	86355	
8	04	12.55	87051	87053
9	05	10.35	87052	
10	02	7.55	87030	87060

CUST

ID	LNAME	FNAME	CITY	ST
01	DANGERFIELD	RODNEY	HARTFORD	CT
02	FIELD	SALLY	FRANKLIN	NY
03	NICHOLSON	JACK	HARTFORD	CT
04	MURRAY	FRED	BOZRAH	CT
05	MADDEN	JOHN	BRATTLEBORO	VT
06	WEST	MAE	PARK PLACE	CA
07	WOODWARD	JOANNE	GETTYSBURG	PA
08	ROGERS	ROY	HAPPY TRAILS	TX
09	RINGWALD	MOLLY	PINSKVILLE	CA
10	ROGERS	FRED	PARK PLACE	CA

I_M

I_NBR	M_NBR
1	4
1	5
2	3
3	1
3	10
4	6
4	3
4	5
4	1
5	1
5	7
7	1
7	3
7	5
8	4
8	7
9	9
9	2
10	4
10	4
10	1
10	5
10	4

MOV

NBR	TITLE	TYPE	RATING	STAR	QTY	PRICE
1	GONE WITH THE WIND	DRAMA		GABLE	4	39.95
2	FRIDAY THE 13TH	HORROR	R	JASON	2	69.95
3	TOP GUN	DRAMA	PG	CRUISE	7	49.95
4	SPLASH	COMEDY	PG13	HANKS	3	29.95
5	101 DALMATIONS	COMEDY	G		3	59.95
6	BODY HEAT	DRAMA	R	TURNER	3	19.95
7	RISKY BUSINESS	COMEDY	R	CRUISE	2	44.55
8	COCOON	SCIFI	PG	AMECHE	2	
9	CROCODILE DUNDEE	COMEDY	PG13	HARRIS	2	69.95
10	TOOTSIE	COMEDY	PG	HOFFMAN	1	29.95

Figure 3.2. The video store tables.

3.3 THE INVOICE (INV) TABLE

The invoice table (Figure 3.3) is used to track our movie rentals and returns, as well as to store the total amount for all of the movies rented, on a single invoice. It consists of five columns.

INV

NBR	CUSTID	TOT	RENT	RETURN
1	01	9.55	87055	*
2	02	13.55	87030	
3	02	21.01	87041	87042
4	01	11.25	87042	87072
5	03	12.55	87045	87051
6	02	9.75	87047	87051
7	04	10.35	86355	
8	04	12.55	87051	87053
9	05	10.35	87052	
10	02	7.55	87030	87060

*Please note that the blank fields in the book will appear as dashes in DB2 or QMF, and question marks in ISQL. They represent NULL or missing values and will be explained in Chapter 5.

COL NAME	TYPE OF INFORMATION STORED	COL SIZE	DATA TYPE
NBR	The invoice number that uniquely identifies each invoice	2	integer
CUSTID	The customer number that uniquely identifies the customer renting the movies	2	alpha/ numeric character
TOT	The total amount for all of the movies rented	up to $100	decimal
RENT	The date the movies were rented (Julian format)*	very large number	integer
RETURN	The date the movies were returned (Julian format)*	very large number	integer

*Julian date format stores dates as YYDDD, where YY represents the year (87,88,89, etc.) and DDD represents the day of the year RELATIVE to January 1. For instance 87001 is January 1, 1987, 88059 represents February 29th, 1988.

Figure 3.3. The Invoice (INV) Table.

3.4 THE CUSTOMER (CUST) TABLE

The customer table (Figure 3.4) holds name and address information on our customers:

CUST

ID	LNAME	FNAME	CITY	ST
01	DANGERFIELD	RODNEY	HARTFORD	CT
02	FIELD	SALLY	FRANKLIN	NY
03	NICHOLSON	JACK	HARTFORD	CT
04	MURRAY	FRED	BOZRAH	CT
05	MADDEN	JOHN	BRATTLEBORO	VT
06	WEST	MAE	PARK PLACE	CA
07	WOODWARD	JOANNE	GETTYSBURG	PA
08	ROGERS	ROY	HAPPY TRAILS	TX
09	RINGWALD	MOLLY	PINKSVILLE	CA
10	ROGERS	FRED	PARK PLACE	CA

COL NAME	TYPE OF INFORMATION STORED	COL SIZE	DATA TYPE
ID	The identification number that uniquely identifies each customer	2	alpha/ numeric character
LNAME	The customer's last name	12	alpha/ numeric character
FNAME	The customer's first name	6	alpha/ numeric character
CITY	The city in which the customer lives	15	alpha/ numeric character
ST	The state in which the customer lives	2	alpha/ numeric character

Figure 3.4. The Customer (CUST) Table.

3.5 THE MOVIE (MOV) TABLE

The movie table (Figure 3.5) holds information of interest to prospective movie renters on all of the films currently in our inventory:

MOV

NBR	TITLE	TYPE	RATING	STAR	QTY	PRICE
1	GONE WITH THE WIND	DRAMA		GABLE	4	39.95
2	FRIDAY THE 13TH	HORROR	R	JASON	2	69.95
3	TOP GUN	DRAMA	PG	CRUISE	7	49.95
4	SPLASH	COMEDY	PG13	HANKS	3	29.95
5	101 DALMATIONS	COMEDY	G		3	59.95
6	BODY HEAT	DRAMA	R	TURNER	3	19.95
7	RISKY BUSINESS	COMEDY	R	CRUISE	2	44.55
8	COCOON	SCIFI	PG	AMECHE	2	
9	CROCODILE DUNDEE	COMEDY	PG13	HARRIS	2	69.95
10	TOOTSIE	COMEDY	PG	HOFFMAN	1	29.95

COL NAME	TYPE OF INFORMATION STORED	COL SIZE	DATA TYPE
NBR	The number that uniquely identifies each movie	2	integer
TITLE	The movie's title	20	alpha/ numeric character
TYPE	The general category of film type describing this movie	6	alpha/ numeric character
RATING	The rating established for the movie by the rating commission	4	alpha/ numeric character
STAR	The name of the star	10	alpha/ numeric character
QTY	The number of copies of this movie that we currently own	2	integer
PRICE	The original buying price of the movie	up to $100	decimal

Figure 3.5. The Movie (MOV) Table.

3.6 THE INVOICE-NUMBER MOVIE-NUMBER (I_M) TABLE

The I_M table is a cross reference table. It cross references our movie numbers (M_NBR) with invoice numbers (I_NBR) for our customers invoice (Figure 3.6). It contains line item information on the movies rented for each invoice. The invoice number is carried, along with the movie number, providing a way of finding out which rented movies correspond to each invoice:

I_M

I_NBR	M_NBR
1	4
1	5
2	3
3	1
3	10
4	6
4	3
4	5
4	1
5	1
5	7
7	1
7	3
7	5
8	4
8	7
9	9
9	2
10	4
10	4
10	1
10	5
10	4

COL NAME	TYPE OF INFORMATION STORED	COL SIZE	DATA TYPE
I_NBR	The invoice number that uniquely identifies each invoice	2	number
M_NBR	The movie number that uniquely identifies each movie	2	number

Figure 3.6. The Invoice-Number Movie-Number (I_M) Table.

3.7 LOADING THE TABLES

Now it's time to load the tables. We will use SQL to do this. Because you have no SQL experience, we will show you how to use the CREATE and INSERT statements. Follow the instructions carefully. Type slowly and check what you type before you enter it. Of course you know to press **ENTER** following each command semicolon.

Before doing Section 3.8, you may want to scan the sections on "Data Definition Statements" and "SQL Update Statements." These cover the CREATE Table and INSERT statements. After you have an idea of what CREATE Table and INSERT do, return here and carefully follow the instruction to enter the tables.

3.8 CREATING THE TABLES

If you scanned the suggested sections, you now know that the CREATE Table and INSERT statements are SQL commands used to set up tables. The CREATE Table statement is used to set up the structure of the table. It defines the table name, the column headings, and the type of information that will appear in the table. The INSERT statement is used to place the data into the tables.

If you follow the instructions below carefully you will create and insert the data for the four tables called INV, CUST, I_M, and MOV. We are assuming that you know how to enter data with SQL. If not, you may need help in getting started.

3.8.1 Create Table MOV

The first table we will create is the MOV table (see Figure 3.5). There are seven column headings to enter plus the table name which is MOV. To create the structure for this table you type the following exactly as shown. We assume you are logged on to SQL and have a SQL prompt on the screen.

Type:

```
CREATE TABLE MOV (NBR SMALLINT,
TITLE CHAR(20),
TYPE CHAR(6),
RATING CHAR(4),
STAR CHAR(10),
QTY SMALLINT,
PRICE DECIMAL(5,2) );
```

What this says is "create a table called MOV with column headings:

NBR (a number value of 1 to 32,760)
TITLE (20 characters)
TYPE (6 characters)
RATING (4 characters)
STAR (10 characters)
QTY (a number value of 1 to 32,760)
PRICE (5 numbers with 2 decimal places)"

When this is complete you should receive a message saying it has been created. Now you must insert the data. This is done row by row. To insert row 1 into the MOV table type the following:

Type:

```
INSERT INTO MOV
  VALUES (1, 'GONE WITH THE WIND', 'DRAMA', NULL, 'GABLE', 4,39.95);
```

Now insert rows 2 through 10.

Type:

```
INSERT INTO MOV
  VALUES (2, 'FRIDAY THE 13TH', 'HORROR', 'R', 'JASON', 2,69.95);
INSERT INTO MOV
  VALUES (3, 'TOP GUN', 'DRAMA', 'PG', 'CRUISE', 7, 49.95);
INSERT INTO MOV
  VALUES (4, 'SPLASH', 'COMEDY', 'PG13', 'HANKS', 3, 29.95);
INSERT INTO MOV
  VALUES (5, '101 DALMATIONS', 'COMEDY', 'G', NULL, 3, 59.95);
INSERT INTO MOV
  VALUES (6, 'BODY HEAT', 'DRAMA', 'R', 'TURNER', 3, 19.95);
INSERT INTO MOV
  VALUES (7, 'RISKY BUSINESS', 'COMEDY', 'R', 'CRUISE', 2, 44.55);
INSERT INTO MOV
  VALUES (8, 'COCOON', 'SCIFI', 'PG', 'AMECHE', 2, NULL);
INSERT INTO MOV
  VALUES (9, 'CROCODILE DUNDEE', 'COMEDY', 'PG13', 'HARRIS', 2,
  69.95);
INSERT INTO MOV
  VALUES (10, 'TOOTSIE', 'COMEDY', 'PG', 'HOFFMAN', 1, 29.95);
```

When you have completed typing the rows into the table, type the following:

```
SELECT * FROM MOV;
```

This will display the MOV table so you can check your accuracy. It should look like this:

NBR	TITLE	TYPE	RATING	STAR	QTY	PRICE
1	GONE WITH THE WIND	DRAMA		GABLE	4	39.95
2	FRIDAY THE 13TH	HORROR	R	JASON	2	69.95
3	TOP GUN	DRAMA	PG	CRUISE	7	49.95
4	SPLASH	COMEDY	PG13	HANKS	3	29.95
5	101 DALMATIONS	COMEDY	G		3	59.95
6	BODY HEAT	DRAMA	R	TURNER	3	19.95
7	RISKY BUSINESS	COMEDY	R	CRUISE	2	44.55
8	COCOON	SCIFI	PG	AMECHE	2	
9	CROCODILE DUNDEE	COMEDY	PG13	HARRIS	2	69.95
10	TOOTSIE	COMEDY	PG	HOFFMAN	1	29.95

If there are errors, you may correct them with the UPDATE command. It's easy, but as you have never done it, follow the instructions.

Assume you entered line 4 as:

4 SPLASH COMEDY PG13 HANKS 3 49.95

The price column should be changed to 29.95. To do this enter the following:

```
UPDATE MOV SET PRICE = 29.95
    WHERE TITLE LIKE 'SPLASH';
```

What the instruction above says is: UPDATE the movie table price field in the row that has the movie called Splash.

If you type it exactly as shown, it will change the price. Edit your table and try to make the corrections. Read the section on UPDATE in Chapter 11 for more details.

3.8.2 Create Table INV

You will use the same procedure you used for the MOV table to create the INV table (see Figure 3.3). Take your time and type carefully

Type:

```
CREATE TABLE INV (NBR SMALLINT,
CUSTID CHAR(2),
TOT DECIMAL (5,2),
RENT INTEGER,
RETURN INTEGER);
```

Now insert the data. Type:

```
INSERT INTO INV
  VALUES (1, '01', 9.55, 87055, NULL);
INSERT INTO INV
  VALUES (2, '02', 13.55, 87030, NULL);
INSERT INTO INV
  VALUES (3, '02', 21.01, 87041, 87042);
INSERT INTO INV
  VALUES (4, '01', 11.25, 87042, 87072);
INSERT INTO INV
  VALUES (5, '03', 12.55, 87045, 87051);
INSERT INTO INV
  VALUES (6, '02', 9.75, 87047, 87051);
INSERT INTO INV
  VALUES (7, '04', 10.35, 86355, NULL);
INSERT INTO INV
  VALUES (8, '04', 12.55, 87051, 87053);
INSERT INTO INV
  VALUES (9, '05', 10.35, 87052, NULL);
INSERT INTO INV
  VALUES (10, '02', 7.55, 87030, 87060);
```

Check your result. Type:

```
SELECT * FROM INV;
```

The table should look like this:

NBR	CUSTID	TOT	RENT	RETURN
1	01	9.55	87055	
2	02	13.55	87030	
3	02	21.01	87041	87042
4	01	11.25	87042	87072
5	03	12.55	87045	87051
6	02	9.75	87047	87051
7	04	10.35	86355	
8	04	12.55	87051	87053
9	05	10.35	87052	
10	02	7.55	87030	87060

3.8.3 Create Table CUST

Now you will create the CUST table (see Figure 3.4).

Type:

```
CREATE TABLE CUST (ID CHAR(2),
LNAME CHAR(12),
FNAME CHAR(6),
CITY VARCHAR(15),
ST CHAR(2)) ;
```

```
INSERT INTO CUST
  VALUES ('09', 'RINGWALD', 'MOLLY', 'PINKSVILLE', 'CA');
INSERT INTO CUST
  VALUES ('10', 'ROGERS', 'FRED', 'PARK PLACE', 'CA');
```

Check your result. Type:

```
SELECT * FROM CUST;
```

Your result should look like this:

```
ID   LNAME         FNAME     CITY          ST
===  ============  ========  ============  ====
01   DANGERFIELD   RODNEY    HARTFORD      CT
02   FIELD         SALLY     FRANKLIN      NY
03   NICHOLSON     JACK      HARTFORD      CT
04   MURRAY        FRED      BOZRAH        CT
05   MADDEN        JOHN      BRATTLEBORO   VT
06   WEST          MAE       PARK PLACE    CA
07   WOODWARD      JOANNE    GETTYSBURG    PA
08   ROGERS        ROY       HAPPY TRAILS  TX
09   RINGWALD      MOLLY     PINKSVILLE    CA
10   ROGERS        FRED      PARK PLACE    CA
```

3.8.4 Create Table I_M

Now you will create the Invoice-Number Movie-Number table (see Figure 3.6).

Type:

```
CREATE TABLE I_M (I_NBR SMALLINT,
M_NBR SMALLINT);
```

Insert the data. Type:

```
INSERT INTO I_M
  VALUES (1, 4);
INSERT INTO I_M
  VALUES (1, 5);
INSERT INTO I_M
  VALUES (2, 3);
INSERT INTO I_M
  VALUES (3, 1);
INSERT INTO I_M
  VALUES (3, 10);
```

```
INSERT INTO I_M
  VALUES (4, 6);
INSERT INTO I_M
  VALUES (4, 3);
INSERT INTO I_M
  VALUES (4,5);
INSERT INTO I_M
  VALUES (4, 1);
INSERT INTO I_M
  VALUES (5, 1);
INSERT INTO I_M
  VALUES (5,7);
INSERT INTO I_M
  VALUES (7, 1);
INSERT INTO I_M
  VALUES (7, 3);
INSERT INTO I_M
  VALUES (7,5);
INSERT INTO I_M
  VALUES (8, 4);
INSERT INTO I_M
  VALUES (8, 7);
INSERT INTO I_M
  VALUES (9, 9);
INSERT INTO I_M
  VALUES (9, 2);
INSERT INTO I_M
  VALUES (10, 4);
INSERT INTO I_M
  VALUES (10, 4);
INSERT INTO I_M
  VALUES (10, 1);
INSERT INTO I_M
  VALUES (10, 5);
INSERT INTO I_M
  VALUES (10, 4);
```

Check your result. Type:

```
SELECT * FROM I_M;
```

```
INSERT INTO I_M
  VALUES (10, 5);
INSERT INTO I_M
  VALUES (10, 4);
```

Check your result. Type:

```
SELECT * FROM I_M;
```

Your result should look like this:

I_NBR	M_NBR
1	4
1	5
2	3
3	1
3	10
4	6
4	3
4	5
4	1
5	1
5	7
7	1
7	3
7	5
8	4
8	7
9	9
9	2
10	4
10	4
10	1
10	5
10	4

If you find errors, you may go back and do line changes by using the UPDATE statement. If all is correct, you may proceed through the book and do the exercises.

If you entered the data correctly, you have learned a great deal about SQL. Most of this book should be easy for you to grasp.

Now exit from SQL to commit and save your tables.

Chapter 4
DATA RETRIEVAL STATEMENTS

The following topics are covered:

If you have entered the data as previously shown, you may do the exercises on your computer. If not, write your answers in the space provided. Check your answers with the answer key. Let's proceed.

4.1 THE SELECT STATEMENT

SELECT retrieves data from tables. It provides a query capability. This means that by executing SELECT statements, information currently in tables will be displayed.

You can think of SELECT as a request to the DBMS for a "snapshot image" of the table information.

SELECT is the most important SQL statement you will learn. With it you can:

- Display some or all of the COLUMNS from tables;
- Display some or all of the ROWS from tables;
- Display calculated values from tables;
- Display statistical information from tables, like averages or sums of column values;
- Combine information from two or more tables.

SELECT has many options and formats. We will be covering most of them.

4.1.1 SELECT Statement Format

The format or layout of SELECT statements is as follows:

```
SELECT <columnname(s)> FROM <tablename>;
```

This is interpreted as:

SELECT (what?)	One or more column names from a table. If you SELECT more than one column, the column names must be separated from each other with commas **(you may optionally add blanks or spaces for readability). Note that there is no comma following the last column name.**
FROM (where?)	A correctly spelled table name.

SELECT and FROM are the two mandatory words in a SELECT query. Other additional words and qualifiers will be covered shortly. In SELECTing data, your main concerns for now will simply be to spell the column and table names correctly. This should not be a problem, because we will be showing them to you on the same page where we ask you to code SELECTs. Again, the structure of the command is:

```
SELECT <columnname(s)> FROM <tablename>;
```

Example 1: Read the Movie table, and display the title and star columns from all the rows.

```
SELECT TITLE, STAR FROM MOV;
```

Example 2: Read the Invoice table, and get the number, rental data, and return date from all the rows.

```
SELECT NBR, RENT, RETURN FROM INV;
```

Note the use of spaces for readability. We recommend you leave a space between every keyword. For example:

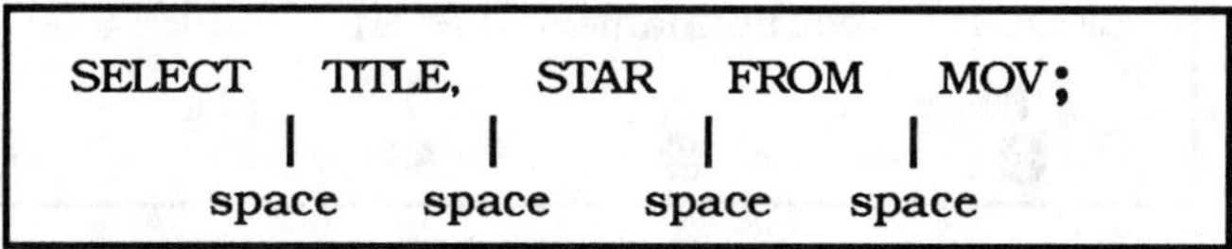

4.1.2 SELECT Statement Results Tables

The output of any successful SELECT statement execution is an unnamed, temporary table known as a results table. Results tables represent a copy of the data as it exists in the database. These results tables can have zero rows, or up to as many rows as the original table — this depends on your query statement.

Example 1:

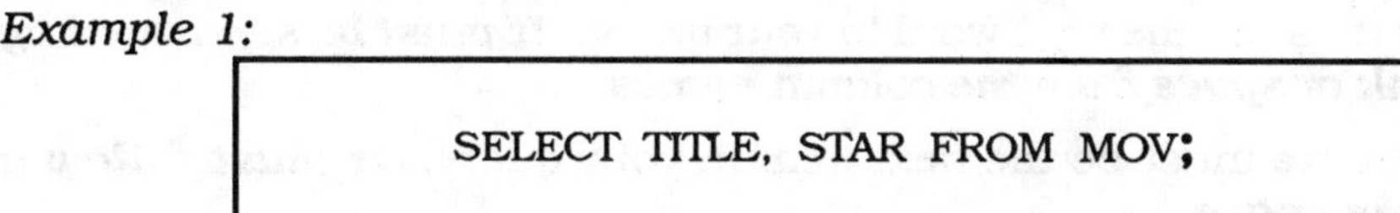

RESULTS TABLE:
The results table will display all titles and stars from the movie table.

TITLE	STAR
GONE WITH THE WIND	GABLE
FRIDAY THE 13TH	JASON
TOP GUN	CRUISE
SPLASH	HANKS
101 DALMATIONS	
BODY HEAT	TURNER
RISKY BUSINESS	CRUISE
COCOON	AMECHE
CROCODILE DUNDEE	HARRIS
TOOTSIE	HOFFMAN

4.2 CODING SELECT STATEMENTS

SQL statements are easy to code when compared to almost any other programming language. There are no rules for column justification (right or left positioning). This means that you don't have to use certain SQL words in certain columns as you do when using some other languages.

There are five parts to a SELECT statement, as seen in Figure 4.1.

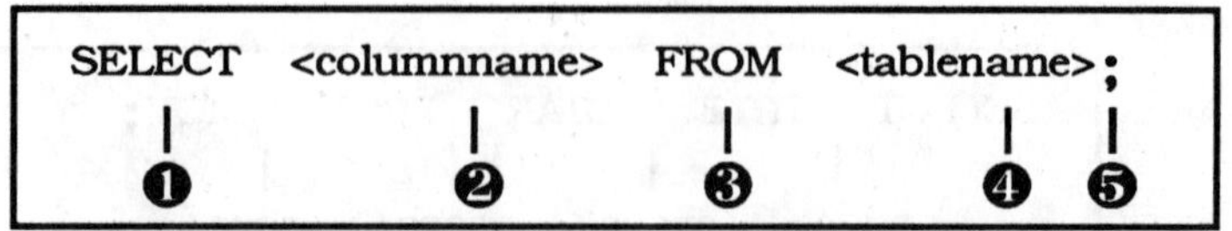

Figure 4.1. SELECT statement.

❶ ***SELECT*** is the first word in your query. ***It must be separated by a blank or space from the column names,*** and, of course, must be spelled correctly.

❷ ***column names*** must be the next items in your query. They must be separated by a comma, exist in the table you are querying, and, of course, must be spelled correctly.

❸ ***FROM*** must be the next word in your query. ***It must be separated by a blank or space from the column names.***

❹ ***tablename*** must be the next item in your query. ***It must follow a blank or space.***

❺ ***semicolon*** lowercase key. The **;** tells SQL that the query is complete and may be executed. If you are using SQL on a mainframe using QMF, you do **not** use a semicolon at the end of the statement.

Even though we have not covered some of the details of the statements shown in the following, look at them carefully. Notice that whether you put the statement on the same line, or break it up onto two lines, the results table is the same. Remember to press **ENTER** after typing each command.

Type: **SELECT LNAME, FNAME FROM CUST;**

RESULTS TABLE:

LNAME	FNAME
DANGERFIELD	RODNEY
FIELD	SALLY
NICHOLSON	JACK
MURRAY	FRED
MADDEN	JOHN
WEST	MAE
WOODWARD	JOANNE
ROGERS	ROY
RINGWALD	MOLLY
ROGERS	FRED

Type: **SELECT LNAME, FNAME FROM CUST;**

RESULTS TABLE:

LNAME	FNAME
DANGERFIELD	RODNEY
FIELD	SALLY
NICHOLSON	JACK
MURRAY	FRED
MADDEN	JOHN
WEST	MAE
WOODWARD	JOANNE
ROGERS	ROY
RINGWALD	MOLLY
ROGERS	FRED

4.3 LAYOUT OF TASK SPECIFICATION SHEET

From here on, you will be completing exercises to achieve mastery in SQL. The exercises will be presented in the format shown in Figure 4.2.

❶ SQL TOPIC. This represents the SQL concept or statement on which we are currently working.

❷ EXERCISE. This represents our business specification written in English.

❸ INPUT TABLE. This represents the table that holds the information that will complete our task.

❹ SQL STATEMENT STRUCTURE OR FORMAT. This describes the general syntax for your statements. In particular, notice the two variable areas for column and table names:

4a. <columnname> represents the column name(s), which will change from request to request.

4b. <tablename> represents the table name, which will change from request to request.

❺ After reading the exercise over two or three times to gain a thorough understanding of the problem to be solved, follow these steps to complete the exercise:

1. Write your answer in the space provided.
2. Type in your statements (watch <columnname> and <tablename> spelling) and execute them.
3. Verify that your answers are correct by looking at both the results table ❻ and the input table. For further verification, look at the answer key. ❼

❶ TOPIC: How to Complete This Book
❷ EXERCISE: This is an example.
❸ INPUT TABLE:

SQL Statement Structure:
FORMAT

❹
```
SELECT <columnname> FROM <tablename>;
```

space space space

❺ STEP 1. ____________________
Write your answer.

STEP 2. Enter your answer into the computer.

If you are using a computer, complete the exercise. Otherwise, check your answer by looking at the answer key.

STEP 3. Check your results.

❻ RESULTS TABLE:

❼ If your screen does not look like this, try again or look at the answer key and execute the solution given.

Figure 4.2. Exercise format.

4.4 SELECTing SINGLE COLUMNS EXERCISES

Because these are your first SQL statements, we'll go through the procedures step by step.

Exercise 4-1 asks you to display (SELECT) all of the movie titles using the movie table. This is a very straightforward request, asking you to SELECT the TITLE column FROM the MOV table. Remember our steps:

4.4.1. Steps Towards Problem Solution

1. Read the exercise carefully.
2. Formulate the solution mentally. Try rephrasing the task using SQL-like syntax (SELECT , FROM, etc.). Check the format box and write your answer in the space provided.
3. Type in your statements and execute them.
4. Verify that your answers are correct by both looking at the results table given and looking up the input table.

The answer you will write down and type is:

```
SELECT   TITLE   FROM   MOV;
```

After you press the **ENTER** key, the data on your screen should look like the results table in Exercise 4-1. If your screen doesn't look like this, check for spelling errors in the <columnname> or <tablename>. Make sure the words are in the correct order as shown in the format box, and that a semicolon(;) appears at the end.

Now do Exercise 4-1.

EXERCISE 4-1

❶ TOPIC: SELECTing Single Columns.

❷ EXERCISE 4-1. Display the movie TITLE(s) from the MOV table.

❸ INPUT TABLE:

MOV

NBR	TITLE	TYPE	RATING	STAR	QTY	PRICE
1	GONE WITH THE WIND	DRAMA		GABLE	4	39.95
2	FRIDAY THE 13TH	HORROR	R	JASON	2	69.95
3	TOP GUN	DRAMA	PG	CRUISE	7	49.95
4	SPLASH	COMEDY	PG13	HANKS	3	29.95
5	101 DALMATIONS	COMEDY	G		3	59.95
6	BODY HEAT	DRAMA	R	TURNER	3	19.95
7	RISKY BUSINESS	COMEDY	R	CRUISE	2	44.55
8	COCOON	SCIFI	PG	AMECHE	2	
9	CROCODILE DUNDEE	COMEDY	PG13	HARRIS	2	69.95
10	TOOTSIE	COMEDY	PG	HOFFMAN	1	29.95

FORMAT

❹

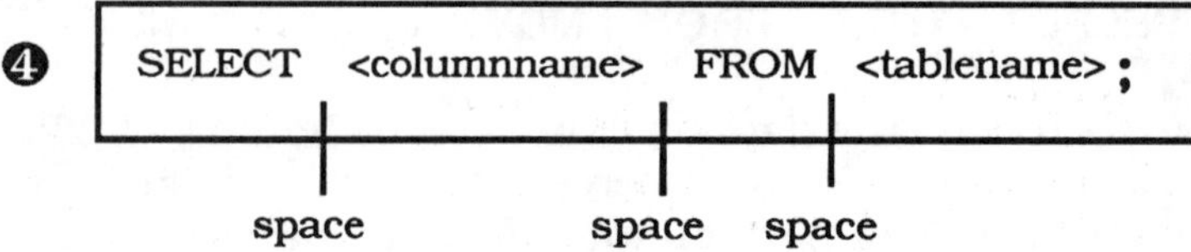

❺ STEP 1. ______________________________

Write your answer.

____

STEP 2. Enter your answer into the computer.

If you are using a computer, complete the exercise. Otherwise, check your answer by looking at the answer key.

STEP 3. Check your results.

❻ RESULTS TABLE:

TITLE
GONE WITH THE WIND
FRIDAY THE 13TH
TOP GUN
SPLASH
101 DALMATIONS
BODY HEAT
RISKY BUSINESS
COCOON
CROCODILE DUNDEE
TOOTSIE

❼ If your screen does not look like this, try again or look at the answer key and execute the solution given there.

Note: **We have given you the answer to Exercise 4-1. From now on, you will write your own answers.**

EXERCISE 4-2

❶ TOPIC: SELECTing Single Columns.
❷ EXERCISE 4-2. Display the CITY column from the CUST table.
❸ INPUT TABLE:

CUST

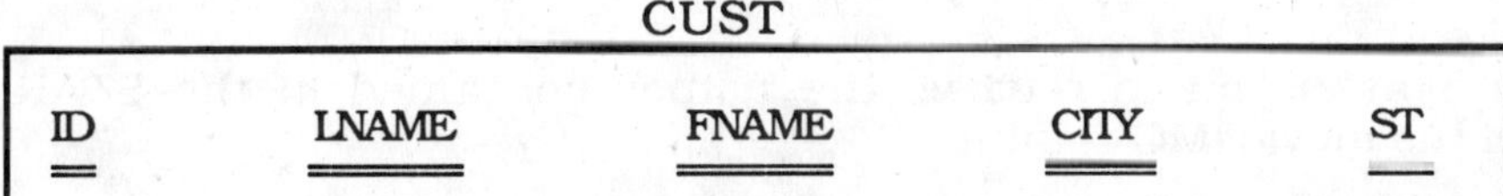

ID	LNAME	FNAME	CITY	ST

FORMAT

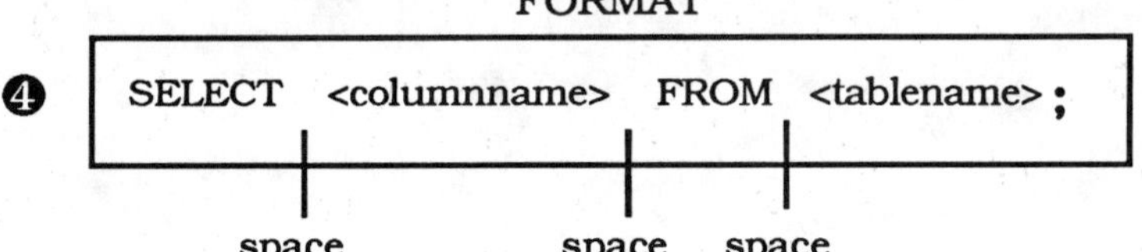

❺ STEP 1. ____

Write your answer.

____

STEP 2. Enter your answer into the computer.

If you are using a computer, complete the exercise. Otherwise, check your answer by looking at the answer key.

STEP 3. Check your results.

❻ RESULTS TABLE:

```
CITY
================
HARTFORD
FRANKLIN
HARTFORD
BOZRAH
BRATTLEBORO
PARK PLACE
GETTYSBURG
HAPPY TRAILS
PINKSVILLE
PARK PLACE
```

❼ If your screen does not look like this, try again or look at the answer key and execute the solution given there.

Exercise 4-3 asks you to retrieve the names contained in the STAR column, on the movie (MOV) table.

EXERCISE 4-3

❶ TOPIC: SELECTing Single Columns.
❷ EXERCISE 4-3. Display all the stars from the MOV table.
❸ INPUT TABLE:

MOV

NBR	TITLE	TYPE	RATING	STAR	QTY	PRICE
1	GONE WITH THE WIND	DRAMA		GABLE	4	39.95
2	FRIDAY THE 13TH	HORROR	R	JASON	2	69.95
3	TOP GUN	DRAMA	PG	CRUISE	7	49.95
4	SPLASH	COMEDY	PG13	HANKS	3	29.95
5	101 DALMATIONS	COMEDY	G		3	59.95
6	BODY HEAT	DRAMA	R	TURNER	3	19.95
7	RISKY BUSINESS	COMEDY	R	CRUISE	2	44.55
8	COCOON	SCIFI	PG	AMECHE	2	
9	CROCODILE DUNDEE	COMEDY	PG13	HARRIS	2	69.95
10	TOOTSIE	COMEDY	PG	HOFFMAN	1	29.95

❹ RESULTS TABLE:

STAR
GABLE
JASON
CRUISE
HANKS
TURNER
CRUISE
AMECHE
HARRIS
HOFFMAN

4.5 SELECT DISTINCT — ELIMINATING DUPLICATE ROWS

In Exercise 4-3, you may have noticed that CRUISE was listed twice in our results table. This kind of redundant information is often unnecessary and can clutter up our reports. SQL does not eliminate duplicate rows from the results table unless explicitly requested. The way to accomplish this is with the DISTINCT modifier.

DISTINCT eliminates duplicate rows from our results table. Its syntax is very straightforward:

```
SELECT DISTINCT <columnname> FROM <tablename>;
```

Just remember that DISTINCT comes after SELECT and before the first column name.

The next three exercises are easy to accomplish and short, so you might want to run the queries with and without DISTINCT to see its effect more clearly.

EXERCISE 4-4

❶ TOPIC: DISTINCT—Eliminating duplicate rows.
❷ EXERCISE 4-4. Display the different stars on the MOV table, and do not duplicate any rows.
❸ INPUT TABLE:

MOV

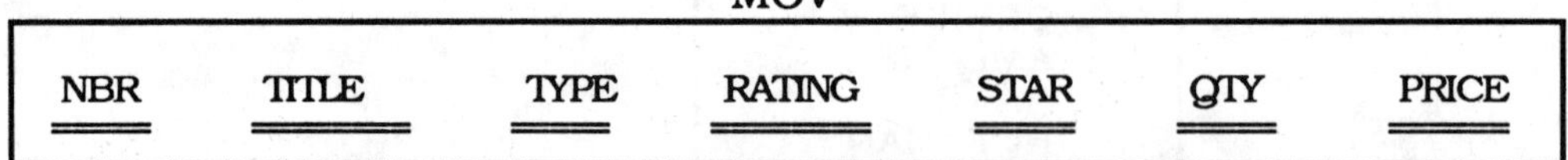

NBR	TITLE	TYPE	RATING	STAR	QTY	PRICE

FORMAT

❹
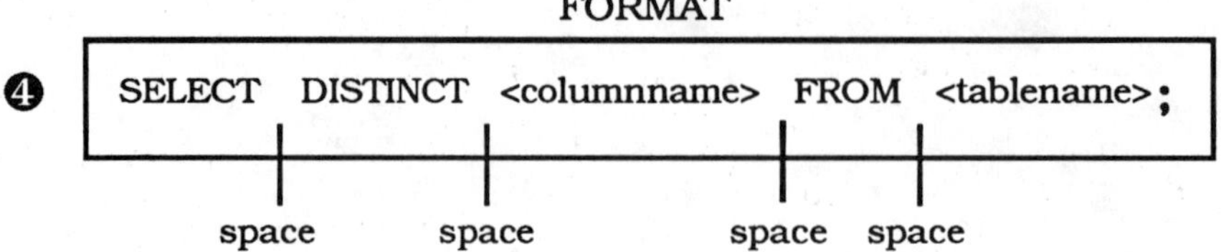

❺ STEP 1. ______________________________________
Write your answer.
____

STEP 2. Enter your answer into the computer.

If you are using a computer, complete the exercise. Otherwise, check your answer by looking at the answer key.

STEP 3. Check your results.

❻ RESULTS TABLE:

STAR
AMECHE
CRUISE
GABLE
HANKS
HARRIS
HOFFMAN
JASON
TURNER

Note that DISTINCT sorted our output here on Star's name. This is purely coincidental. Later on we will learn how to sort our Results Tables in SQL.

EXERCISE 4-5

❶ TOPIC: DISTINCT—Eliminating duplicate rows.
❷ EXERCISE 4-5. Display the different movie types on the MOV table. Remove duplicate rows from your output.
❸ INPUT TABLE:

MOV

NBR	TITLE	TYPE	RATING	STAR	QTY	PRICE

❹ RESULTS TABLE:

TYPE
COMEDY
DRAMA
HORROR
SCIFI

Note: **Mainframe users may see a results table with these rows displayed in a different order.**

4.6 SELECTING MULTIPLE COLUMNS FROM A TABLE

SELECTing multiple columns is as easy as retrieving single columns. Just type the column names after your SELECT statement (but before the FROM clause) and separate each column name with a comma. You can optionally add spaces after each comma to make your query more readable.

The columns will be displayed from left to right, in the same order in which they appear after SELECT.

Note that DISTINCT applies to all of the columns selected in the query, so that if you request more than one column, one results table row must duplicate another results row in every column before it is eliminated by DISTINCT. Exercise 4-7 will show you this more clearly.

Try these exercises. Refer to the table listings shown to get the correct spelling for the columns requested.

EXERCISE 4-6

❶ TOPIC: SELECTing Multiple Columns.

❷ EXERCISE 4-6. Display all the titles, quantities, and prices on the MOV table.

❸ INPUT TABLE:

MOV

NBR	TITLE	TYPE	RATING	STAR	QTY	PRICE

FORMAT

❹
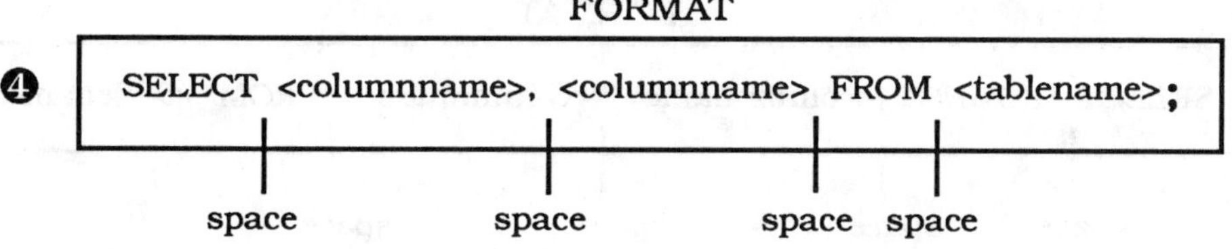

❺ STEP 1. ______________________________

Write your answer.

____

STEP 2. Enter your answer into the computer.

❻ RESULTS TABLE:

TITLE	QTY	PRICE
GONE WITH THE WIND	4	39.95
FRIDAY THE 13TH	2	69.95
TOP GUN	7	49.95
SPLASH	3	29.95
101 DALMATIONS	3	59.95
BODY HEAT	3	19.95
RISKY BUSINESS	2	44.55
COCOON	2	
CROCODILE DUNDEE	2	69.95
TOOTSIE	1	29.95

EXERCISE 4-7

❶ TOPIC: DISTINCT—Selecting multiple columns.

❷ EXERCISE 4-7. Select all of the different combinations of movie types and ratings.

❸ INPUT TABLE:

MOV

NBR	TITLE	TYPE	RATING	STAR	QTY	PRICE

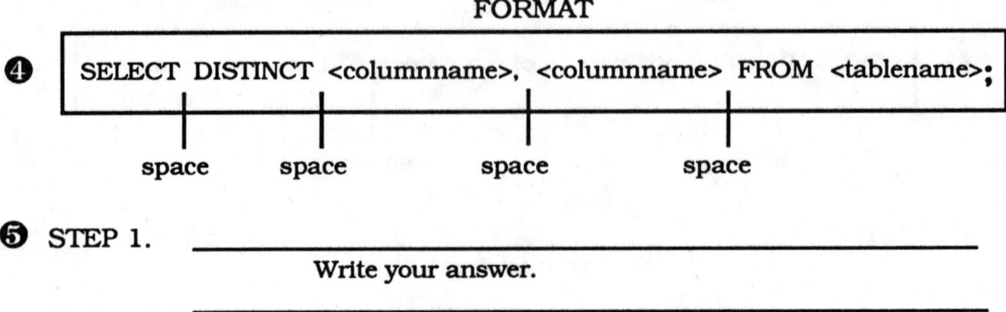

❺ STEP 1. ______________________________

Write your answer.

____

❻ RESULTS TABLE:

TYPE	RATING
DRAMA	
HORROR	R
DRAMA	PG
COMEDY	PG13
COMEDY	G
DRAMA	R
COMEDY	R
SCIFI	PG
COMEDY	PG

Note: **DISTINCT eliminated only one movie which duplicated "comedy" and "PG-13."**

4.7 SELECT * — DISPLAYING ALL COLUMNS FROM A TABLE

SELECT * is a shorthand method of telling SQL that you want all of the columns in a table displayed. Its format is very simple (see the FORMAT box below). It can save time in your SQL coding efforts.

SELECT * is convenient, and we will be using it quite a bit throughout the next few lessons to allow you to concentrate on learning new SQL concepts, and to spend less time typing.

EXERCISE 4-8

❶ TOPIC: SELECTing All Columns.

❷ EXERCISE 4-8. Display all the rows and columns in the INV table.

❸ INPUT TABLE:

INV

NBR	CUSTID	TOT	RENT	RETURN
1	01	9.55	87055	
2	02	13.55	87030	
3	02	21.01	87041	87042
4	01	11.25	87042	87072
5	03	12.55	87045	87051
6	02	9.75	87047	87051
7	04	10.35	86355	
8	04	12.55	87051	87053
9	05	10.35	87052	
10	02	7.55	87030	87060

FORMAT

❹
```
SELECT * FROM <tablename> ;
```

❺ STEP 1. ______________________________

Write your answer.

____

STEP 2. Enter your answer into the computer.

If you are using a computer, complete the exercise. Otherwise, check your answer by looking at the answer key.

STEP 3. Check your results.

❻ RESULTS TABLE:

NBR	CUSTID	TOT	RENT	RETURN
1	01	9.55	87055	
2	02	13.55	87030	
3	02	21.01	87041	87042
4	01	11.25	87042	87072
5	03	12.55	87045	87051
6	02	9.75	87047	87051
7	04	10.35	86355	
8	04	10.35	87051	87053
9	05	10.35	87052	
10	02	7.55	87030	87060

4.8 SUMMARY

So far we have covered three ways to SELECT data:

1. SELECT <columnname>, <columnname>, <columnname>, etc.
2. SELECT DISTINCT <columnname>, etc.
3. SELECT * , etc.

Next you will learn how to tell SQL which particular rows to select.

4.9 INTERPRETING SQL STATEMENTS

In these examples, try to write out an English translation, or picture the results of executing the SQL statements shown. The fourth example uses the WHERE clause, which we haven't covered yet. What do you think its effect will be on the SELECT statement?

1. SELECT DISTINCT I_NBR FROM I_M;
2. SELECT DISTINCT CUSTID, RENT FROM INV;
3. SELECT FNAME, LNAME, CITY FROM CUST;
4. SELECT * FROM MOV WHERE TYPE = 'DRAMA';

1. ______________________________

   ____

2. ______________________________

   ____

3. ______________________________

   ____

4. ______________________________

   ____

4.10 ANSWER KEY

There may be more than one answer to an exercise. However, the one given emphasizes the topic discussed.

EXERCISE 4-1	SELECT TITLE FROM MOV;
EXERCISE 4-2	SELECT CITY FROM CUST;
EXERCISE 4-3	SELECT STAR FROM MOV;
EXERCISE 4-4	SELECT DISTINCT STAR FROM MOV;
EXERCISE 4-5	SELECT DISTINCT TYPE FROM MOV;
EXERCISE 4-6	SELECT TITLE, QTY, PRICE FROM MOV;
EXERCISE 4-7	SELECT DISTINCT TYPE, RATING FROM MOV;
EXERCISE 4-8	SELECT * FROM INV;

4.10.1 Interpreting SQL statements

1. Check to see that no duplicate invoice numbers exist on the I_M Table.

2. Find out who rented movies — remove duplicates.

3. Display our customers' first and last names and their cities.

4. Display all of the rows and columns for DRAMA movies.

Chapter 5
CONDITIONAL RETRIEVAL: SELECT WHERE

Here you will learn to:

1. Select specific rows from a table;
2. Control the order in which columns are displayed;
3. Select rows that satisfy all conditions;
4. Select rows that satisfy ONE of several conditions;
5. Select rows that do not meet specific conditions.

The following topics are covered:

5.1 WHERE—CONDITIONAL SELECTION

Up to now, we've been SELECTing all of the rows (or all of the non-redundant rows) from our tables. Suppose you only wanted to see certain rows containing specific values; for example, invoices totaling over $10, or customers living in the state of Vermont, or movies with a G rating (see Figure 5.1). To do this, you would need to place a WHERE clause in your SELECT statement.

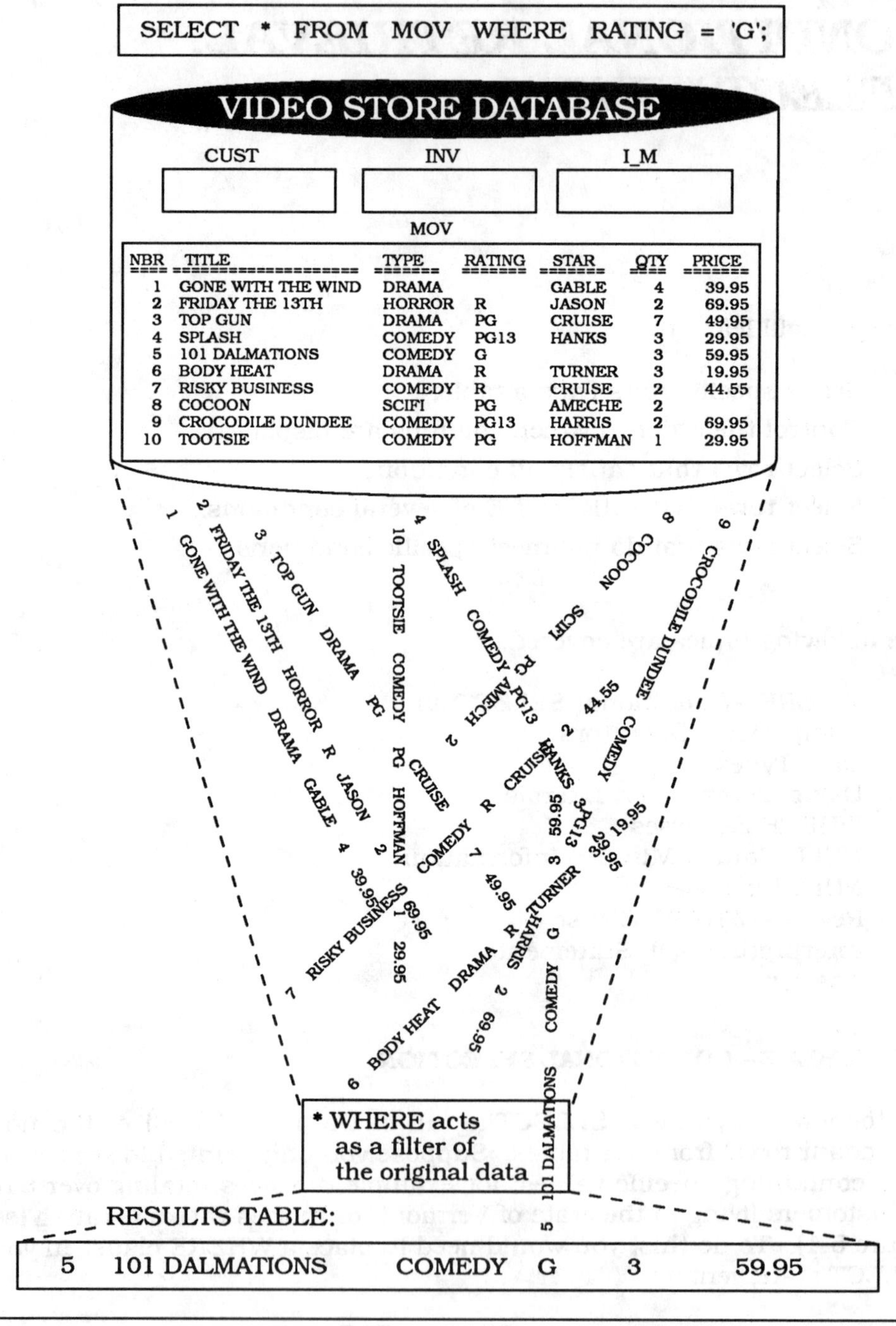

Figure 5.1. SELECTing certain rows (WHERE).

WHERE limits SQL's row SELECTion to only those rows that satisfy a condition that you specify. Figure 5.2 shows a SELECT statement with a WHERE clause in it. There are four parts to the WHERE clause:

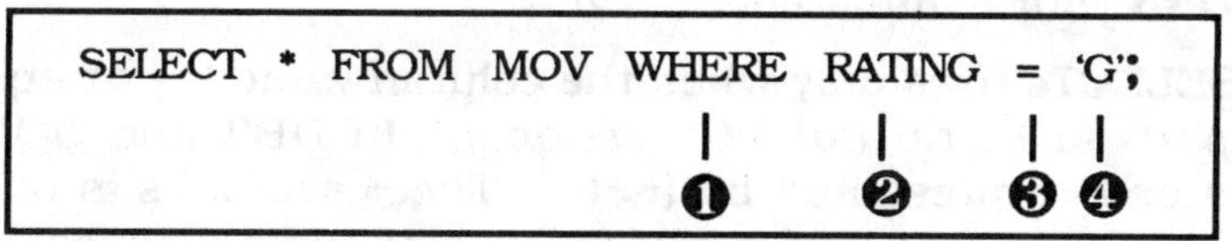

Figure 5.2. The WHERE clause.

❶ ***WHERE*** This is a SQL keyword like SELECT and FROM. It always follows "FROM <tablename>" in a SELECT statement. WHERE itself, is always followed by a condition. The condition is made up of a column name, a comparison operator, and (for now at least), a literal value.

❷ ***Columnname*** A valid column name follows WHERE.

❸ ***Comparison operator*** One of six valid SQL comparison operators follows the column name (see Section 5.2).

❹ ***Literal value*** A literal value (also called a comparison literal) follows the comparison operator. Later we shall see that another column name or an Arithmetic Expression can also be used.

In Figure 5.2, SQL uses the WHERE clause to compare the values stored in the RATING column on the MOV table with the letter 'G'. Only rows with a 'G' rating will be displayed on the results table.

Type in the statement in Figure 5.2 and execute it to see for yourself WHERE's effect on a SELECT statement. **Note the single quotation marks enclosing the literal value 'G'.**

5.2 COMPARISON OPERATORS

As is shown in Figure 5.3, SQL gives us six *comparison operators.*

1. (=) SELECTs rows only when the column value is **equal** to your comparison literal.
2. (<) SELECTs rows only when the column value is **less than** your comparison literal.
3. (>) SELECTs rows only when the column value is **greater than** your comparison literal.

4. **(<=)** SELECTs rows only when the column value is **less than or equal to** your comparison literal.
5. **(>=)** SELECTs rows only when the column value is **greater than or equal to** your comparison literal.
6. **(!=)** SELECTs rows only when the column value is **not equal** to your comparison literal (for PC versions). In DB2 and SQL/DS, this operator is represented by **(¬=)** . The ¬ symbol sits on top of the number 6 on most IBM keyboards. You can also specify not equal as < >.

Operator	Meaning
=	Equal to
<	Less than
>	Greater than
<=	Less than or equal to
>=	Greater than or equal to
¬=	Not equal to*

*NOTE: For PC versions, NOT EQUAL TO is typed as: !=

Example 1:

```
SELECT  *
  FROM  INV
    WHERE  CUSTID  =  '01';
```

Example 2:

```
SELECT  *
  FROM  INV
    WHERE  RENT  >=  87012;
```

Example 3:

```
SELECT  *
  FROM  MOV
    WHERE  RATING  ¬=  'R';
```

Figure 5.3. Comparison operators.

These operators can be used for making comparisons with both character and numeric data. Figure 5.3 shows some examples using comparison operators. Try them out by typing them in, and running them.

- Example 1 will display all the information on invoices belonging to customer ID '01'.
- Example 2 will display all the information for invoices where the rental date is equal to or later than January 12, 1987.
- Example 3 will display all the information for movies where the rating is not equal to an 'R'.

There are other combinations you can use like ¬> (not greater than) and ¬< (not less than). I don't suggest you use these because they can be confusing to read, and you can always make the same comparison more clearly another way.

5.3 DATA TYPES

Every column in a table is given a *data type* when the table is defined. These data types describe the kind of information that can be stored in a column. There are two main categories or classifications of data types: character and numeric. The literals you use to compare column values with must match the kind of data stored in the column.

A column defined with a character data type is allowed to store all types of characters including:

- **Letters** — upper and lower case;
- **Numbers** — they are perceived by the system as letters, and you are not allowed to use them in mathematical expressions computed values (we'll be covering these later on);
- **Special characters** — like (@ # $ % ¬ * & ? / : " ; \ + = < >) etc.

A column defined with a numeric data type is allowed to store only numeric data. Further, you have to tell the system how to represent your numbers:

- SMALLINT— allows only whole numbers to be stored with values less than approximately 32,700;
- INTEGER— allows only whole numbers to be stored with values less than approximately 2,500,000,000 (2 and 1/2 billion);
- DECIMAL— allows decimal numbers to be stored with up to 15 digits of numeric precision.

All numeric columns store positive and negative numbers. To represent negatives, we simply place the minus sign (–) in front of the number:

- –1
- –1.3
- –10

Look over the following table descriptions in Figure 5.4. They show the data types of the columns in our video store database. This information is necessary to correctly code your WHERE clauses for the next set of tasks.

MOV

COL NAME	TYPE OF INFORMATION STORED	COL SIZE	DATA TYPE
NBR	The movie number that uniquely identifies each movie	2	integer
TITLE	The movie's title	20	alpha/numeric character
TYPE	The general category of film type describing this movie	6	alpha/numeric character
RATING	The rating established for this movie by the rating commission	4	alpha/numeric character
STAR	The name of the star	10	alpha/numeric character
QTY	The number of copies of this movie currently that we own	2	integer
PRICE	The original buying price of the movie	up to $100	decimal

CUST

COL NAME	TYPE OF INFORMATION STORED	COL SIZE	DATA TYPE
ID	The identification number that uniquely identifies each customer	2	alpha/numeric character
LNAME	The customer's last name	12	alpha/numeric character
FNAME	The customer's first name	6	alpha/numeric character
CITY	The city in which the customer lives	15	alpha/numeric character
ST	The state in which the customer lives	2	alpha/numeric character

Figure 5.4. Video store table descriptions (*continued on next page*).

INV

COL NAME	TYPE OF INFORMATION STORED	COL SIZE	DATA TYPE
NBR	The invoice number that uniquely identifies each invoice	2	integer
CUSTID	The customer number that uniquely identifies the customer renting the movies	2	alpha/numeric character
TOT	The total amount for all of the movies rented	up to $100	decimal
RENT	The date the movies were rented (in Julian format YYDDD)	very large	integer
RETURN	The date the movies were returned (in Julian format YYDDD)	very large	integer

I_M

COL NAME	TYPE OF INFORMATION STORED	COL SIZE	DATA TYPE
I_NBR	The invoice number that uniquely identifies each invoice	2	integer
M_NBR	The movie number that uniquely identifies each movie	2	integer

Figure 5.4. *(Continued)*

5.4 USING COMPARISON LITERALS

You need to consider data types when using comparison literals in your WHERE clause:

1. *Character Comparison.* When you are comparing a literal value with a character column you will need to enclose the value in single quotation marks as shown in Figure 5.5.
2. *Numeric Comparison.* When you are comparing a literal value with a numeric column you do not enclose the value in quotation marks (see Figure 5.5).

3. *Upper/Lower Case Sensitivity.* DB2, QMF and PC database systems may not convert your comparison literal to upper case if you enter a statement in lower case. This means that the rating 'pg13' will not compare equal to 'PG13'. If you are using one of these systems, you should set your keyboard up with all capital letters (either toggle on a "caps lock" key, or "caps on" key). SQL/DS converts your WHERE clause literal to upper case automatically.

Two other things to keep in mind:

1. With character comparisons, you will have to type the literal value *Exactly* — character for character — as it is stored on the table in order for SQL to find it. For instance:

   WHERE RATING = 'PG 13'; or WHERE RATING = 'PG-13';

   will not find movies stored on the MOV table with PG13 ratings.

   WHERE RATING = 'PG13';

   will find those movies.

2. Numeric comparisons behave a little differently. To retrieve invoices with a total value (TOT) equal to $10, you may code:

   WHERE TOT = 10; or WHERE TOT = 10.00;

   Both of these WHERE clauses will return the same rows. The algebraic values have to be equivalent for successful SELECTion, but that's all.

* Comparison with character data type:

```
SELECT  *
  FROM  MOV
     WHERE  RATING  =  'PG13';
```

— Character literal embedded between single quotes (' ').

* Comparison with numeric data type (whole numbers):

```
SELECT  *
  FROM  INV
     WHERE  RENT  >=  87200;
```

— Constant NOT embedded between single quotes.
— Any numbers (0–9) and a (+ or -) sign may be used.

```
SELECT  *
  FROM  MOV
     WHERE  RATING  ¬=  'R';
```

Figure 5.5. Using comparison literals.

5.5 WHERE EXERCISES

In all of these exercises, use SELECT * as opposed to typing in the different column names (we are learning how to conditionally SELECT, not how to type).

Exercise 5-1 will use the greater than (>) operator. Don't be tempted to type a dollar sign — that is not considered numeric data by SQL.

EXERCISE 5-1

❶ TOPIC: WHERE exercises — SELECTing Multiple Columns using search conditions.

❷ EXERCISE 5-1. Display all the information from the invoice table for movie rentals totaling over $10.00.

❸ INPUT TABLE:

INV

NBR	CUSTID	TOT	RENT	RETURN

FORMAT

❹
```
SELECT * FROM <tablename>
    WHERE <columnname> <operator> <value>;
```

❺ STEP 1. ____________________

Write your answer.

____

STEP 2. Enter your answer into the computer.

❻ RESULTS TABLE:

NBR	CUSTID	TOT	RENT	RETURN
2	02	13.55	87030	
3	02	21.01	87041	87042
4	01	11.25	87042	87072
5	03	12.55	87045	87051
7	04	10.35	86355	
8	04	12.55	87051	87053
9	05	10.35	87052	

EXERCISE 5-2

Here you will use the movie table and do a character comparison, looking for comedies (refer to the actual table information in the MOV table to see how the value "COMEDY" is stored in the TYPE column).

❶ TOPIC: WHERE exercises — SELECTing Multiple Columns using search conditions.

❷ EXERCISE 5-2. Display all of the information about comedies on the MOV table.

❸ INPUT TABLE:

MOV

NBR	TITLE	TYPE	RATING	STAR	QTY	PRICE

❹ RESULTS TABLE:

NBR	TITLE	TYPE	RATING	STAR	QTY	PRICE
4	SPLASH	COMEDY	PG13	HANKS	3	29.95
5	101 DALMATIONS	COMEDY	G		3	59.95
7	RISKY BUSINESS	COMEDY	R	CRUISE	2	44.55
9	CROCODILE DUNDEE	COMEDY	PG13	HARRIS	2	69.95
10	TOOTSIE	COMEDY	PG	HOFFMAN	1	29.95

Hint: **Don't forget the single quotation marks, and the upper/lower case sensitivity.**

EXERCISE 5-3

Here you should use the not equal symbol (¬= on mainframe systems and != on PCs).

❶ TOPIC: WHERE exercises — SELECTing Multiple Columns using search conditions.

❷ EXERCISE 5-3. Display all of the information on movies that are not 'R' rated.

❸ INPUT TABLE:

MOV

NBR	TITLE	TYPE	RATING	STAR	QTY	PRICE

❹ STEP 1. ______________________________
Write your answer.
____

STEP 2. Enter your answer into the computer.

❺ RESULTS TABLE:

NBR	TITLE	TYPE	RATING	STAR	QTY	PRICE
3	TOP GUN	DRAMA	PG	CRUISE	7	49.95
4	SPLASH	COMEDY	PG13	HANKS	3	29.95
5	101 DALMATIONS	COMEDY	G		3	59.95
8	COCOON	SCIFI	PG	AMECHE	2	
9	CROCODILE DUNDEE	COMEDY	PG13	HARRIS	2	69.95
10	TOOTSIE	COMEDY	PG	HOFFMAN	1	29.95

5.6 NULL VALUES — MISSING INFORMATION

NULL values represent another data type in SQL. They signify the absence of any value (in other words, missing information), for a particular field in the table.

NULLs are displayed as dashes (-) in DB2 and QMF, as Blanks ( ) in SQLBase, and as question marks (?) in SQL/DS. If you are using DB2, QMF, or SQL/DS, you might have noticed that the results tables in this workbook show NULLs as blanks. That's because I've been using SQLBase to display my results.

NULL values are not the same as zeros, low-values, spaces, or any other traditional data processing values, and they are treated specially by SQL. Generally, fields that contain NULL values are not processed by SQL, they are ignored. Remember the last example where we wanted to retrieve all of the movies without an R rating? Were you wondering why "Gone With the Wind" wasn't SELECTed? This was because its rating was a NULL value (it had no rating value), and was ignored by SQL in its search. Fields that contain NULL values will only satisfy a special WHERE condition clause containing the keyword NULL.

You retrieve rows searching for NULL values in a column by specifying the phrase WHERE <columnname> IS NULL. No other syntax is allowed (such as WHERE <columnname> = NULL, or WHERE <columnname> NULL, etc.). To retrieve movies with unknown stars (i.e., 101 Dalmations), you would code:

```
SELECT * FROM MOV WHERE STAR IS NULL;
```

At this point, you're probably wondering why NULL values exist at all. Why not just fill empty fields with zeros, or spaces? The answer is that SQL uses this special data type in order to correctly process statistical queries using SQL functions such as average (AVG), or to find the highest and lowest values in a column (MAX, MIN).

5.7 NULL EXERCISES

EXERCISE 5-4

This task asks you to retrieve "Gone With the Wind" from the MOV table by requesting movies with unknown ratings.

❶ TOPIC: NULL exercises — Selecting all Columns — NULL value conditions.

❷ EXERCISE 5-4. Display all rows for movies with unknown ratings (where the RATING is a NULL value).

❸ INPUT TABLE:

MOV

NBR	TITLE	TYPE	RATING	STAR	QTY	PRICE

FORMAT

❹

```
SELECT * FROM <tablename>
    WHERE <columnname>   IS NULL;
```

❺ STEP 1. ______________________________

Write your answer.

____

STEP 2. Enter your answer into the computer.

❻ RESULTS TABLE:

NBR	TITLE	TYPE	RATING	STAR	QTY	PRICE
1	GONE WITH THE WIND	DRAMA		GABLE	4	39.95

EXERCISE 5-5

❶ TOPIC: Character Comparison — more practice exercises.

❷ EXERCISE 5-5. Display all the information for customers living in TEXAS.

❸ INPUT TABLE:

CUST

ID	LNAME	FNAME	CITY	ST

FORMAT

❹
```
SELECT * FROM <tablename>
     WHERE <columnname> <operator> <value>;
```

❺ STEP 1. ______________________________

Write your answer.

____

STEP 2. Enter your answer into the computer.

❻ RESULTS TABLE:

ID	LNAME	FNAME	CITY	ST
08	ROGERS	ROY	HAPPY TRAILS	TX

❼ If your screen does not look like this, try again or look at the answer key and execute the solution given there.

EXERCISE 5-6

❶ TOPIC: No rows selected.
❷ EXERCISE 5-6. Display all the I_M table's rows for movie number 8.
❸ INPUT TABLE:

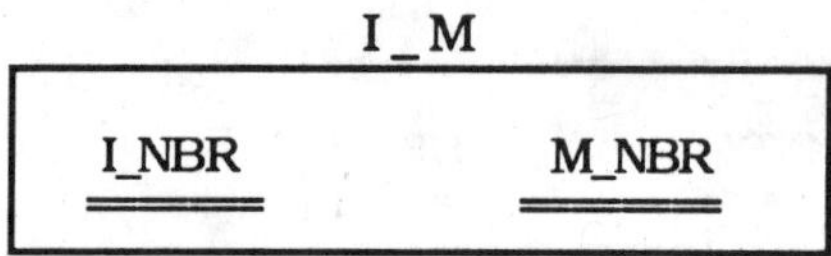

❹ RESULTS TABLE:

There are no rows selected.

5.8 REVIEW — WHERE CLAUSE

Character comparison. When you are comparing a literal value with a character column, you will need to enclose the value in single quotation marks.

Numeric Comparison. When you are comparing a literal value with a numeric column, you may not enclose the value in quotation marks.

Upper/Lower Case Sensitivity. DB2, QMF, and SQLBase may not convert your comparison literal to upper case if you enter a statement in lower case. This means that the rating 'pg13' will not compare as equal to 'PG13'. If you are using one of these systems, you should set your keyboard up with all capital letters (either toggle on a "caps lock" key, or "caps on" key). SQL/DS converts your WHERE clause literal to upper case automatically.

Two other things to keep in mind:

1. With character comparisons, you will have to type the literal value *exactly* — character for character — as it is stored on the table, in order for SQL to find it. For instance:

   WHERE RATING = 'PG 13'; or WHERE RATING = 'PG-13';

   will not find movies stored on the MOV table with PG13 ratings.

   WHERE RATING = 'PG13';

   will find those movies.

2. Numeric comparisons behave a little differently. To retrieve invoices with a total value (TOT) equal to $10, you can use

   WHERE TOT = 10; or WHERE TOT = 10.00;

5.9 INTERPRETING SQL STATEMENTS

In these examples, try to write an English translation, or to picture the results of executing the SQL statements shown. If you want to see how you did, type in the examples and try running them.

1. SELECT LNAME FROM CUST WHERE ID = '03';
2. SELECT RETURN FROM INV WHERE RENT = 87010;
3. SELECT STAR, TITLE, PRICE FROM MOV WHERE STAR != 'CRUISE';
4. SELECT LNAME FROM CUST WHERE ST = 'VT' OR ST = 'NY';

5.10 ANSWER KEY

There may be more than one answer to an exercise. However, the one given emphasizes the topic discussed.

```
EXERCISE 5-1    SELECT * FROM INV WHERE TOT > 10;
EXERCISE 5-2    SELECT * FROM MOV WHERE TYPE = 'COMEDY';
EXERCISE 5-3    SELECT * FROM MOV WHERE RATING != 'R';
EXERCISE 5-4    SELECT * FROM MOV WHERE RATING IS NULL;
EXERCISE 5-5    SELECT * FROM CUST WHERE ST = 'TX';
EXERCISE 5-6    SELECT * FROM I_M WHERE M_NBR = 8;
```

Chapter 6
SORTING THE RESULTS TABLE: ORDER BY

Now you will learn how to sort the rows of the results table, in ascending or descending order, using multiple sort criteria.

The following topics are covered:

6.1 ORDER BY — SIMPLE DATA ARRANGEMENT

In general, the output rows from your query (the results table), are not guaranteed to be in any particular order. This may not be the most informative way to present your data. You can control the display order of the SELECTed rows by adding an ORDER BY clause to your SELECT statement. ORDER BY will sort the rows in your results table on the column values you specify.

You have two options when coding ORDER BY:

1. ORDER BY <columnname> ASC — Sort and display in ascending order; (This is the default arrangement — which means that SQL will automatically order from lowest to highest unless you specify DESC.)

2. ORDER BY <columnname> DESC — Sort and display in descending order; (To do this, you must explicitly code DESC after the column name.)

You may also sort rows on multiple columns; this permits a major to minor sort ordering to be established (simply specify ORDER BY MAJCOL, MINCOL1, MINCOL2, etc., making sure to separate the columnnames with commas). For example, you may want to sort by TYPE within PRICE. PRICE is the major column. The command would be:

```
SELECT * FROM MOV ORDER BY PRICE, TYPE;
```

Figure 6.1 presents three ORDER BY examples:

1. Orders the customer table on the last name column in ascending sequence;

2. Orders the customer table on the state column in descending sequence (notice that when your SELECT statement has no WHERE clause, ORDER BY follows the FROM <tablename>. When your query has a WHERE clause, ORDER BY follows the WHERE condition.);

3. Orders the customer table on the last name/first name columns in ascending sequence (notice that Fred and Roy Rogers are presented in reverse sequence. This is because of the minor sort ordering on FNAME.)

1. SELECT * FROM CUST ORDER BY LNAME ;

ID	LNAME	FNAME	CITY	ST
01	DANGERFIELD	RODNEY	HARTFORD	CT
02	FIELD	SALLY	FRANKLIN	NY
05	MADDEN	JOHN	BRATTLEBORO	VT
04	MURRAY	FRED	BOZRAH	CT
03	NICHOLSON	JACK	HARTFORD	CT
09	RINGWALD	MOLLY	PINKSVILLE	CA
08	ROGERS	ROY	HAPPY TRAILS	TX
10	ROGERS	FRED	PARK PLACE	CA
06	WEST	MAE	PARK PLACE	CA
07	WOODWARD	JOANNE	GETTYSBURG	PA

2. SELECT * FROM CUST ORDER BY ST DESC ;

ID	LNAME	FNAME	CITY	ST
05	MADDEN	JOHN	BRATTLEBORO	VT
08	ROGERS	ROY	HAPPY TRAILS	TX
07	WOODWARD	JOANNE	GETTYSBURG	PA
02	FIELD	SALLY	FRANKLIN	NY
01	DANGERFIELD	RODNEY	HARTFORD	CT
04	MURRAY	FRED	BOZRAH	CT
06	WEST	MAE	PARK PLACE	CA
09	RINGWALD	MOLLY	PINKSVILLE	CA
03	NICHOLSON	JACK	HARTFORD	CT
10	ROGERS	FRED	PARK PLACE	CA

3. SELECT * FROM CUST ORDER BY LNAME, FNAME ;

ID	LNAME	FNAME	CITY	ST
01	DANGERFIELD	RODNEY	HARTFORD	CT
02	FIELD	SALLY	FRANKLIN	NY
05	MADDEN	JOHN	BRATTLEBORO	VT
04	MURRAY	FRED	BOZRAH	CT
03	NICHOLSON	JACK	HARTFORD	CT
09	RINGWALD	MOLLY	PINKSVILLE	CA
10	ROGERS	FRED	PARK PLACE	CA
08	ROGERS	ROY	HAPPY TRAILS	TX
06	WEST	MAE	PARK PLACE	CA
07	WOODWARD	JOANNE	GETTYSBURG	PA

Figure 6.1. ORDER BY examples.

6.2 ORDER BY—SORTING SEQUENCE

ORDER BY will sort your rows based on what is known as the collating sequence (sorting sequence) of the computer hardware you are using. There are two standard collating sequences: EBCDIC, and ASCII. Mainframe computers use EBCDIC* collating sequence, and personal computers use ASCII** .

If you are using a mainframe system, when you sort on a character data column the collating sequence is as follows:

Lowest — Special characters (blanks, *&^%#!@#^& etc.)

Higher — Letters (the alphabet)

Highest — Numbers (the integers 0 thru 9)

Numeric column data will sort in algebraic value order.
NULL values collate higher than everything else (all of the above).

If you are using a personal computer, when you sort on a character data column, the collating sequence is as follows:

Highest — Other special characters

Higher — Letters (the alphabet)

High— Numbers (the integers 0 thru 9)

Lowest — Some Special characters

Numeric column data will sort in algebraic value order.
NULL values collate lower than everything else (all of the above).

Figure 6.2 demonstrates the difference in the way that mainframe and PC systems sort the same data. We will be presenting our results table examples based on PC system (ASCII) collating sequence.

---

* EBCDIC stands for Extended Binary Code Decimal Interchange Code — aren't you glad you asked?

** ASCII stands for American Standard Code for Information Interchange.

MAINFRAME COLLATING SEQUENCE

SQL > SELECT * FROM INV ORDER BY TITLE;

NBR	TITLE	TYPE	RATING	STAR	QTY	PRICE
6	BODY HEAT	DRAMA	R	TURNER	3	19.95
8	COCOON	SCIFI	PG	AMECHE	2	
9	CROCODILE DUNDEE	COMEDY	PG13	HARRIS	2	69.95
2	FRIDAY THE 13TH	HORROR	R	JASON	2	69.95
1	GONE WITH THE WIND	DRAMA		GABLE	4	39.95
7	RISKY BUSINESS	COMEDY	R	CRUISE	2	44.55
4	SPLASH	COMEDY	PG13	HANKS	3	29.95
10	TOOTSIE	COMEDY	PG	HOFFMAN	1	29.95
3	TOP GUN	DRAMA	PG	CRUISE	7	49.95
5	101 DALMATIONS	COMEDY	G		3	59.95

PC COLLATING SEQUENCE

SQL > SELECT * FROM INV ORDER BY TITLE;

NBR	TITLE	TYPE	RATING	STAR	QTY	PRICE
5	101 DALMATIONS	COMEDY	G		3	59.95
6	BODY HEAT	DRAMA	R	TURNER	3	19.95
8	COCOON	SCIFI	PG	AMECHE	2	
9	CROCODILE DUNDEE	COMEDY	PG13	HARRIS	2	69.95
2	FRIDAY THE 13TH	HORROR	R	JASON	2	69.95
1	GONE WITH THE WIND	DRAMA		GABLE	4	39.95
7	RISKY BUSINESS	COMEDY	R	CRUISE	2	44.55
4	SPLASH	COMEDY	PG13	HANKS	3	29.95
10	TOOTSIE	COMEDY	PG	HOFFMAN	1	29.95
3	TOP GUN	DRAMA	PG	CRUISE	7	49.95

Figure 6.2. Mainframe versus PC collating sequence.

(Continued on next page)

MAINFRAME COLLATING SEQUENCE — NULL VALUES

SQL > SELECT * FROM MOV ORDER BY RETURN;

NBR	CUSTID	TOT	RENT	RETURN
3	02	21.01	87041	87042
5	03	12.55	87045	87051
6	02	9.75	87047	87051
8	04	12.55	87051	87053
10	02	7.55	87030	87060
4	01	11.25	87042	87072
1	01	9.55	87055	
2	02	13.55	87030	
7	04	10.35	86355	
9	05	10.35	87052	

PC COLLATING SEQUENCE — NULL VALUES

SQL > SELECT * FROM MOV ORDER BY RETURN;

NBR	CUSTID	TOT	RENT	RETURN
1	01	9.55	87055	
2	02	13.55	87030	
7	04	10.35	86355	
9	05	10.35	87052	
3	02	21.01	87041	87042
5	03	12.55	87045	87051
6	02	9.75	87047	87051
8	04	12.55	87051	87053
10	02	7.55	87030	87060
4	01	11.25	87042	87072

Figure 6.2. *(Continued)*

6.3 ORDER BY—STATEMENT FORMAT

There are two different ways to specify the ORDER BY clause:

1. by column name
2. by the relative columnnbr (its position within the SELECT clause)

1. ORDERing BY columnname

```
SELECT NBR, QTY FROM TABLE ORDER BY NBR, QTY DESC;
```

2. ORDERing BY columnnbr

You "ORDER BY columnnbr" simply by specifying the relative position your sort column is in after the SELECT verb.

```
SELECT NBR, QTY FROM TABLE ORDER BY 1,2 DESC;
```

This relative ordering can be used anytime, but it is necessary when you want to order by things like calculated values and merged rows — these will be covered later.

6.3.1 ASC/DESC—Reminder

SQL assumes a default sort order of ASCending if you don't explicitly code DESC. This means you can leave off the ASC specifier if you want your rows presented in ascending order.

6.4 ORDER BY EXERCISES

In the exercises for ORDER BY, you can assume two things:

1. Unless you specifically stated DESC for the sort order, the default is ASC.
2. The results tables will be sorted in ASCII collating sequence. **If you are using DB2, QMF, or SQL/DS you may get a different sort output result when sorting on character columns that mix numbers and letters. This may be something to consider in a cooperative processing environment, where the same query may be run on a PC, mini-computer and mainframe. The *same* query executed against the *same* data will return an entirely different report!**

EXERCISE 6-1

❶ TOPIC: ORDER BY.

❷ EXERCISE 6-1. Display a list of all invoices sorted by TOT DESCending.

❸ INPUT TABLE:

INV

NBR	CUSTID	TOT	RENT	RETURN

FORMAT

❹

```
SELECT * FROM <tablename> WHERE <condition>
ORDER BY <the column to sequence> <ASC/DESC>;
```

The answer to this one is

```
SELECT * FROM INV ORDER BY TOT DESC;
```

❺ STEP 1. ____

Write your answer.

____

STEP 2. Enter your answer into the computer.

❻ RESULTS TABLE:

NBR	CUSTID	TOT	RENT	RETURN
3	02	21.01	87041	87042
2	02	13.55	87030	
5	03	12.55	87045	87051
8	04	12.55	87051	87053
4	01	11.25	87042	87072
7	04	10.35	86355	
9	05	10.35	87052	
6	02	9.75	87047	87051
1	01	9.55	87055	
10	02	7.55	87030	87060

EXERCISE 6-2

❶ TOPIC: ORDER BY.

❷ EXERCISE 6-2. Display a list of all movies, with a PRICE over $20, sorted by PRICE.

❸ INPUT TABLE:

MOV

NBR	TITLE	TYPE	RATING	STAR	QTY	PRICE

❹ RESULTS TABLE:

NBR	TITLE	TYPE	RATING	STAR	QTY	PRICE
4	SPLASH	COMEDY	PG13	HANKS	3	29.95
10	TOOTSIE	COMEDY	PG	HOFFMAN	1	29.95
1	GONE WITH THE WIND	DRAMA		GABLE	4	39.95
7	RISKY BUSINESS	COMEDY	R	CRUISE	2	44.55
3	TOP GUN	DRAMA	PG	CRUISE	7	49.95
5	101 DALMATIONS	COMEDY	G		3	59.95
2	FRIDAY THE 13TH	HORROR	R	JASON	2	69.95
9	CROCODILE DUNDEE	COMEDY	PG13	HARRIS	2	69.95

EXERCISE 6-3

❶ TOPIC: ORDER BY.

❷ EXERCISE 6-3. Display a list of all the movies sorted by RATING within TYPE.

❸ INPUT TABLE:

MOV

NBR	TITLE	TYPE	RATING	STAR	QTY	PRICE
1	GONE WITH THE WIND	DRAMA		GABLE	4	39.95
2	FRIDAY THE 13TH	HORROR	R	JASON	2	69.95
3	TOP GUN	DRAMA	PG	CRUISE	7	49.95
4	SPLASH	COMEDY	PG13	HANKS	3	29.95
5	101 DALMATIONS	COMEDY	G		3	59.95
6	BODY HEAT	DRAMA	R	TURNER	3	19.95
7	RISKY BUSINESS	COMEDY	R	CRUISE	2	44.55
8	COCOON	SCIFI	PG	AMECHE	2	
9	CROCODILE DUNDEE	COMEDY	PG13	HARRIS	2	69.95
10	TOOTSIE	COMEDY	PG	HOFFMAN	1	29.95

FORMAT

❹

```
SELECT * FROM <tablename>
WHERE <condition>
ORDER BY <majcol, mincol>   <ASC/DESC>;
```

❺ STEP 1. ______________________________

Write your answer.

____

STEP 2. Enter your answer into the computer.

❻ RESULTS TABLE:

NBR	TITLE	TYPE	RATING	STAR	QTY	PRICE
5	101 DALMATIONS	COMEDY	G		3	59.95
10	TOOTSIE	COMEDY	PG	HOFFMAN	1	29.95
4	SPLASH	COMEDY	PG13	HANKS	3	29.95
9	CROCODILE DUNDEE	COMEDY	PG13	HARRIS	2	69.95
7	RISKY BUSINESS	COMEDY	R	CRUISE	2	44.55
1	GONE WITH THE WIND	DRAMA		GABLE	4	39.95
3	TOP GUN	DRAMA	PG	CRUISE	2	49.95
6	BODY HEAT	DRAMA	R	TURNER	3	19.95
2	FRIDAY THE 13TH	HORROR	R	JASON	2	69.95
8	COCOON	SCIFI	PG	AMECHE	2	

***Note:* The NULL value in the rating field collates lower than PG within the DRAMA type on PC systems.**

EXERCISE 6-4

❶ TOPIC: ORDER BY.

❷ EXERCISE 6-4. Display all the information on customers, sort by city within state.

❸ INPUT TABLE:

CUST

ID	LNAME	FNAME	CITY	ST

❹ RESULTS TABLE:

ID	LNAME	FNAME	CITY	ST
03	NICHOLSON	JACK	HARTFORD	CT
06	WEST	MAE	PARK PLACE	CA
10	ROGERS	FRED	PARK PLACE	CA
09	RINGWALD	MOLLY	PINKSVILLE	CA
04	MURRAY	FRED	BOZRAH	CT
01	DANGERFIELD	RODNEY	HARTFORD	CT
02	FIELD	SALLY	FRANKLIN	NY
07	WOODWARD	JOANNE	GETTYSBURG	PA
08	ROGERS	ROY	HAPPY TRAILS	TX
05	MADDEN	JOHN	BRATTLEBORO	VT

6.5 ORDER BY—INTERPRETING SQL

Write out an English translation for the following SELECT statements. Try running the examples shown if you would like to see the results on your computer.

```
1.   SELECT * FROM CUST ORDER BY ST, CITY, LNAME DESC;
2.   SELECT TITLE, TYPE, RATING FROM MOV
          WHERE TYPE = 'DRAMA' ORDER BY RATING;
```

Use this space to write the answer.

1. ____

   ____

2. ____

   ____

6.6 ANSWER KEY

There may be more than one answer to an exercise. However, this one given emphasizes the topic discussed.

EXERCISE 6-1	SELECT * FROM INV ORDER BY TOT DESC;
EXERCISE 6-2	SELECT * FROM MOV WHERE PRICE > 20 ORDER BY PRICE;
EXERCISE 6-3	SELECT * FROM MOV ORDER BY TYPE, RATING;
EXERCISE 6-4	SELECT * FROM CUST ORDER BY ST, CITY;

6.6.1. Interpreting SQL Statements

1. Create a customer report sorted on Last name descending within City ascending within State.

2. Display a List of our DRAMAS (title, type, and rating only) selected by Rating ascending.

Chapter 7
COMBINING SEARCH CONDITIONS: AND, OR, NOT

Here you will learn how to combine search conditions connecting them using AND and OR and negating them using NOT.

The following topics will be covered:

7.1 AND AND OR — MULTIPLE SEARCH CONDITIONS

At times, it becomes necessary to specify more than one search criteria in a WHERE clause to retrieve the necessary rows from a table. For instance, suppose you want to see all of the invoice data for invoice numbers 1 and 5. To satisfy this request you could run two separate queries, or you could simply combine the two conditions in a single WHERE clause, like this:

```
SELECT * FROM INV WHERE NBR = 1 OR NBR = 5;
```

This SELECT statement retrieves the invoice data for invoices 1 and 5 — go ahead and try it.

You use AND and OR connectors when a single search condition is not sufficient to pinpoint the desired rows.

7.1.1 OR

OR is used when we want to retrieve rows that satisfy one of several conditions. OR means that if either search condition is true, then the row is to be selected. In Figure 7.1, we can see that when OR is used to connect conditions, the effect is to double the opportunities for SQL to SELECT rows.

```
SELECT * FROM MOV WHERE TYPE = 'COMEDY' OR RATING = 'R';
```

NBR	TITLE	TYPE	RATING	STAR	QTY	PRICE
2	FRIDAY THE 13TH	HORROR	R	JASON	2	69.95
4	SPLASH	COMEDY	PG13	HANKS	3	29.95
5	101 DALMATIONS	COMEDY	G		3	59.95
6	BODY HEAT	DRAMA	R	TURNER	3	19.95
7	RISKY BUSINESS	COMEDY	R	CRUISE	2	44.55
9	CROCODILE DUNDEE	COMEDY	PG13	HARRIS	2	69.95
10	TOOTSIE	COMEDY	PG	HOFFMAN	1	29.95

Figure 7.1. OR — Multiple search conditions.

7.1.2 AND

AND is used when we want to retrieve only rows that satisfy all of the conditions in our WHERE clause. AND means that both search conditions must be true for a given row to be retrieved. In Figure 7.2, we can see that when AND is used to connect conditions, the effect is to apply more stringent requirements on row SELECTion.

```
SELECT * FROM MOV WHERE TYPE = 'COMEDY' AND RATING = 'R';
```

NBR	TITLE	TYPE	RATING	STAR	QTY	PRICE
7	RISKY BUSINESS	COMEDY	R	CRUISE	2	44.55

Figure 7.2. AND — Multiple search conditions.

We'll begin learning how to combine multiple search conditions with the AND connector.

7.1.3 Multiple Search Condition — AND

You use AND to filter out layers of rows in a search condition, and really pinpoint what you are after. This is shown in Figure 7.3.

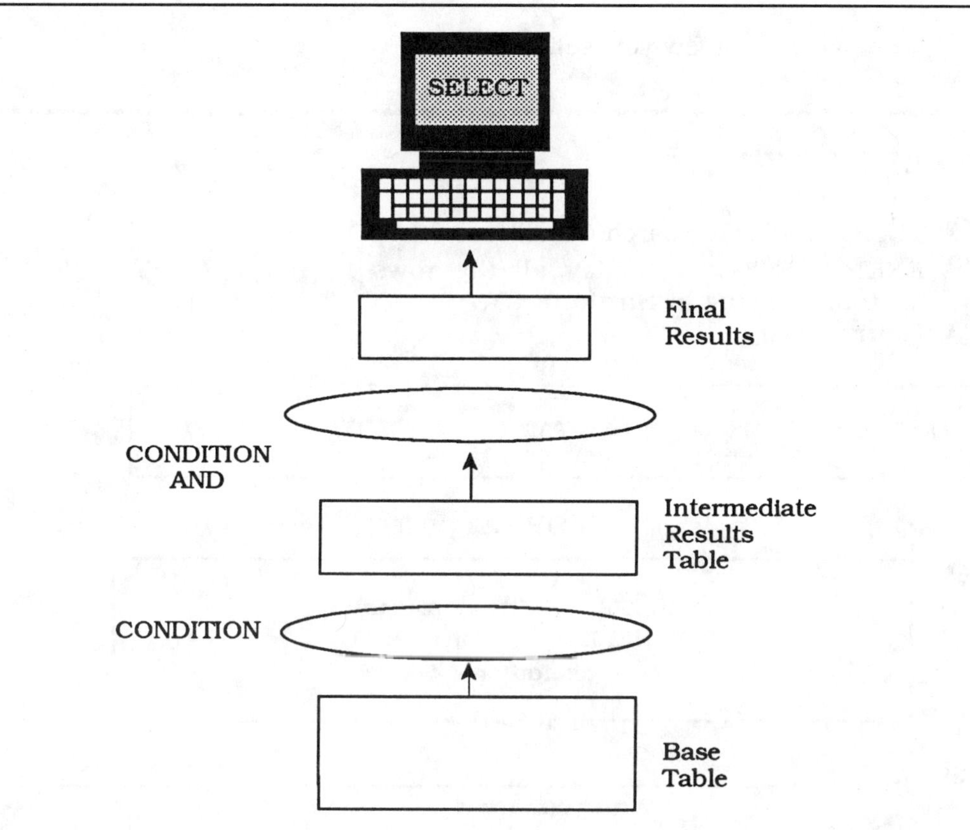

Figure 7.3. AND — Filters data.

There is one thing to look out for when you are coding an AND clause. Because AND decreases the opportunities for row SELECTion, you can end up with too few rows, or no rows in your results table. If this is what you are after, fine, but be careful about coding a WHERE clause such as:

WHERE NBR = 1 AND NBR = 5;

This will return zero rows, because the NBR column cannot contain both a 1 AND 5 for any single row. You would use the OR connector to retrieve rows for NBRs 1 and 5.

Ready to try a few yourself?

EXERCISE 7-1

❶ TOPIC: Multiple Search Conditions — AND.

❷ EXERCISE 7-1. Display all the rows from the CUST table for customers living in Hartford, CT.

❸ INPUT TABLE:

CUST

ID	LNAME	FNAME	CITY	ST

FORMAT

❹

```
SELECT * FROM <tablename>
WHERE <condition1> AND
<conditionN>;
```

❺ STEP 1. ______________________________

Write your answer.

____

STEP 2. Enter your answer into the computer.

If you are using a computer, complete the exercise. Otherwise, check your answer by looking at the answer key.

STEP 3. Check your results.

❻ RESULTS TABLE:

ID	LNAME	FNAME	CITY	ST
01	DANGERFIELD	RODNEY	HARTFORD	CT

EXERCISE 7-2

This exercise asks you to get invoice information on invoices where the total amount is greater than $10, and the return date has not been filled in (IS NULL).

❶ TOPIC: Using Multiple Search Conditions.
❷ EXERCISE 7-2. Display all of the information on INVoices with TOT amounts over $10.00, that have not been returned (RETURN contains NULL values).
❸ INPUT TABLE:

INV

NBR	CUSTID	TOT	RENT	RETURN

❹ RESULTS TABLE:

NBR	CUSTID	TOT	RENT	RETURN
2	02	13.55	87030	
7	04	10.35	86355	
9	05	10.35	87052	

7.1.4 Multiple Search Condition — OR

While AND is used to curtail SQL's row SELECTion, OR is used to increase chances for row SELECTion. With OR, if either or both conditions are met, the row is SELECTed.

An important point that should be mentioned about OR syntax is that you must specify the full condition for all conditions connected by OR. For example, you might think that SQL could interpret a request to retrieve information on movie numbers 1 and 4 such as:

```
SELECT * FROM MOV WHERE NBR = 1 OR 4;
```

This is not so. You will be given a syntax error if you try to execute this statement. To retrieve information on movie numbers 1 and 4, you must code the WHERE clause like this:

```
SELECT * FROM MOV WHERE NBR = 1 OR NBR = 4;
```

Ready to try a few yourself?

EXERCISE 7-3

This exercise asks you to get all of our R-rated movies AND any drama. Notice the semantic problem we have here. In English, when we use the connector AND, it sometimes translates into a SQL OR. The importance of clear communication in business programming specifications can't be overstated when you start working with SQL.

❶ TOPIC: Using Multiple Search Conditions.

❷ EXERCISE 7-3. Display a list of all the 'R' rated movies and all of the DRAMAs.

❸ INPUT TABLE:

MOV

NBR	TITLE	TYPE	RATING	STAR	QTY	PRICE

FORMAT

❹

```
SELECT * FROM <tablename>
WHERE <condition1> OR
<conditionN>;
```

❺ STEP 1. ______________________________

Write your answer.

____

STEP 2. Enter your answer into the computer.

❻ RESULTS TABLE:

NBR	TITLE	TYPE	RATING	STAR	QTY	PRICE
1	GONE WITH THE WIND	DRAMA		GABLE	4	39.95
2	FRIDAY THE 13TH	HORROR	R	JASON	2	69.95
3	TOP GUN	DRAMA	PG	CRUISE	7	49.95
6	BODY HEAT	DRAMA	R	TURNER	3	19.95
7	RISKY BUSINESS	COMEDY	R	CRUISE	2	44.55

EXERCISE 7-4

❶ TOPIC: Using Multiple Search Conditions — A problem with double negatives.

❷ EXERCISE 7-4. List all of the movies that are not 'R' or 'G' rated (RATING).

❸ INPUT TABLE:

MOV

NBR	TITLE	TYPE	RATING	STAR	QTY	PRICE

FORMAT

❹

```
SELECT * FROM <tablename>
WHERE <condition1> OR/AND
<conditionN>;
```

❺ STEP 1. ______________________________
Write your answer.
____

STEP 2. Enter your answer into the computer.

❻ RESULTS TABLE:

NBR	TITLE	TYPE	RATING	STAR	QTY	PRICE
3	TOP GUN	DRAMA	PG	CRUISE	7	49.95
4	SPLASH	COMEDY	PG13	HANKS	3	29.95
8	COCOON	SCIFI	PG	AMECHE	2	
9	CROCODILE DUNDEE	COMEDY	PG13	HARRIS	2	69.95
10	TOOTSIE	COMEDY	PG	HOFFMAN	1	29.95

Hint: **If you use OR, won't the rating always be unequal to one or the other? In other words, see if the Exercise can be translated from not 'R' *or* 'G' rated to not 'R' *and* not 'G' rated.**

7.2 AND-OR HIERARCHY OF EVALUATION

You can combine AND and OR in a WHERE clause to create some rather complex queries. These may be essential in SELECTing the right data, but you must first understand SQL's hierarchy of AND-OR evaluation — in what order SQL interprets search conditions that combine AND-OR.

Just like other procedural programming languages, SQL uses an evaluation hierarchy or strategy in choosing which combinations of WHERE conditions to link together. Basically, the rule is that any two conditions linked by AND will be paired up and evaluated before conditions linked by OR. Think of this process as SQL taking care of the AND conditions before it gets around to the OR conditions.

In Example 1 below, if you want to retrieve the movies rented to customers 01 and 02 that have not been returned, the SELECT statement will not work properly, because SQL first pairs up the conditions of CUSTID being equal to 02 and RETURN being NULL, applying the NULL condition to customer 02's invoices only. Try this query out to see its effect.

Example 1:

```
SELECT * FROM INV
WHERE CUSTID = '01'
OR CUSTID = '02'
AND RETURN IS NULL;
```

To get the proper result, pair up conditions using parentheses. Try the Example 2 SELECT statement and see how it differs from the first.

Example 2:

```
SELECT * FROM INV
WHERE (CUSTID = '01' OR CUSTID = '02')
AND RETURN IS NULL;
```

Without parentheses, SQL pairs up conditions joined by AND then conditions joined by OR — from left to right. With parentheses, SQL pairs up conditions

- first within the parentheses,
- second joined by AND, and
- third joined by OR

from left to right.

Be sure your parentheses are balanced, that is, with an equal number of them on the right as on the left.

EXERCISE 7-5

❶ TOPIC: Using Multiple Search Conditions.

❷ EXERCISE 7-5. List all the COMEDY AND 'PG13' rated movies where the price is over $30.00.

❸ INPUT TABLE:

MOV

NBR	TITLE	TYPE	RATING	STAR	QTY	PRICE
1	GONE WITH THE WIND	DRAMA		GABLE	4	39.95
2	FRIDAY THE 13TH	HORROR	R	JASON	2	69.95
3	TOP GUN	DRAMA	PG	CRUISE	7	49.95
4	SPLASH	COMEDY	PG13	HANKS	3	29.95
5	101 DALMATIONS	COMEDY	G		3	59.95
6	BODY HEAT	DRAMA	R	TURNER	3	19.95
7	RISKY BUSINESS	COMEDY	R	CRUISE	2	44.55
8	COCOON	SCIFI	PG	AMECHE	2	
9	CROCODILE DUNDEE	COMEDY	PG13	HARRIS	2	69.95
10	TOOTSIE	COMEDY	PG	HOFFMAN	1	29.95

FORMAT

❹

```
SELECT * FROM <tablename>
WHERE <condition1> AND/OR <condition2>
AND/OR <condition3>;
```

❺ STEP 1. ______________________________

Write your answer.

____

STEP 2. Enter your answer into the computer.

❻ RESULTS TABLE:

NBR	TITLE	TYPE	RATING	STAR	QTY	PRICE
5	101 DALMATIONS	COMEDY	G		4	59.95
7	RISKY BUSINESS	COMEDY	R	CRUISE	3	44.55
9	CROCODILE DUNDEE	COMEDY	PG13	HARRIS	3	69.95

Hint: **Don't forget the parentheses.**

EXERCISE 7-6

❶ TOPIC: Using Multiple Search Conditions.

❷ EXERCISE 7-6. List all invoice information on invoices where the TOTAL is over $10, or that were rented prior to 87052, and that have not been returned.

❸ INPUT TABLE:

INV

NBR	CUSTID	TOT	RENT	RETURN

❹ RESULTS TABLE:

NBR	CUSTID	TOT	RENT	RETURN
2	02	13.55	87030	
7	04	10.35	86355	
9	05	10.35	87052	

7.3 NOT — NEGATIVE SELECTION

Just as the negative of a photo gives a reverse image of a picture, you can negate your SELECT criteria by using NOT to obtain a reverse image of what data to select from a table. This is done by placing the SQL keyword NOT in front of your condition, for example:

```
SELECT * FROM MOV WHERE PRICE IS NOT NULL;
```

In effect, the request is for SELECTion of all rows from the movie table EXCEPT FOR ROWS WITH AN UNKNOWN VALUE IN THE PRICE FIELD.

When working with the other relational operators, the NOT syntax looks a little confusing, because the NOT is placed in front of the entire condition to get all of the noncomedy movies.

```
SELECT * FROM MOV WHERE NOT TYPE = 'COMEDY';
```

If you tried to use,

```
SELECT * FROM MOV WHERE TYPE NOT = 'COMEDY';
```

your statement would not run.

To make it clear to what the NOT applies, you should use parentheses. It is particularly important to do so when you are mixing combinations of NOT and AND and OR. SQL evaluates NOT before AND before OR, unless you specify differently by enclosing your search conditions in parentheses.

For example, if you want to find all of the movies except for HORROR, COMEDY, and DRAMA, the WHERE clause

```
WHERE NOT TYPE = 'HORROR' OR TYPE = 'COMEDY' OR TYPE =
'DRAMA';
```

gets paired like this:

```
WHERE NOT (TYPE = 'HORROR' OR TYPE = 'COMEDY') OR TYPE =
'DRAMA';
```

This will not bring back HORROR and COMEDY movies, but will get DRAMAS. It should have been coded this way instead:

```
WHERE NOT (TYPE = 'HORROR' OR TYPE = 'COMEDY' OR TYPE =
'DRAMA');
```

To focus on the correct evaluation, study the examples and try typing them in and executing them.

Example 1: — incorrect

```
SELECT * FROM MOV WHERE NOT TYPE = 'HORROR' OR
TYPE = 'COMEDY' OR TYPE = 'DRAMA';
```

Example 2: — correct

```
SELECT * FROM MOV WHERE NOT (TYPE = 'HORROR' OR
TYPE = 'COMEDY' OR TYPE = 'DRAMA');
```

EXERCISE 7-7

❶ TOPIC: NOT — Negative SELECTion.

❷ EXERCISE 7-7. List all information on invoices that have been returned (RETURN).

❸ INPUT TABLE:

INV

NBR	CUSTID	TOT	RENT	RETURN
1	01	9.55	87055	
2	02	13.55	87030	
3	02	21.01	87041	87042
4	01	11.25	87042	87072
5	03	12.55	87045	87051
6	02	9.75	87047	87051
7	04	10.35	86355	
8	04	12.55	87051	87053
9	05	10.35	87052	
10	02	7.55	87030	87060

❹ RESULTS TABLE:

NBR	CUSTID	TOT	RENT	RETURN
3	02	21.01	87041	87042
4	01	11.25	87042	87072
5	03	12.55	87045	87051
6	02	9.75	87047	87051
8	04	12.55	87051	87053
10	02	7.55	87030	87060

Hint: **Remember the syntax for negating NULL searches.**

EXERCISE 7-8

❶ TOPIC: Multiple Search Condition — NOT.

❷ EXERCISE 7-8. List all movie information, except for HORROR movies or those rated 'R' (TYPE, RATING).

❸ INPUT TABLE:

MOV

NBR	TITLE	TYPE	RATING	STAR	QTY	PRICE

❹ RESULTS TABLE:

NBR	TITLE	TYPE	RATING	STAR	QTY	PRICE
3	TOP GUN	DRAMA	PG	CRUISE	7	49.95
4	SPLASH	COMEDY	PG13	HANKS	3	29.95
5	101 DALMATIONS	COMEDY	G		3	59.95
8	COCOON	SCIFI	PG	AMECHE	2	
9	CROCODILE DUNDEE	COMEDY	PG13	HARRIS	2	69.95
10	TOOTSIE	COMEDY	PG	HOFFMAN	1	29.95

Hint: **Think carefully about how to connect the conditions in the parentheses.**

7.4 REVIEW MULTIPLE SEARCH CONDITIONS

Before you begin this review exercise, a word on verifying the correctness of your queries. When you begin to write more involved and complex queries containing AND and OR, or to use some of the more advanced techniques in this book, you must be able to predict the results of your statement, and verify its correctness. This is due partly to the "all–or–nothing" nature of SQL programming. Unlike traditional programming languages, where intermediate results can be seen, and errors can be trapped easily, SQL either gives you what you wanted or not. When results are unexpected, it can be difficult to find out why. It's up to you to understand how to code your request. None of the steps between entering the SQL statement, and SQL's execution of it (in the form of a results table, or lack of one), may be seen.

Until SQL query trace facilities are available, let's concentrate on what can be done to lessen the chances of failure.

1. "Know Thy Data" — use SELECT * (if feasible) or some means of ascertaining what is currently in the table.
2. Test queries thoroughly with small tables, where the results of your requests can be easily verified.
3. Build queries in steps — for instance, start with
   a. SELECT * FROM MOV WHERE PRICE IS NULL;
      — then add to the WHERE clause.
   b. SELECT * FROM MOV WHERE QTY < 2 OR PRICE IS NULL;
      — then add the final portion of the WHERE clause.
   c. SELECT * FROM MOV WHERE (QTY < 2 AND PRICE < 30) OR PRICE IS NULL;

This way, you can verify each portion of the code as it is executed.

7.5 INTERPRETING SQL

Write out an English translation for the following SELECT statements. Try running the examples shown if you would like to see the results on your computer.

1. SELECT LNAME, FNAME FROM CUST WHERE NOT ID = '01' AND NOT ST = 'CA';
2. SELECT QTY, NBR FROM MOV WHERE RATING = 'PG' OR RATING = 'PG13' AND QTY < 8;
3. SELECT * FROM I_M WHERE M_NBR = 1 OR (I_NBR = 10 AND M_NBR = 4);

7.6 ANSWER KEY

There may be more than one answer to an exercise. However, the one given emphasizes the topic discussed.

EXERCISE 7-1 SELECT * FROM CUST WHERE ST = 'CT' AND CITY = 'HARTFORD';

EXERCISE 7-2 SELECT * FROM INV WHERE TOT > 10 AND RETURN IS NULL;

EXERCISE 7-3 SELECT * FROM MOV WHERE RATING = 'R' OR TYPE = 'DRAMA';

EXERCISE 7-4 SELECT * FROM MOV WHERE RATING != 'R' AND RATING != 'G';

EXERCISE 7-5 SELECT * FROM MOV WHERE (TYPE = 'COMEDY' OR RATING = 'PG13') AND PRICE > 30;

EXERCISE 7-6 SELECT * FROM INV WHERE (TOT > 10 OR RENT < 87052) AND RETURN IS NULL;

EXERCISE 7-7 SELECT * FROM INV WHERE RETURN IS NOT NULL;

EXERCISE 7-8 SELECT * FROM MOV WHERE NOT TYPE = 'HORROR' AND NOT RATING = 'R';

OR

SELECT * FROM MOV WHERE TYPE != 'HORROR' AND RATING != 'R';

OR

SELECT * FROM MOV WHERE NOT (TYPE = 'HORROR' OR RATING = "R");

Note: **Mainframe not equal symbol is ¬=.**

Chapter 8
OTHER RELATIONAL OPERATORS: BETWEEN, IN, LIKE

Here you will learn how to:

1. Select data that is within a range using the BETWEEN operator;
2. Select data that matches values contained in a list using the IN operator;
3. Select data that is similar to a set of search characters using the LIKE operator;
4. Manipulate and display the results of arithmetic calculations on your data using arithmetic expressions in your SELECT statement.

The following topics will be covered:

8.1 Other Relational Operators
8.2 Other Relational Operators — Examples
8.3 BETWEEN — Selecting Rows Where Column Values are Within a Range
8.4 BETWEEN Exercises
8.5 IN—Selecting Rows Where Column Values are in a List
8.6 LIKE — Selecting Rows Where Column Values are Approximately Equal
8.7 LIKE Exercises
8.8 Computed Values
8.9 Computational Syntax
8.10 Computed Values — Exercises
8.11 Answer Key

8.1 OTHER RELATIONAL OPERATORS

SQL has three other comparison operators to use in the WHERE clause (recall the previous ones =, <, <=, etc.). These new operators are BETWEEN, IN, and LIKE. With these operators we can:

1. limit coding efforts — by using a kind of "SQL shorthand"
2. limit coding errors — less coding leads to fewer coding errors
3. retrieve rows, based on nonstandard SELECTion criteria. In other words, we can accomplish certain tasks for which the standard comparison operators are inadequate.

SQL allows arithmetic calculations of column values, using mathematical expressions. These new tools are easy to learn and use. They are also necessary in nonprocedural languages, in order to satisfy many standard business processing requests. Section 8.2 shows examples of these relational operators and SQL arithmetic expressions. Look them over before continuing.

8.2 OTHER RELATIONAL OPERATORS — Examples

Figures 8.1 through 8.4 illustrate other relational operators.

```
SELECT * FROM MOV WHERE QTY BETWEEN 3 AND 5;
```

NBR	TITLE	TYPE	RATING	STAR	QTY	PRICE
====	====================	=========	========	========	====	======
1	GONE WITH THE WIND	DRAMA		GABLE	4	39.95
4	SPLASH	COMEDY	PG13	HANKS	3	29.95
5	101 DALMATIONS	COMEDY	G		3	59.95
6	BODY HEAT	DRAMA	R	TURNER	3	19.95

Figure 8.1. BETWEEN — Selects rows when column values lie within a specified range.

```
SELECT * FROM MOV WHERE QTY IN (3, 4, 5);
```

NBR	TITLE	TYPE	RATING	STAR	QTY	PRICE
1	GONE WITH THE WIND	DRAMA		GABLE	4	39.95
4	SPLASH	COMEDY	PG13	HANKS	3	29.95
5	101 DALMATIONS	COMEDY	G		3	59.95
6	BODY HEAT	DRAMA	R	TURNER	3	19.95

Figure 8.2. IN — Selects rows when column values equal the values specified in a list.

```
SELECT * FROM MOV WHERE STAR LIKE '%ABLE%';
```

NBR	TITLE	TYPE	RATING	STAR	QTY	PRICE
1	FONE WITH THE WIND	DRAMA		GABLE	4	39.95

Figure 8.3. LIKE — Selects rows when column values match a specified character search pattern.

Arithmetic expressions can be included in SELECT statements to perform simple math functions on column values.

Figure 8.4 says "select the quantity column and the quantity column multiplied (*) by five, from the movie table where all quantity values times five are greater than 10."

```
SELECT QTY,QTY * 5 FROM MOV WHERE QTY * 5 > 10;
```

QTY	QTY*5
4	20
7	35
3	15
3	15
3	15

Figure 8.4. Selects columns and performs math functions using an arithmetic expression.

8.3 BETWEEN — SELECTING ROWS WHERE COLUMN VALUES ARE WITHIN A RANGE

The BETWEEN comparison operator lets you retrieve rows where the columns contain values within a specified lower and upper limit. For example, if you wanted to see all invoices where the total amount is between $12 and $16, you could write a query like this:

```
SELECT * FROM INV WHERE TOT < = 12 AND TOT > = 16;
```

or you could use the BETWEEN operator:

```
SELECT * FROM INV WHERE TOT BETWEEN 12 AND 16;
```

which would produce the same result. I think you'll agree that the example using BETWEEN is both more understandable and more convenient. So for the types of queries that involve inclusive search logic (greater than or equal to a low value, and less than or equal to a high value) I suggest using BETWEEN.

8.3.1 Using BETWEEN

1. BETWEEN is part of the WHERE clause
2. The structure is:

```
WHERE <column_name> BETWEEN x AND y
```

- The lower value (x) must be coded first
- The reserved word AND must be next
- The higher value (y) must be coded last

3. BETWEEN can be used with character, numeric and date datatypes (you must enclose character and date x and y values with single quotes. BETWEEN is inclusive in scope (equivalent to greater than or equal to a low value, and less than or equal to a high value.
4. You can search for rows outside of a range of values by combining NOT with BETWEEN (ex. list all invoices where the total is less than $12 and greater than $16:

```
SELECT * FROM INV WHERE TOT NOT BETWEEN 12 AND 16;
```

8.4 BETWEEN EXERCISES

EXERCISE 8-1

This exercise asks you to use BETWEEN to retrieve rows from the INV table where the total amount on the invoice is >= $10, and <= $15.

❶ TOPIC: Selecting Multiple Columns (BETWEEN Two Values).

❷ EXERCISE 8-1. Display all of the information on INVoices where the TOTal amount is between $15 and $10, inclusive.

❸ INPUT TABLE:

INV

NBR	CUSTID	TOT	RENT	RETURN

FORMAT

❹

```
SELECT * FROM <tablename>
WHERE <columnname> BETWEEN <lowvalue> AND <highvalue>;
```

❺ STEP 1. ______________________________

Write your answer.

____

STEP 2. Enter your answer into the computer.

❻ RESULTS TABLE:

NBR	CUSTID	TOT	RENT	RETURN
2	02	13.55	87030	
4	01	11.25	87042	87072
5	03	12.55	87045	87051
7	04	10.35	86355	
8	04	12.55	87051	87053
9	05	10.35	87052	

EXERCISE 8-2

This exercise requests a list of customers by last name. This is an example of using BETWEEN on character data. Don't forget to enclose values in single quotation marks.

❶ TOPIC: Selecting Multiple Columns (BETWEEN Two Values).

❷ EXERCISE 8-2. Display a list of all CUSTomer information, on customers whose LNAMEs are between Nicholson and Woodward alphabetically.

❸ INPUT TABLE:

CUST

ID	LNAME	FNAME	CITY	ST

❹ RESULTS TABLE:

ID	LNAME	FNAME	CITY	ST
03	NICHOLSON	JACK	HARTFORD	CT
06	WEST	MAE	PARK PLACE	CA
07	WOODWARD	JOANNE	GETTYSBURG	PA
08	ROGERS	ROY	HAPPY TRAILS	RX
09	RINGWALD	MOLLY	PINKSVILLE	CA
10	ROGERS	FRED	PARK PLACE	CA

EXERCISE 8-3

Using NOT with BETWEEN is also simple. Code the NOT before the BETWEEN keyword.

❶ TOPIC: Selecting Multiple Columns (BETWEEN Two Values).

❷ EXERCISE 8-3. Display a list of all the invoice information for invoices where the RENT date is not between January 1, 1987 and February 20, 1987 (JANUARY 1, 1987 = 87001, FEBRUARY 20, 1987 = 87051).

❸ INPUT TABLE:

INV

NBR	CUSTID	TOT	RENT	RETURN

FORMAT

❹
```
SELECT * FROM <tablename> WHERE <columnname>
   NOT BETWEEN <lowvalue> AND <highvalue>;
```

❺ STEP 1. ______________________________
Write your answer.
____

STEP 2. Enter your answer into the computer.

❻ RESULTS TABLE:

NBR	CUSTID	TOT	RENT	RETURN
1	01	9.55	87055	
7	04	10.35	86355	
9	05	10.35	87052	

8.5 IN — SELECTING ROWS WHERE COLUMN VALUES ARE IN A LIST

The IN comparison operator lets you retrieve rows where the columns contain fields that match specified values in a list. For example, if you wanted to see all invoices for customers 01, 04 and 06, you could write a query like this:

```
SELECT * FROM INV WHERE CUSTID='01' OR CUSTID='04' OR CUSTID='06';
```

or you could use the IN operator like this;

```
SELECT * FROM INV WHERE CUSTID IN ('01', '04', '06');
```

which would produce the same result. IN simplifies the query, and makes this typical logic less clumsy and error-prone. It is a shorthand substitution for coding multiple OR conditions.

```
column equal value1 OR
      column equal value2 OR
            column equal value 3 . . .
```

8.5.1 Using IN

1. IN is part of the WHERE clause
2. The structure is WHERE <column_name> IN (<list_of_values>) where the list_of_values can be numbers, characters, or dates
3. The list_of_values is enclosed in parentheses
4. Individual values in the list are separated by commas (ex. WHERE PRICE IN (20, 50, 35)
5. Individual values in the list can be in any order
6. You can search for rows containing values NOT in a list by combining NOT and IN (ex. list all invoices for customers other than 01, 04, 06: SELECT * FROM INV WHERE CUSTID NOT IN ('01', '04', '06');

EXERCISE 8-4

❶ TOPIC: Selecting Multiple Columns (IN a list of values).

❷ EXERCISE 8-4. Display a list of all customers from Vermont (VT), New York (NY), and Connecticut (CT).

❸ INPUT TABLE:

CUST

ID	LNAME	FNAME	CITY	ST
01	DANGERFIELD	RODNEY	HARTFORD	CT
02	FIELD	SALLY	FRANKLIN	NY
03	NICHOLSON	JACK	HARTFORD	CT
04	MURRAY	FRED	BOZRAH	CT
05	MADDEN	JOHN	BRATTLEBORO	VT
06	WEST	MAE	PARK PLACE	CA
07	WOODWARD	JOANNE	GETTYSBURG	PA
08	ROGERS	ROY	HAPPY TRAILS	TX
09	RINGWALD	MOLLY	PINKSVILLE	CA
10	ROGERS	FRED	PARK PLACE	CA

FORMAT

❹
```
SELECT * FROM <tablename>
WHERE <columnname> IN (<value1>, <value2>, <valueN>);
```

❺ STEP 1. ____

Write your answer.

____

STEP 2. Enter your answer into the computer.

❻ RESULTS TABLE:

ID	LNAME	FNAME	CITY	ST
01	DANGERFIELD	RODNEY	HARTFORD	CT
02	FIELD	SALLY	FRANKLIN	NY
04	MURRAY	FRED	BOZRAH	CT
05	MADDEN	JOHN	BRATTLEBORO	VT

EXERCISE 8-5

IN combination with NOT. Use NOT IN to complete this task. (Hint: the structure for this is similar to that used for NOT BETWEEN.)

❶ TOPIC: Selecting Multiple Columns (IN a list of Values).
❷ EXERCISE 8-5. List all of the invoices, except for customers 02, 03, and 06.
❸ INPUT TABLE:

INV

NBR	CUSTID	TOT	RENT	RETURN

FORMAT

❹
```
SELECT <columnname> FROM <tablename>
WHERE <columnname> NOT IN
(<value1>, <value2>, <valueN>);
```

❺ STEP 1. ______________________________

Write your answer.

____

STEP 2. Enter your answer into the computer.

❻ RESULTS TABLE:

NBR	CUSTID	TOT	RENT	RETURN
1	01	9.55	87055	
4	01	11.25	87042	87072
7	04	10.35	86355	
8	04	12.55	87051	87053
9	05	10.35	87052	

8.6 LIKE — SELECTING ROWS WHERE COLUMN VALUES ARE APPROXIMATELY EQUAL

The LIKE comparison operator lets you retrieve rows where the columns contain values that resemble a specified search string. For example, if you wanted to see all movies where the title contained the word THE somewhere in it you could write a query like this:

```
SELECT * FROM MOV WHERE TITLE LIKE '%THE%';
```

Or if you wanted to see a list of all customers that had the letter O in the second position of their last name, you could write a query like this:

```
SELECT * FROM CUST WHERE LNAME LIKE '_O%';
```

LIKE allows you to be less exact in your WHERE clause specification. It is useful for selecting rows whose columns:

- have approximate but not exact values
- have values you are unsure of
- have values that would be time-consuming to identify more specifically

8.6.1 Using LIKE

1. LIKE is part of the WHERE clause
2. LIKE can only be used with character columns
3. The structure is

```
WHERE <column_name> LIKE '<search_string>)
```

where the search_string can be any combination of letters, numbers, special characters (!@#$^&*_+=) and the SQL wildcard characters percent sign (%) and underscore (_).

4. The percent sign stands for "any sequence of, and number of characters or no characters at all". In the first example above, we were searching for the word THE in our movie titles using the search string '%THE%'. The percent signs tells DB2 and SQL/DS to retrieve rows where the characters THE appear anywhere in the title. This search string would have returned the following movie rows:

```
THE THING
GONE WITH THE WIND
THE
OVER THERE
ANOTHER BIZARRE HOLLYWOOD FILM
```

5. The underscore stands for "any single character <u>at that particular position in the column</u>" — one and only one character. In a previous example:

```
SELECT * FROM CUST WHERE LNAME LIKE '_0%';
```

we were searching through a fictitious table for any customers whose last names had an O in the second position '_O%'. The underscore tells DB2 and SQL/DS to retrieve rows where the character O appears in position two of the LNAME column. This search string would have returned the following customer rows:

```
column position
123456789012345678901234567890

SO
AORTA
1ORTEGA
 O'LEARY (note blank in column one)
```

It would not have returned the following customer rows:

```
OZ
OLIVIER
SMOTHERS BROTHERS
```

6. Besides the wildcard characters, all other characters in the search string stand for themselves.
7. The search_string is enclosed in single quotes.
8. You can mix the underscore and percent sign in the same search string.
9. The wildcard characters can be positioned anywhere in the search string in any combination.
10. You can search for rows containing values NOT in the search string by combining NOT and LIKE (ex. list all customers whose last names do not begin with the letter A: SELECT * FROM CUST WHERE LNAME NOT LIKE 'A%';

8.6.2 LIKE With Fixed Character Columns on Mainframe Systems

When columns are defined as fixed length character columns, DB2 and SQL/DS pads each field in the table with spaces if the actual character data in the row is less than the maximum defined size of the column. Figure 8.5 below shows our CUST CITY column and the trailing spaces in most rows (trailing spaces are displayed with the small letter b).

```
CITY
==================
HARTFORDbbbb
FRANKLINbbbb
HARTFORDbbbb
BOZRAHbbbbbb
BRATTLEBOROb
PARK PLACEbb
GETTYSBURGbb
HAPPY TRAILS
PINKSVILLEbb
PARK PLACEbb
```

Figure 8.5. CITY column with trailing spaces.

If you wanted to return all rows where the CITY ended in ACE you might be tempted to write a query like this:

```
SELECT * FROM CUST WHERE CITY LIKE '%ACE';
```

expecting it to return PARK PLACE. This won't work. Why not?

The trailing blanks at the end of the PARK PLACE field are considered significant characters by SQL and you would have had to type:

```
WHERE CITY LIKE '%PLACE
```

with two "hard spaces (blanks)" after the word PLACE in order to account for them. The probability of trailing blanks in fixed character fields must be considered when you are using LIKE to test fixed length character data:

- At the end of a field
- Separated by blanks (Exercise 8.9)

If you are using this course or programming on a micro or mini-computer, you should not have to worry about this situation. Even if your system accepts fixed length columns as valid definitions, most micro and mini-computer relational database systems store the data as variable length and fields are not padded with trailing blanks.

Here are two more examples before you try the exercises.

Example 1:

```
SELECT * FROM MOV WHERE RATING LIKE 'PG _ _';
```

```
NBR  TITLE                CLASS     RATING   STAR        QTY  PRICE
==== ==================== ========= ======== =========== ==== ======
   4 SPLASH               COMEDY    PG13     HANKS          3  29.95
   9 CROCODILE DUNDEE     COMEDY    PG13     HARRIS         2  69.95
```

Example 2 shows a search string, using the % search character, that will find either PG or PG13 rated movies.

Example 2:

```
SELECT * FROM MOV WHERE RATING LIKE "PG%";
```

NBR	TITLE	TYPE	RATING	STAR	QTY	PRICE
3	TOP GUN	DRAMA	PG	CRUISE	7	49.95
4	SPLASH	COMEDY	PG13	HANKS	3	29.95
8	COCOON	SCIFI	PG	AMECHE	2	
9	CROCODILE DUNDEE	COMEDY	PG13	HARRIS	2	69.95
10	TOOTSIE	COMEDY	PG	HOFFMAN	1	29.95

8.7 LIKE EXERCISES

LIKE — Search Characters

% (percent sign) — indicates a string of unknown characters — that matches any value of any length, including no value at all

_ (underscore) — indicates any SINGLE unknown character position

EXERCISE 8-6

This exercise requires you to use the % search character.

LIKE — Search Characters
% (percent sign) — indicates a string of unknown characters — that matches any value of any length, including no value at all
_ (underscore) — indicates any SINGLE unknown character position

❶ TOPIC: Selecting Multiple Columns (LIKE search string pattern)

❷ EXERCISE 8-6. Display all the customers with last names that include FIELD anywhere in them.

❸ INPUT TABLE:

CUST

ID	LNAME	FNAME	CITY	ST

FORMAT

❹

```
SELECT <columnname> FROM <tablename>
WHERE <columnname> LIKE <'stringpattern'>;
```

❺ STEP 1. ______________________________

Write your answer.

____

STEP 2. Enter your answer into the computer.

❻ RESULTS TABLE:

ID	LNAME	FNAME	CITY	ST
01	DANGERFIELD	RODNEY	HARTFORD	CT
02	FIELD	SALLY	FRANKLIN	NY

EXERCISE 8-7

This exercise is a tricky one. It requires you to combine the use of _ and % search characters. Can you see that if you simply wrote:

. . . WHERE CITY LIKE '% %'

you could pick up rows with blanks either before the city, or rows where the city column was entirely blank filled?

LIKE — Search Characters
% (percent sign) — indicates a string of unknown characters — that matches any value of any length, including no value at all
_ (underscore) — indicates any SINGLE unknown character position

❶ TOPIC: Selecting Multiple Columns (LIKE a search string pattern)
❷ EXERCISE 8-7. List all of the CUSTomers that live in a CITY that has as its name two words separated by a space.
❸ INPUT TABLE:

CUST

ID	LNAME	FNAME	CITY	ST
==	=======	=======	=====	===

FORMAT

❹
```
SELECT <columnname> FROM <tablename>
WHERE <columnname> LIKE <'stringpattern'>;
```

❺ STEP 1. ______________________________

Write your answer.

____

STEP 2. Enter your answer into the computer.

❻ RESULTS TABLE:

ID	LNAME	FNAME	CITY	ST
06	WEST	MAE	PARK PLACE	CA
08	ROGERS	ROY	HAPPY TRAILS	TX
10	ROGERS	FRED	PARK PLACE	CA

Hint: **You can use any combination of % and — characters in your search string.**

8.8 COMPUTED VALUES

SQL can SELECT not only numeric values already in a table, but also results that can be calculated using those values. For instance, you can write:

```
SELECT NBR, QTY * PRICE FROM MOV;
```

This is asking SQL to display NBR and (QTY times PRICE) from the MOV table to show the total investment we have in each movie. To SELECT computed values, you use arithmetic expressions. These are expressions made up of column names, literal numeric values, and arithmetic operators. There are two ways to use arithmetic expressions in SELECT statements:

1. When you write an expression after the SELECT clause (as shown above), SQL computes the value for you and places it in your results table.
2. You can also use arithmetic expressions in a WHERE clause to retrieve rows based on a "calculated search". This is shown in Figure 8.6.

* You can use arithmetic expressions in SELECT statements, just as you would column names

```
SELECT TITLE, QTY * PRICE
 FROM MOV
  WHERE TITLE = 'COCOON';
```

* You can also use arithmetic expressions in a WHERE clause.

```
SELECT TITLE, QTY * PRICE
 FROM MOV
  WHERE QTY * PRICE > 100;
```

* You may also add a descriptive column to your query, by enclosing a literal in quotation marks

```
SELECT TITLE, 'QUANT * PRICE = ', QTY * PRICE
 FROM MOV
  WHERE QTY * PRICE > 100;
```

Figure 8.6. Computed values.

Notice in the examples shown that you can mix literal values and column names in the arithmetic expression. Try these out before reading any further.

```
column     * column     SELECT QTY    * PRCE    FROM MOV;
column     * literal    SELECT QTY    * 10      FROM MOV;
literal    * column     SELECT 10     * QTY     FROM MOV;
literal    * literal    SELECT 10     * 10      FROM MOV;
character  * literal
                        SELECT 'ROW NBR', NBR   FROM MOV;
```

There are several considerations for using computed values in SQL queries:

1. The columns used in your computations must be defined as numeric.

2. You might have to be aware of arithmetic overflow. Numeric columns can only hold a finite range of values. If, when calculating values, you go beyond the range allowed by the column, you will get an arithmetic overflow error. This can sometimes be avoided by multiplying your calculation by 1.0 which will allow you to obtain 15 decimal places of value.

Example:

```
SELECT QTY * PRICE * 1.0
```

3. You only have the basic four arithmetic operators from which to choose (addition, subtraction, multiplication, and division). Note the symbols below:

   * - multiply
   / - divide
   + - add
   – - subtract

   These operators can be used on any combination of columns and numeric values.

4. You might want to add descriptive columns to your queries to make them more readable.

Example:

```
SELECT 'QTY * PRICE =', QTY * PRICE FROM MOV;
```

```
QTY * PRICE =      QTY * PRICE
===============    =============
QTY * PRICE =              159.8
QTY * PRICE =              139.9
QTY * PRICE =             349.65
QTY * PRICE =              89.85
QTY * PRICE =             179.85
QTY * PRICE =              59.85
QTY * PRICE =               89.1
QTY * PRICE =
QTY * PRICE =              139.9
QTY * PRICE =              29.95
```

5. You must use the relative column position option of the ORDER BY clause to ORDER your results on an arithmetic expression. In the SELECT statement below, we cannot ORDER BY QTY * PRICE, we must specify the calculated value's relative column position (2).

Example:

```
SELECT NBR, QTY * PRICE FROM MOV ORDER BY
2 DESC;
```

```
     NBR        QTY * PRICE
========      =============
       3             349.65
       5             179.85
       1              159.8
       2              139.9
       9              139.9
       4              89.85
       7               89.1
       6              59.85
      10              29.95
       8
```

6. Null values are ignored in arithmetic operations of all kinds. For example, the outcome of multiplying a column with a NULL value in it by anything, will be NULL.

7. Different DBMSs will display arithmetic expressions with different levels of decimal precision. If you are using a mainframe product your actual results table may be slightly different from what is shown. However, the numeric values should be equivalent.

Example:

```
SELECT RENT, RETURN, RETURN-RENT FROM INV;
```

RENT	RETURN	RETURN-RENT
87055		
87030		
87041	87042	1
87042	87072	30
87045	87051	6
87047	87051	4
86355		
87051	87053	2
87052		
87030	87060	30

8.9. COMPUTATIONAL SYNTAX

- Include expressions after the SELECT clause, like any other SELECTed value.
- Only use columns in arithmetic expressions that are defined as numeric data types.
- Keep in mind the following SQL hierarchy for arithmetic expression evaluation.

The SQL default evaluation order is shown here. This is very important to remember.

SQL DEFAULT EVALUATION ORDER
All multiplication and/or division is done first — from left to right. Any remaining addition and/or subtraction is done last — from left to right.

Study Figure 8.7 and learn the computation operators.

When combining arithmetic expressions, you can change SQL's default evaluation order by using parentheses (these also serve as good query documentation). See Example 2.

Example 1:

```
SELECT QTY * PRICE + 10    FROM MOV    WHERE NBR = 3;
```

Example 2:

```
SELECT QTY * (PRICE + 10)    FROM MOV    WHERE NBR = 3;
```

Operator	Meaning
*	Multiplication
/	Division
+	Addition
-	Subtraction

* Evaluation rules for arithmetic operators

* and / + and -	Multiplication and division will be evaluated before addition and subtraction

* Data items with NULL values will always return a NULL result

* Suggestion — include NOT NULL in the WHERE clause of your SELECT statement

```
SELECT QTY * PRICE
 FROM MOV WHERE QTY IS NOT NULL
  AND PRICE IS NOT NULL;
```

Figure 8.7. Computation operators.

8.10 COMPUTED VALUES EXERCISES

EXERCISE 8-8

❶ TOPIC: Selecting Multiple Columns (Calculated Values).

❷ EXERCISE 8-8. Build a report listing the movie number, current value and replacement value for each movie in the MOV table. Calculate the replacement value for all movies as QTY * PRICE * 1.15.

❸ INPUT TABLE:

MOV

NBR	TITLE	TYPE	RATING	STAR	QTY	PRICE
1	GONE WITH THE WIND	DRAMA		GABLE	4	39.95
2	FRIDAY THE 13TH	HORROR	R	JASON	2	69.95
3	TOP GUN	DRAMA	PG	CRUISE	7	49.95
4	SPLASH	COMEDY	PG13	HANKS	3	29.95
5	101 DALMATIONS	COMEDY	G		3	59.95
6	BODY HEAT	DRAMA	R	TURNER	3	19.95
7	RISKY BUSINESS	COMEDY	R	CRUISE	2	44.55
8	COCOON	SCIFI	PG	AMECHE	2	
9	CROCODILE DUNDEE	COMEDY	PG13	HARRIS	2	69.95
10	TOOTSIE	COMEDY	PG	HOFFMAN	1	29.95

FORMAT

❹

```
SELECT   <columnname>, <arithmetic expression>
             FROM  <tablename>
             WHERE <condition>;
```

❺ STEP 1. ________________________________

Write your answer.

____

STEP 2. Enter your answer into the computer.

❻ RESULTS TABLE:

NBR	QTY * PRICE	QTY * PRICE * 1.15
1	159.8	183.77
2	139.9	160.885
3	349.65	402.0975
4	89.85	103.3275
5	179.85	206.8275
6	59.85	68.8275
7	89.1	102.465
8		
9	139.9	160.885
10	29.95	34.4425

Note: **Mainframe users will see different column headings and formatting. The values should be equivalent.**

EXERCISE 8-9

This exercise requires you to use columns for both operands of your arithmetic expression.

❶ TOPIC: Selecting Multiple Columns (Calculated Values).

❷ EXERCISE 8-9. Build a report containing (INVoice NBR, RENT, RETURN, RETURN minus RENT) to list how long our movies are being kept.

❸ INPUT TABLE:

INV

NBR	CUSTID	TOT	RENT	RETURN

❹ RESULTS TABLE:

NBR	RENT	RETURN	RETURN–RENT
1	87055		
2	87030		
3	87041	87042	1
4	87042	87072	30
5	87045	87051	6
6	87047	87051	4
7	86355		
8	87051	87053	2
9	87052		
10	87030	87060	30

Note: **Mainframe terminal users will not see the RETURN–RENT heading.**

8.11 ANSWER KEY

There may be more than one answer to an exercise. However, the one given emphasizes the chapter topic.

EXERCISE 8-1 SELECT * FROM INV WHERE TOT BETWEEN 10 AND 15;
EXERCISE 8-2 SELECT * FROM CUST WHERE LNAME BETWEEN 'NICHOLSON' AND 'WOODWARD';
EXERCISE 8-3 SELECT * FROM INV WHERE RENT NOT BETWEEN 87001 AND 87051;
EXERCISE 8-4 SELECT * FROM CUST WHERE ST IN ('VT', 'NY', 'CT');
EXERCISE 8-5 SELECT * FROM INV WHERE CUSTID NOT IN ('O2', '03', '06');
EXERCISE 8-6 SELECT * FROM CUST WHERE LNAME LIKE '%FIELD%';
EXERCISE 8-7 SELECT * FROM CUST WHERE CITY LIKE '_% %_';
EXERCISE 8-8 SELECT NBR, QTY * PRICE, QTY * PRICE * 1.15 FROM MOV;
EXERCISE 8-9 SELECT NBR, RENT, RETURN, RETURN–RENT FROM INV;

Chapter 9
JOINING TABLES

Here you will learn how to:

1. Select data that is contained on more than one table with a single select statement.
2. Combine the information from three and four tables in a single query.

The following topics will be covered:

9.1 JOIN — RETRIEVING DATA FROM TWO OR MORE TABLES

Suppose we wanted a report listing the customer name (LNAME), invoice number (NBR), and return date (RETURN), for all of the invoices. Notice in Fig. 9.1 that the columns we need for the report are carried on two separate tables. LNAME is on the CUST table, and NBR and RETURN are on the INV table. We do this by "joining" them together using columns that carry corresponding information. In this case, customer identification number — which is called ID on the CUST table, and CUSTID on the INV table — contains values which correspond between the two tables. This allows us to merge CUST and INV as illustrated in Figure 9.1.

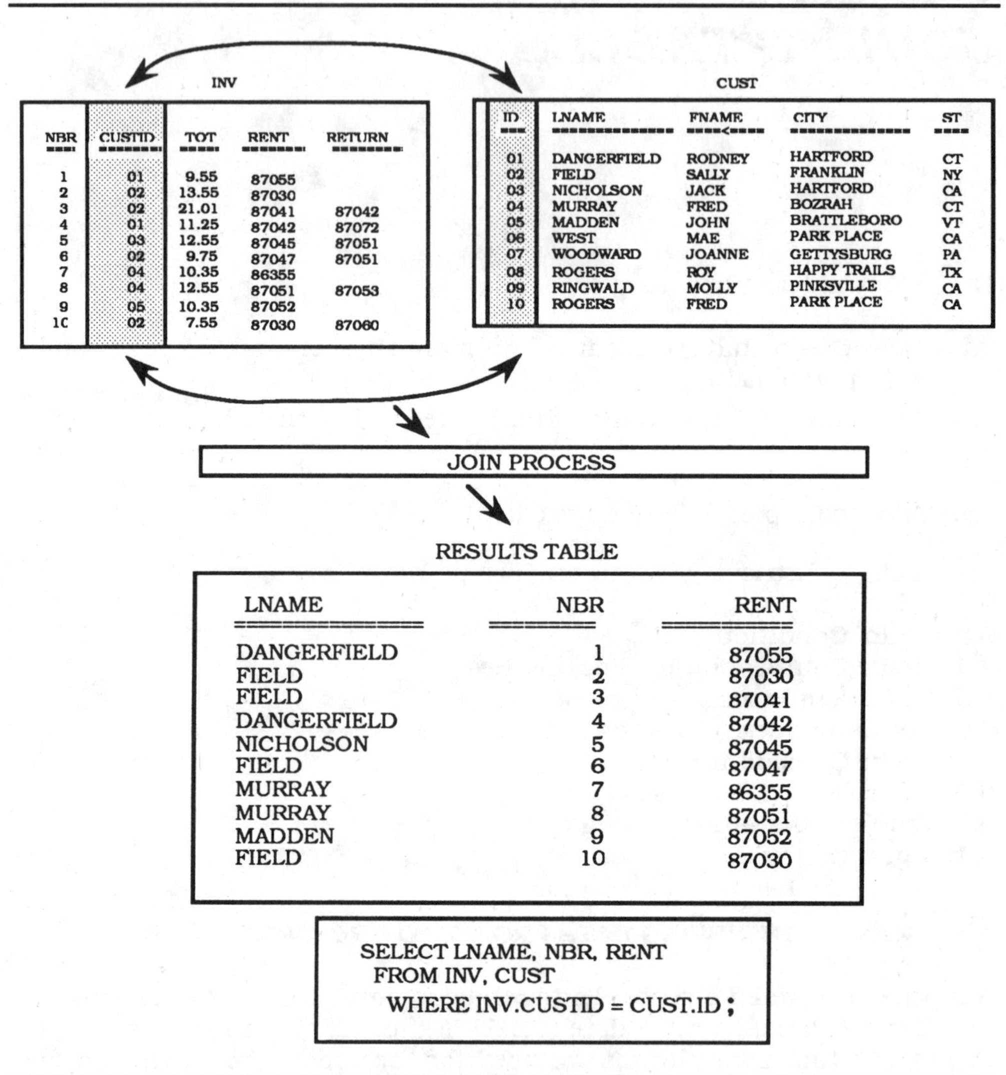

INV

NBR	CUSTID	TOT	RENT	RETURN
1	01	9.55	87055	
2	02	13.55	87030	
3	02	21.01	87041	87042
4	01	11.25	87042	87072
5	03	12.55	87045	87051
6	02	9.75	87047	87051
7	04	10.35	86355	
8	04	12.55	87051	87053
9	05	10.35	87052	
1C	02	7.55	87030	87060

CUST

ID	LNAME	FNAME	CITY	ST
01	DANGERFIELD	RODNEY	HARTFORD	CT
02	FIELD	SALLY	FRANKLIN	NY
03	NICHOLSON	JACK	HARTFORD	CA
04	MURRAY	FRED	BOZRAH	CT
05	MADDEN	JOHN	BRATTLEBORO	VT
06	WEST	MAE	PARK PLACE	CA
07	WOODWARD	JOANNE	GETTYSBURG	PA
08	ROGERS	ROY	HAPPY TRAILS	TX
09	RINGWALD	MOLLY	PINKSVILLE	CA
10	ROGERS	FRED	PARK PLACE	CA

RESULTS TABLE

LNAME	NBR	RENT
DANGERFIELD	1	87055
FIELD	2	87030
FIELD	3	87041
DANGERFIELD	4	87042
NICHOLSON	5	87045
FIELD	6	87047
MURRAY	7	86355
MURRAY	8	87051
MADDEN	9	87052
FIELD	10	87030

```
SELECT LNAME, NBR, RENT
 FROM INV, CUST
   WHERE INV.CUSTID = CUST.ID ;
```

Figure. 9.1. Selecting from more than one table.

This technique is called joining tables, because the corresponding fields from two or more different tables are actually "joined together" to form a single row in the results table. The ability to join rows and combine data from two or more tables is one of the most powerful features of relational systems.

With SQL's joining ability, we can store all the information about customers on the CUST table, and all the information about invoices on the INV table. We can then combine the information when necessary.

And when is it necessary to join tables? There are two situations which require you to join tables:

1. When the data you need for a single report is on two or more tables.

2. When you need to select data from one table based on a condition that exists in another table.

Study these two examples:

1. **Merge columns from two or more tables.** If for instance, as in the example above, we wanted a list of the customer LNAMEs and their corresponding invoice NBRs, we would use the following statement:

```
SELECT LNAME, NBR FROM CUST, INV WHERE ID = CUSTID;
```

LNAME	NBR
DANGERFIELD	1
FIELD	2
FIELD	3
DANGERFIELD	4
NICHOLSON	5
FIELD	6
MURRAY	7
MURRAY	8
MADDEN	9
FIELD	10

2. **SELECT data from one table, based on a condition applied to another table.** For instance, if we wanted a list of the customers who have videos yet to be returned, we would use the following statement:

```
SELECT LNAME, ID FROM CUST, INV WHERE
ID = CUSTID AND RETURN IS NULL;
```

LNAME	ID
DANGERFIELD	01
FIELD	02
MURRAY	04
MADDEN	05

9.2 JOIN GRAPHIC

Figure 9.2 demonstrates the join of the INV and CUST tables to produce the report (the results table) from a statement similar to Example 1.

Notice that the customer identification number is common to both tables. It is used as the joining column in the WHERE condition, to "link" the tables together.

Notice that customer IDs 01, 02, 03, 04 and 05 on the CUST table have corresponding entries on the INV table. That is why they are the only customers that are in the results table. Values on CUST that are not on INV, or values on INV that are not on CUST are not SELECTed.

The key to understanding the subtleties of joining tables lies in the WHERE condition that links the tables together. This WHERE condition is called (appropriately enough) a "join condition". Let's look at it a little more closely.

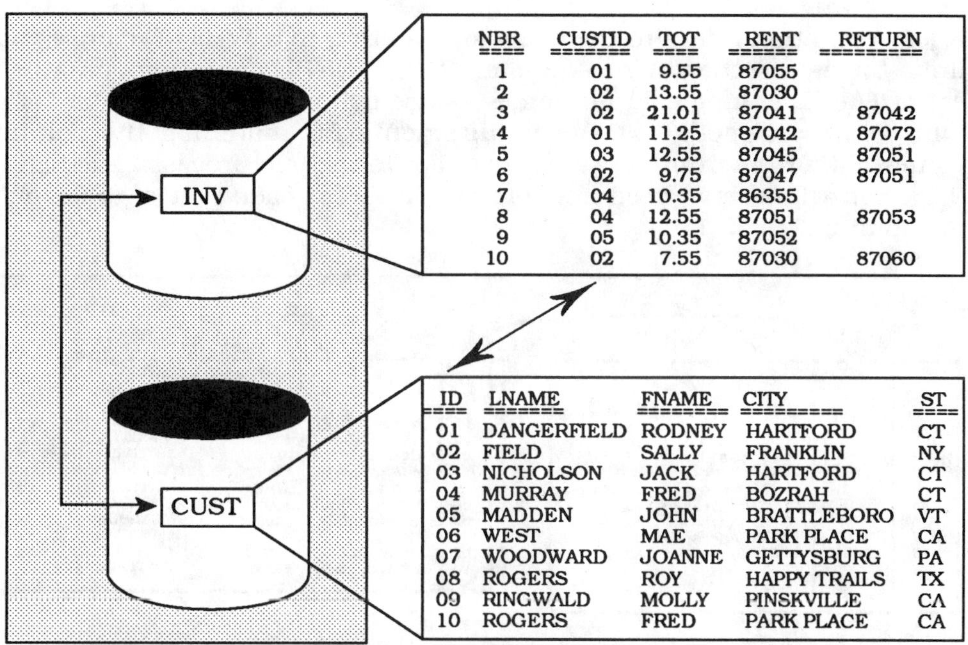

NBR	CUSTID	TOT	RENT	RETURN
1	01	9.55	87055	
2	02	13.55	87030	
3	02	21.01	87041	87042
4	01	11.25	87042	87072
5	03	12.55	87045	87051
6	02	9.75	87047	87051
7	04	10.35	86355	
8	04	12.55	87051	87053
9	05	10.35	87052	
10	02	7.55	87030	87060

ID	LNAME	FNAME	CITY	ST
01	DANGERFIELD	RODNEY	HARTFORD	CT
02	FIELD	SALLY	FRANKLIN	NY
03	NICHOLSON	JACK	HARTFORD	CT
04	MURRAY	FRED	BOZRAH	CT
05	MADDEN	JOHN	BRATTLEBORO	VT
06	WEST	MAE	PARK PLACE	CA
07	WOODWARD	JOANNE	GETTYSBURG	PA
08	ROGERS	ROY	HAPPY TRAILS	TX
09	RINGWALD	MOLLY	PINSKVILLE	CA
10	ROGERS	FRED	PARK PLACE	CA

Join SELECT statement

```
SELECT LNAME, NBR, CUSTID
 FROM INV, CUST
  WHERE ID = CUSTID;
```

RESULTS TABLE

LNAME	CUSTID	NBR
DANGERFIELD	01	1
DANGERFIELD	01	4
FIELD	02	2
FIELD	02	3
FIELD	02	6
FIELD	02	10
NICHOLSON	03	5
MURRAY	04	7
MURRAY	04	8
MADDEN	05	9

Figure 9.2. Join (another view).

9.3 JOIN CONDITION

A join condition refers to the specific condition in a WHERE clause that links tables together in a table join. This WHERE condition contains a set of TABLE linking column names — one column from each table. The join condition spells out the requirements for matching the column values to tie rows from two tables together in the join process. See Figure 9.3. You will always need one join condition for each pair of tables you are joining.

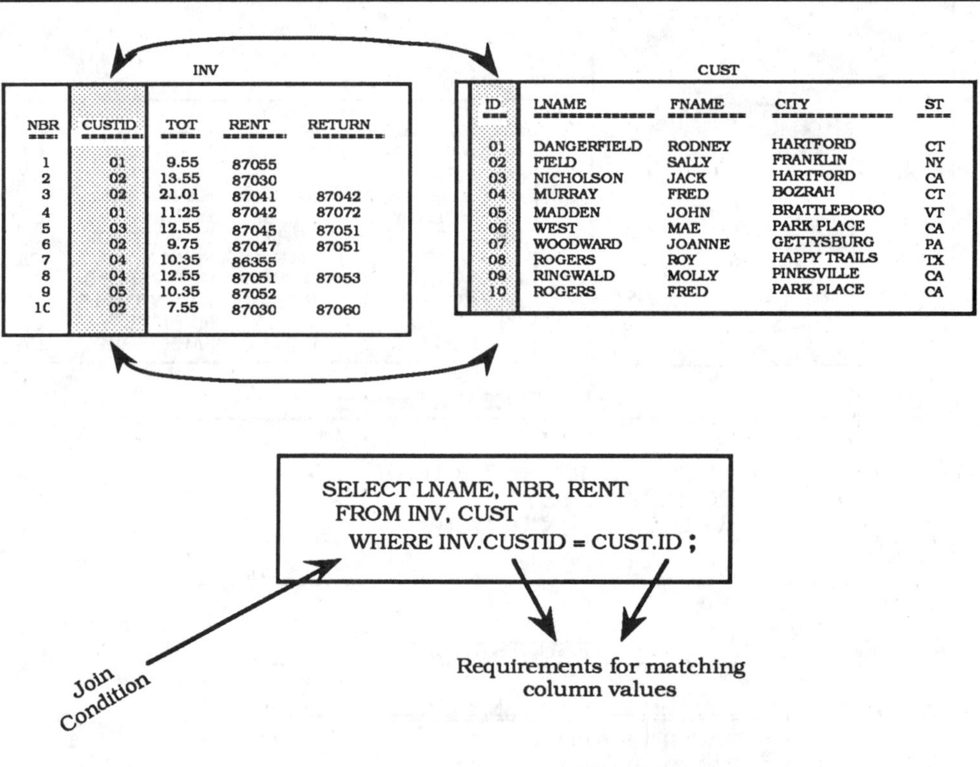

NBR	CUSTID	TOT	RENT	RETURN
1	01	9.55	87055	
2	02	13.55	87030	
3	02	21.01	87041	87042
4	01	11.25	87042	87072
5	03	12.55	87045	87051
6	02	9.75	87047	87051
7	04	10.35	86355	
8	04	12.55	87051	87053
9	05	10.35	87052	
10	02	7.55	87030	87060

ID	LNAME	FNAME	CITY	ST
01	DANGERFIELD	RODNEY	HARTFORD	CT
02	FIELD	SALLY	FRANKLIN	NY
03	NICHOLSON	JACK	HARTFORD	CA
04	MURRAY	FRED	BOZRAH	CT
05	MADDEN	JOHN	BRATTLEBORO	VT
06	WEST	MAE	PARK PLACE	CA
07	WOODWARD	JOANNE	GETTYSBURG	PA
08	ROGERS	ROY	HAPPY TRAILS	TX
09	RINGWALD	MOLLY	PINKSVILLE	CA
10	ROGERS	FRED	PARK PLACE	CA

Figure 9.3. Join condition.

9.4 FOUR JOIN CONDITION POSSIBILITIES

When you join one table with another, there are four possibilities that can unfold, depending on what happens when the Join Condition is executed (how the values from one table-column match the values from the other table-column.) They are illustrated in Figure 9.5, but you can type them in and execute them to verify the results:

1. Values in the columns we are joining on, appear one time in each table. In this case, one row will be created in the results table for each matching value.

```
SELECT LNAME, CUSTID, NBR FROM INV, CUST
WHERE CUSTID = ID AND CUSTID IN ('03', '05')
```

LNAME	CUSTID	NBR
NICHOLSON	03	5
MADDEN	05	9

Figure 9.4a. One to one correspondence.

2. Values in the columns we are joining on do not appear in one or the other table. In this case, no entry is placed in the results table. This type of exception data is sometimes important to display.

```
SELECT LNAME, CUSTID, NBR FROM INV, CUST
WHERE CUSTID = ID AND ID IN ('06', '07');

0 ROWS SELECTED
```

Figure 9.4b. One to zero correspondence.

3. Values in the columns we are joining on appear multiple times in one or the other table. In this case, several rows will be SELECTed into our results table. In fact, the number of rows that is equal to the higher number of occurrences found in either table will be SELECTed.

```
SELECT LNAME, CUSTID, NBR FROM INV, CUST
WHERE CUSTID = ID;
```

LNAME	CUSTID	NBR
DANGERFIELD	01	1
DANGERFIELD	01	4
FIELD	02	2
FIELD	02	3
FIELD	02	6
FIELD	02	10
NICHOLSON	03	5
MURRAY	04	7
MURRAY	04	8
MADDEN	05	9

Figure 9.4c. One to many correspondence.

4. Values in the columns we are joining on appear multiple times in both tables. This happens infrequently in most systems. In the case of multiple corresponding values, (M * M Rows) will be returned (i.e. value 'ACC' appears two times in a Department Table DEPT column, and value 'ACC' appears six times in an Employee Table DEPT column. The number of results rows for the 'ACC' Dept would be the product of the number of times the value is found in the Department Table and the number of times the value is found in the Employee Table. Twelve rows for 'ACC' would be returned.).

Figure 9.4d. Many to many corresponding values.

Understanding how SQL executes a Join will help make clearer how these four conditions can come about.

9.5 SQL JOIN PROCESS — CONCEPTUAL*

Whenever SQL finds two or more table names in a FROM clause, it builds a "holding" table, which is a temporary results table. This table is stored in the computer's memory and may take a great deal of space, as well as time, to build.

This table is known as a Cartesian Product Table, and you can picture it as a table that SQL creates, containing all possible row combinations from the tables you are joining:

---

* This section describes how SQL joins tables on a "conceptual" level. Physically, DB2 and SQL/DS generally choose more efficient joining processes.

Cartesian Product Table — unconditional join of Table 1 to Table 2.

Cartesian Product Table

	TABLE 1	TABLE 2
Combination 1	row 1	row 1
	row 1	row 2
	row 1	row 3
	row 1	row 4
	row 1	row n
Combination 2	row 2	row 1
	row 2	row 2
	row 2	row 3
	row 2	row 4
	row 2	row n
Combination n	row n	row 1
	row n	row 2
	row n	row 3
	row n	row 4
	row n	row n

Figures 9.5a and 9.5b illustrate this process in greater detail.

After the Cartesian Product Table is built, SQL applies the WHERE clause to it, and filters out all rows that do not meet the join condition (WHERE CUST.ID = INV.CUSTID, as shown in Figure 9.5c).

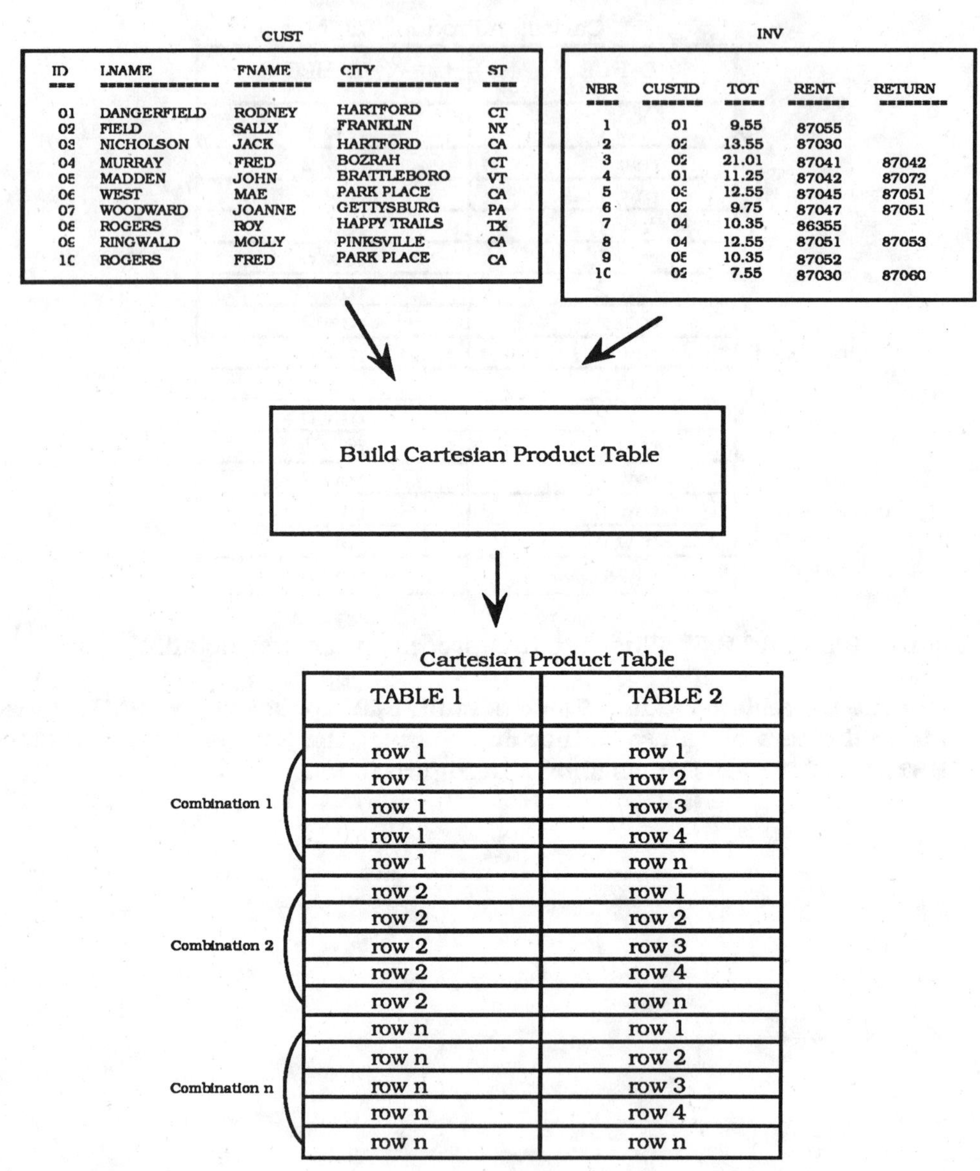

CUST

ID	LNAME	FNAME	CITY	ST
01	DANGERFIELD	RODNEY	HARTFORD	CT
02	FIELD	SALLY	FRANKLIN	NY
03	NICHOLSON	JACK	HARTFORD	CA
04	MURRAY	FRED	BOZRAH	CT
05	MADDEN	JOHN	BRATTLEBORO	VT
06	WEST	MAE	PARK PLACE	CA
07	WOODWARD	JOANNE	GETTYSBURG	PA
08	ROGERS	ROY	HAPPY TRAILS	TX
09	RINGWALD	MOLLY	PINKSVILLE	CA
10	ROGERS	FRED	PARK PLACE	CA

INV

NBR	CUSTID	TOT	RENT	RETURN
1	01	9.55	87055	
2	02	13.55	87030	
3	02	21.01	87041	87042
4	01	11.25	87042	87072
5	03	12.55	87045	87051
6	02	9.75	87047	87051
7	04	10.35	86355	
8	04	12.55	87051	87053
9	05	10.35	87052	
10	02	7.55	87030	87060

TABLE 1	TABLE 2
row 1	row 1
row 1	row 2
row 1	row 3
row 1	row 4
row 1	row n
row 2	row 1
row 2	row 2
row 2	row 3
row 2	row 4
row 2	row n
row n	row 1
row n	row 2
row n	row 3
row n	row 4
row n	row n

Figure 9.5a. Join process — Building the Cartesian Product Table.

```
SQL > SELECT LNAME, NBR, RENT FROM INV, CUST ;
```

	LNAME	NBR	RENT
Combination 1	DANGERFIELD	1	87055
	DANGERFIELD	2	87030
	DANGERFIELD	3	87041
	DANGERFIELD	4	87042
	DANGERFIELD	5	87045
	DANGERFIELD	6	87047
	DANGERFIELD	7	86355
	DANGERFIELD	8	87051
	DANGERFIELD	9	87052
	DANGERFIELD	10	87030
Combination 2	FIELD	1	87055
	FIELD	2	87030
	FIELD	3	87041
	FIELD	4	87042
	FIELD	5	87045
	FIELD	6	87047
	FIELD	7	86355
	FIELD	8	87051
	FIELD	9	87052
	FIELD	10	87030
Combination 3	NICHOLSON	1	87055
	NICHOLSON	2	87030
	NICHOLSON	3	87041
	NICHOLSON	4	87042
	NICHOLSON	5	87045
	NICHOLSON	6	87047
	NICHOLSON	7	86355
	NICHOLSON	8	87051
	NICHOLSON	9	87052
	NICHOLSON	10	87030
Combination 4	MURRAY	1	87055
	MURRAY	2	87030
	MURRAY	3	87041
	MURRAY	4	87042
	MURRAY	5	87045
	MURRAY	6	87047
	MURRAY	7	86355
	MURRAY	8	87051
	MURRAY	9	87052
	MURRAY	10	87030
Combination 5	MADDEN	1	87055
	MADDEN	2	87030
	MADDEN	3	87041
	MADDEN	4	87042
	MADDEN	5	87045
	MADDEN	6	87047
	MADDEN	7	86355
	MADDEN	8	87051
	MADDEN	9	87052
	MADDEN	10	87030
Combination 6	WEST	1	87055
	WEST	2	87030
	WEST	3	87041
	WEST	4	87042
	WEST	5	87045

Figure 9.5b. Join Process — Cartesian Product of INV and CUST.

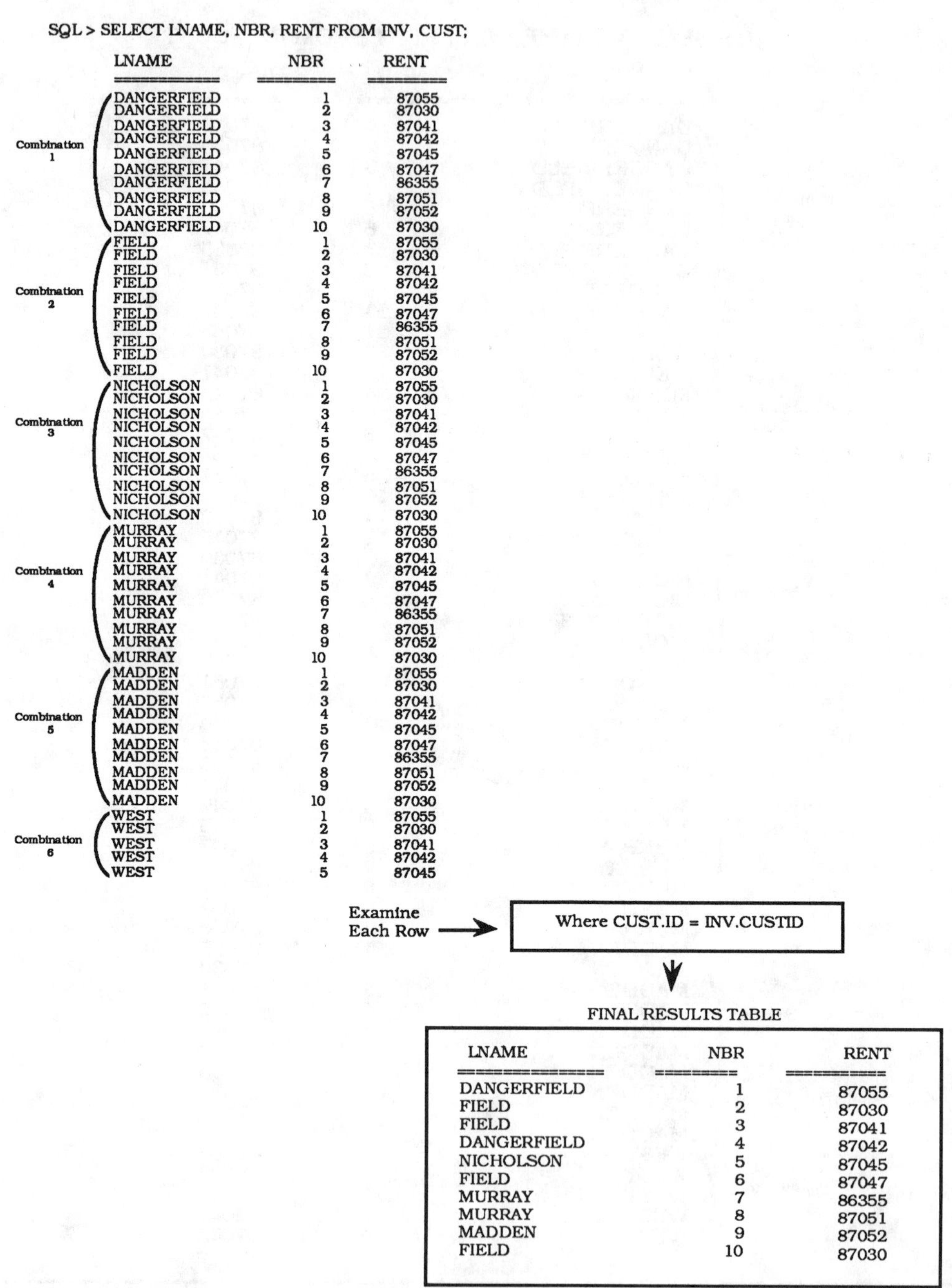

SQL > SELECT LNAME, NBR, RENT FROM INV, CUST;

	LNAME	NBR	RENT
Combination 1	DANGERFIELD	1	87055
	DANGERFIELD	2	87030
	DANGERFIELD	3	87041
	DANGERFIELD	4	87042
	DANGERFIELD	5	87045
	DANGERFIELD	6	87047
	DANGERFIELD	7	86355
	DANGERFIELD	8	87051
	DANGERFIELD	9	87052
	DANGERFIELD	10	87030
Combination 2	FIELD	1	87055
	FIELD	2	87030
	FIELD	3	87041
	FIELD	4	87042
	FIELD	5	87045
	FIELD	6	87047
	FIELD	7	86355
	FIELD	8	87051
	FIELD	9	87052
	FIELD	10	87030
Combination 3	NICHOLSON	1	87055
	NICHOLSON	2	87030
	NICHOLSON	3	87041
	NICHOLSON	4	87042
	NICHOLSON	5	87045
	NICHOLSON	6	87047
	NICHOLSON	7	86355
	NICHOLSON	8	87051
	NICHOLSON	9	87052
	NICHOLSON	10	87030
Combination 4	MURRAY	1	87055
	MURRAY	2	87030
	MURRAY	3	87041
	MURRAY	4	87042
	MURRAY	5	87045
	MURRAY	6	87047
	MURRAY	7	86355
	MURRAY	8	87051
	MURRAY	9	87052
	MURRAY	10	87030
Combination 5	MADDEN	1	87055
	MADDEN	2	87030
	MADDEN	3	87041
	MADDEN	4	87042
	MADDEN	5	87045
	MADDEN	6	87047
	MADDEN	7	86355
	MADDEN	8	87051
	MADDEN	9	87052
	MADDEN	10	87030
Combination 6	WEST	1	87055
	WEST	2	87030
	WEST	3	87041
	WEST	4	87042
	WEST	5	87045

Examine Each Row → Where CUST.ID = INV.CUSTID

FINAL RESULTS TABLE

LNAME	NBR	RENT
DANGERFIELD	1	87055
FIELD	2	87030
FIELD	3	87041
DANGERFIELD	4	87042
NICHOLSON	5	87045
FIELD	6	87047
MURRAY	7	86355
MURRAY	8	87051
MADDEN	9	87052
FIELD	10	87030

Figure 9.5c. Join condition filters out rows not meeting the condition.

To summarize this two-phase process:

1. SQL creates a large, temporary table that contains all possible row combinations from the tables you are joining — the Cartesian Product of the two tables.
2. SQL applies the WHERE clause, like a filter on the table, to sift out rows that do not meet the WHERE condition. These rows become the results table.

As you might imagine, Cartesian Product Tables can be rather large. In spite of the fact that they are only temporary structures, the amount of disk space and internal SQL processing that occurs places exhaustive demands on the system. Joins of large tables (over 100,000 rows) can produce major performance problems.

To avoid getting Cartesian Product Tables back as Results Tables, you must remember to include a Join condition in your WHERE clause that references the correct table linking columns.

9.6 TABLE-LINKING COLUMNS

Figure 9.6a shows the table-linking columns for our video database. These columns hold the "intertable values" necessary to correctly join two or more tables together. Look at the actual column information in Figure 9.6b and study the values stored in the fields that are specified as the table-linking columns.

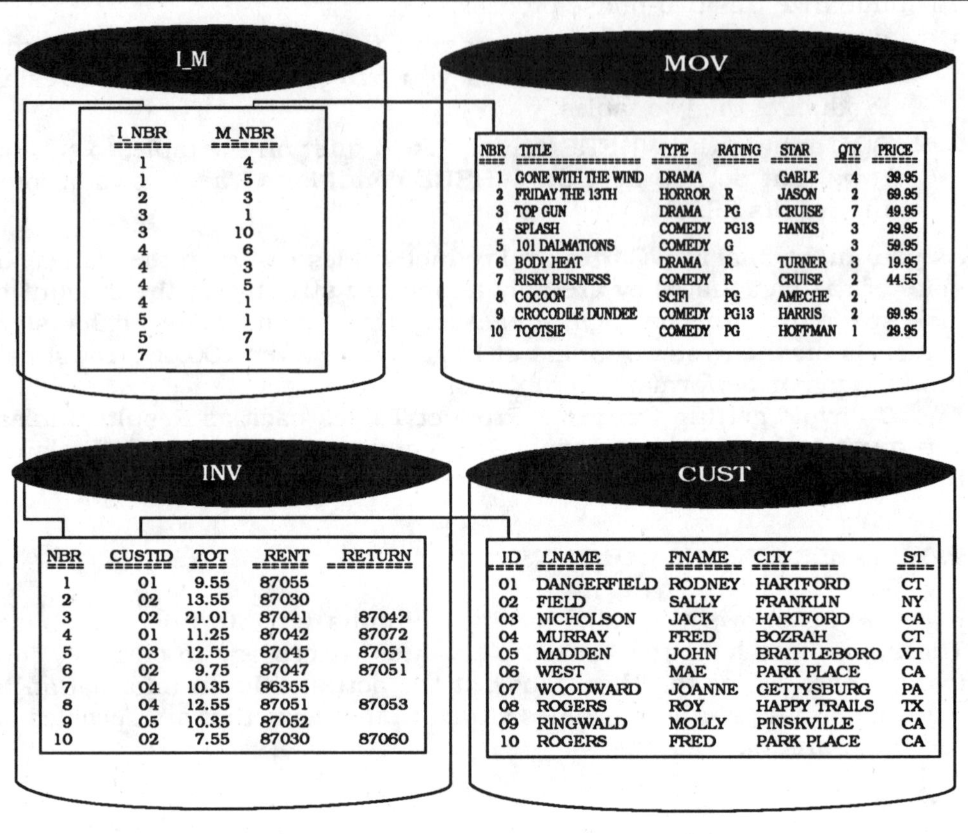

I_M

I_NBR	M_NBR
1	4
1	5
2	3
3	1
3	10
4	6
4	3
4	5
4	1
5	1
5	7
7	1

MOV

NBR	TITLE	TYPE	RATING	STAR	QTY	PRICE
1	GONE WITH THE WIND	DRAMA		GABLE	4	39.95
2	FRIDAY THE 13TH	HORROR	R	JASON	2	69.95
3	TOP GUN	DRAMA	PG	CRUISE	7	49.95
4	SPLASH	COMEDY	PG13	HANKS	3	29.95
5	101 DALMATIONS	COMEDY	G		3	59.95
6	BODY HEAT	DRAMA	R	TURNER	3	19.95
7	RISKY BUSINESS	COMEDY	R	CRUISE	2	44.55
8	COCOON	SCIFI	PG	AMECHE	2	
9	CROCODILE DUNDEE	COMEDY	PG13	HARRIS	2	69.95
10	TOOTSIE	COMEDY	PG	HOFFMAN	1	29.95

INV

NBR	CUSTID	TOT	RENT	RETURN
1	01	9.55	87055	
2	02	13.55	87030	
3	02	21.01	87041	87042
4	01	11.25	87042	87072
5	03	12.55	87045	87051
6	02	9.75	87047	87051
7	04	10.35	86355	
8	04	12.55	87051	87053
9	05	10.35	87052	
10	02	7.55	87030	87060

CUST

ID	LNAME	FNAME	CITY	ST
01	DANGERFIELD	RODNEY	HARTFORD	CT
02	FIELD	SALLY	FRANKLIN	NY
03	NICHOLSON	JACK	HARTFORD	CA
04	MURRAY	FRED	BOZRAH	CT
05	MADDEN	JOHN	BRATTLEBORO	VT
06	WEST	MAE	PARK PLACE	CA
07	WOODWARD	JOANNE	GETTYSBURG	PA
08	ROGERS	ROY	HAPPY TRAILS	TX
09	RINGWALD	MOLLY	PINSKVILLE	CA
10	ROGERS	FRED	PARK PLACE	CA

Figure 9.6a. The video store database.

These table-linking columns are sometimes referred to as key columns. They are described in the relational database world as being either Primary or Foreign keys, depending upon their significance to the table itself, and their significance to other tables in the database.

Primary Key — the column(s) that would be considered the most important column(s) in the table and uniquely identify each row (ID from the CUST table, NBR from the MOV and INV tables).

Foreign Key — the column that stores values corresponding to another table's primary key column (CUSTID from the INV table is an example of a foreign key, which relates to the ID column from the CUST table).

To join tables correctly, we link one table's foreign key to another table's primary key by means of the WHERE clause, as in Figure 9.6b.

TABLE I	JOIN TABLE II	PRIMARY KEY	FOREIGN KEY	JOIN CONDITION (WHERE CLAUSE)
CUST	INV	ID	CUSTID	WHERE ID = CUSTID
INV	I_M	NBR	I_NBR	WHERE NBR = I_NBR
I_M	INV	(The primary key consists of *both* columns	NBR	WHRE NBR = I_NBR
I_M	MOV		NBR	WHERE NBR = M_NBR
MOV	I_M	NBR	M_NBR	WHERE NBR = M_NBR

Figure 9.6b. Join condition using primary and foreign keys.

9.7 JOIN CONSIDERATIONS

Figure 9.7 summarizes the joining concepts.

Concept 1, the limit of joining fifteen tables in a single SELECT statement, would have to be considered a bit impractical, particularly in light of what we've just learned about join performance. IBM recommends a limit of no more than seven tables joined in a single statement. Four tables is even more sensible — especially if you are using medium- to large-sized tables (over 50,000 rows).

1. May join up to 15 tables in one SELECT

2. If there are identical column names referred to in the join, those column names must be prefixed by a table name to ensure uniqueness (INV.CUSTID, MOV.NBR, I_M.M_NBR)

3. Joins will retrieve all possible combinations of rows that satisfy the Join condition.

4. A Join condition that is part of the WHERE clause that creates the link between the tables necessary to restrict SELECTion to useful rows

   ```
   WHERE ID = CUSTID
   ```

5. The WHERE clause may specify multiple selection criteria:

   ```
   WHERE ID = CUSTID
   AND RETURN < 87301
   ```

6. All other selection criteria tested for in the WHERE clause must be logically AND'd together with the Tablejoin condition, or you will generate a cartesian product results table.

7. The data types for the columns in the Join condition should be compatible (either both numeric or both character)

8. Rows with a NULL value in the Joining column will never be SELECTed

9. There are four possible Join results

   1 → 1 = 1 row returned
   1 → 0 = 0 rows returned
   1 → n = n rows returned
   n → n = n*n rows returned

Figure 9.7. Join considerations.

In Concept 2, if we are joining two tables on a column where the column name is duplicated in both tables (for instance PERSONNEL and DEPT tables both referring to the Employee number column as EMPNO), we have to qualify the EMPNO columnname by prefixing it with the table name and a period. This will make our column name references clear to SQL.

9.8 SIMPLE JOIN EXERCISES

Exercise 9-1 asks that you join the MOV and I_M tables to display movie TITLE and NBR and I_M columns for all the movies that have been rented.* This is a simple join, with no extra conditions. Let's do it together, step by step.

1. First check back in Section 9.5 to find the Linking Columns between these two tables.

2. These two columns will make up your WHERE clause. Write the WHERE clause out here.

   WHERE ____;

3. Now we need the SELECT. Which columns are we requesting?

   SELECT __________, __________, __________, __________,

4. We need two tables in our FROM statement. Separate their names with a comma:

   FROM __________, __________

---

* The fact that an entry appears in I_M for a movie means that it has been rented (It is a line item on some invoice).

EXERCISE 9-1

Finally, we are ready to assemble the pieces and try it out.

❶ TOPIC: Intermediate SELECT Examples.

❷ EXERCISE 9-1. Join the MOV and I_M tables on the movie NBR values. Display the movie TITLE and NBR from the MOV table, and all of the columns from the I_M table.

❸ INPUT TABLE:

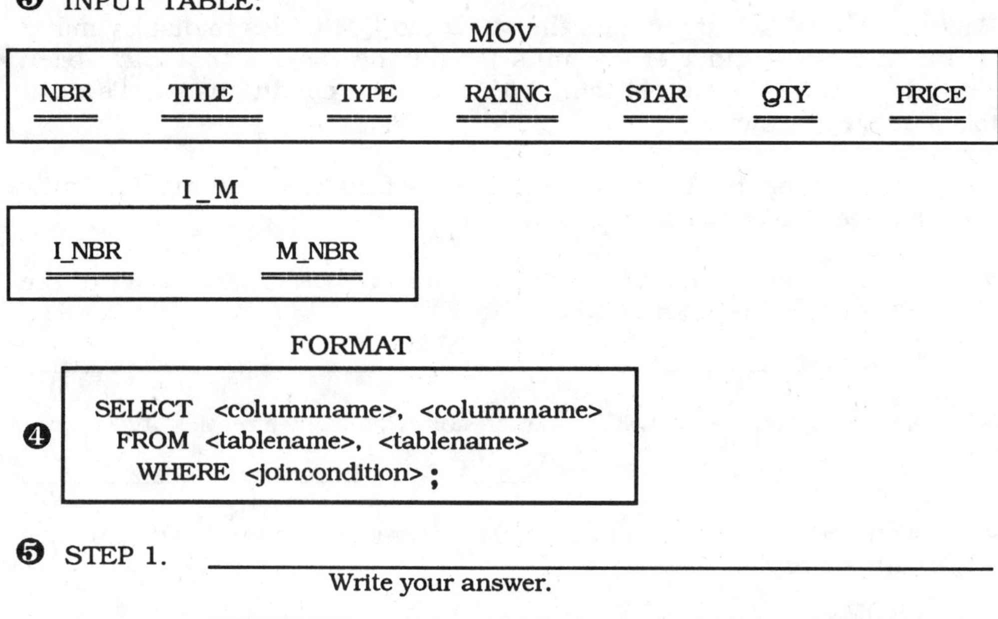

❺ STEP 1. ______________________________

Write your answer.

____

STEP 2. Enter your answer into the computer.

❻ RESULTS TABLE:

TITLE	NBR	I_NBR	M_NBR
GONE WITH THE WIND	1	3	1
GONE WITH THE WIND	1	4	1
GONE WITH THE WIND	1	5	1
GONE WITH THE WIND	1	7	1
GONE WITH THE WIND	1	10	1
FRIDAY THE 13TH	2	9	2
TOP GUN	3	2	3
TOP GUN	3	4	3
TOP GUN	3	7	3
SPLASH	4	1	4
SPLASH	4	8	4
SPLASH	4	10	4
SPLASH	4	10	4
SPLASH	4	10	4
101 DALMATIONS	5	1	5
101 DALMATIONS	5	4	5
101 DALMATIONS	5	7	5
101 DALMATIONS	5	10	5
TOP GUN	3	4	3
TOP GUN	3	7	3
SPLASH	4	1	4
SPLASH	4	8	4
SPLASH	4	10	4
SPLASH	4	10	4
SPLASH	4	10	4
101 DALMATIONS	5	1	5
101 DALMATIONS	5	4	5
101 DALMATIONS	5	7	5
101 DALMATIONS	5	10	5
BODY HEAT	6	4	6
RISKY BUSINESS	7	5	7
RISKY BUSINESS	7	8	7
CROCODILE DUNDEE	9	9	9
TOOTSIE	10	3	10

Note: **If you have trouble with this exercise, carefully review and try it again until you understand the concepts. Redo the exercise until it's clear.**

EXERCISE 9-2

❶ TOPIC: Intermediate Join Examples.

❷ EXERCISE 9-2. Display all invoice information and line item information relating to that invoice. Join the INV and I_M tables on the invoice NBR values. Display ALL of the columns and ALL of the rows on both tables.

❸ INPUT TABLE:

INV

NBR	CUSTID	TOT	RENT	RETURN

I_M

I_NBR	M_NBR

FORMAT

❹
```
SELECT * FROM <tablename>
  <tablename>;
```

❺ STEP 1. ______________________________

Write your answer.

____

STEP 2. Enter your answer into the computer.

❻ RESULTS TABLE:

NBR	CUSTID	TOT	RENT	RETURN	I_NBR	IM_NBR
1	01	9.55	87055		1	4
1	01	9.55	87055		1	5
2	02	13.55	87030		2	3
3	02	21.01	87041	87042	3	1
3	02	21.01	87041	87042	3	10
4	01	11.25	87042	87072	4	6
4	01	11.25	87042	87072	4	3
4	01	11.25	87042	87072	4	5
4	01	11.25	87042	87072	4	1
5	03	12.55	87045	87051	5	1
5	03	12.55	87045	87051	5	7
7	04	10.35	86355		7	1
7	04	10.35	86355		7	3
7	04	10.35	86355		7	5
8	04	12.55	87051	87053	8	4
8	04	12.55	87051	87053	8	7
9	05	10.35	87052		9	9
9	05	10.35	87052		9	2
10	02	7.55	87030	87060	10	4
10	02	7.55	87030	87060	10	4
10	02	7.55	87030	87060	10	1
10	02	7.55	87030	87060	10	5
10	02	7.55	87030	87060	10	4

EXERCISE 9-3

This exercise asks you to join the CUSTomer and INVoice tables, and to sort the results table. ORDER BY always comes at the end of a SELECT statement.

❶ TOPIC: Intermediate SELECT Examples (EQUI-JOIN with ORDER BY).

❷ EXERCISE 9-3. Show a list of our customers and all their invoices. Join the INV and CUST tables on the customer ID values. Display the LNAME and ID columns from the CUST table, and the NBR and CUSTID columns from the INV table. Sort the output on customer LNAME ascending.

❸ INPUT TABLE:

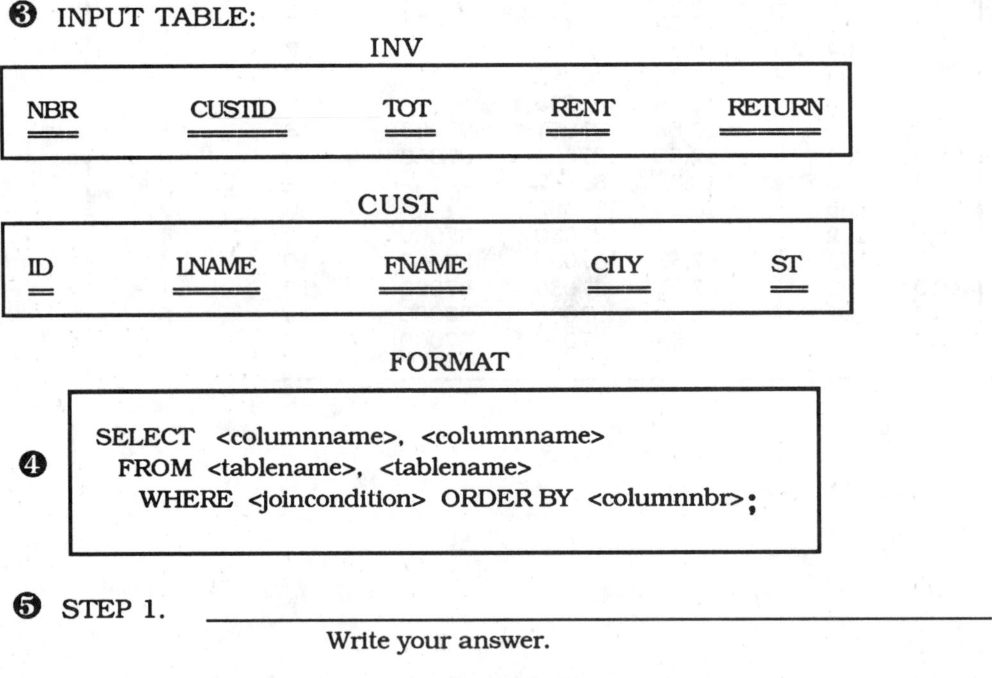

INV

NBR	CUSTID	TOT	RENT	RETURN

CUST

ID	LNAME	FNAME	CITY	ST

FORMAT

❹

```
SELECT <columnname>, <columnname>
  FROM <tablename>, <tablename>
    WHERE <joincondition> ORDER BY <columnnbr>;
```

❺ STEP 1. ______________________________________

Write your answer.

____

STEP 2. Enter your answer into the computer.

❻ RESULTS TABLE:

LNAME	ID	NBR	CUSTID
========	====	======	========
DANGERFIELD	01	1	01
DANGERFIELD	01	4	01
FIELD	02	2	02
FIELD	02	3	02
FIELD	02	6	02
FIELD	01	10	02
NICHOLSON	03	5	03
MURRAY	04	7	04
MURRAY	04	8	04
MADDEN	05	9	05

Note: **We did not get a list of all our customers. Why?**

EXERCISE 9-4

Exercise 9-4 adds an additional SELECTion condition. This is no harder than the multiple conditions we tried earlier. When you add extra conditions to a table join, it doesn't matter if you place them before or after the Join Condition.

Join
Condition ______________________________

Additional WHERE
Condition ______________________________

SELECT __________, __________, __________

FROM __________, __________

❶ TOPIC: Intermediate SELECT Examples (JOIN with additional WHERE condition)

❷ EXERCISE 9-4. Join the INV and CUST tables. Display the INV information for movies rented by Rodney Dangerfield.

❸ INPUT TABLE:

INV

NBR	CUSTID	TOT	RENT	RETURN

CUST

ID	LNAME	FNAME	CITY	ST

FORMAT

❹
```
SELECT <columnname>, <columnname>
  FROM <tablename>, <tablename>
    WHERE <joincondition> AND/OR <other condition>;
```

Hint: **A short way to SELECT all columns from a table is to place a period followed by an asterisk following the tablename. For example:**

SELECT INV.* will select all columns from the INV table.

❺ STEP 1. ______________________________

Write your answer.

____

STEP 2. Enter your answer into the computer.

❻ RESULTS TABLE:

INV.NBR	INV.CUSTID	INV.TOT	INV.RENT	INV.RETURN	ID	LNAME	FNAME
1	01	9.55	87055	99999	01	DANGERFIELD	RODNEY
4	01	11.25	87042	87072	01	DANGERFIELD	RODNEY

EXERCISE 9-5

This exercise also includes an additional WHERE condition.

❶ TOPIC: Intermediate SELECT Examples (JOIN with additional WHERE condition).

❷ EXERCISE 9-5. Join the INV and CUST tables. Display all of the last names of customers with movies to be returned.

❸ INPUT TABLE:

INV

NBR	CUSTID	TOT	RENT	RETURN
1	01	9.55	87055	
2	02	13.55	87030	
3	02	21.01	87041	87042
4	01	11.25	87042	87072
5	03	12.55	87045	87051
6	02	9.75	87047	87051
7	04	10.35	86355	
8	04	12.55	87051	87053
9	05	10.35	87052	
10	02	7.55	87030	87060

CUST

ID	LNAME	FNAME	CITY	ST
03	NICHOLSON	JACK	HARTFORD	CT
06	WEST	MAE	PARK PLACE	CA
10	ROGERS	FRED	PARK PLACE	CA
09	RINGWALD	MOLLY	PINKSVILLE	CA
04	MURRAY	FRED	BOZRAH	CT
01	DANGERFIELD	RODNEY	HARTFORD	CT
02	FIELD	SALLY	FRANKLIN	NY
07	WOODWARD	JOANNE	GETTYSBURG	PA
08	ROGERS	ROY	HAPPY TRAILS	TX
05	MADDEN	JOHN	BRATTLEBORO	VT

FORMAT

❹
```
SELECT  <columnname>, <columnname>
  FROM <tablename>, <tablename>
    WHERE <joincondition> AND/OR <other condition>;
```

❺ STEP 1. ______________________________________

Write your answer.

____

STEP 2. Enter your answer into the computer.

❻ RESULTS TABLE:

```
LNAME
=============
DANGERFIELD
FIELD
MURRAY
MADDEN
```

EXERCISE 9-6

The exercise requires an additional WHERE condition.

❶ TOPIC: Intermediate SELECT Examples (JOIN with additional WHERE condition).

❷ EXERCISE 9-6. Join the MOV and I_M tables. Display a list of movie information on 'PG13' comedy films that have been rented. Remove duplicate rows.

❸ INPUT TABLE:

MOV

NBR	TITLE	TYPE	RATING	STAR	QTY	PRICE
1	GONE WITH THE WIND	DRAMA		GABLE	4	39.95
2	FRIDAY THE 13TH	HORROR	R	JASON	2	69.95
3	TOP GUN	DRAMA	PG	CRUISE	7	49.95
4	SPLASH	COMEDY	PG13	HANKS	3	29.95
5	101 DALMATIONS	COMEDY	G		3	59.95
6	BODY HEAT	DRAMA	R	TURNER	3	19.95
7	RISKY BUSINESS	COMEDY	R	CRUISE	2	44.55
8	COCOON	SCIFI	PG	AMECHE	2	
9	CROCODILE DUNDEE	COMEDY	PG13	HARRIS	2	69.95
10	TOOTSIE	COMEDY	PG	HOFFMAN	1	29.95

I_M

I_NBR	M_NBR
1	4
1	5
2	3
3	1
3	10
4	6
4	3
4	5
4	1
5	1
5	7
7	1
7	3
7	5
8	4
8	7
9	9
9	2
10	4
10	4
10	1
10	5
10	4

FORMAT

❹
```
SELECT  <columnname>, <columnname>
  FROM <tablename>, <tablename>
    WHERE <joincondition>
      AND/OR <other condition>;
```

❺ STEP 1. ______________________________

Write your answer.

____

STEP 2. Enter your answer into the computer.

❻ RESULTS TABLE:

MOV.NBR	MOV.TITLE	MOV.TYPE	MOV.RATING	MOV.STAR	MOV.QTY	MOV.PRICE
4	SPLASH	COMEDY	PG13	HANKS	3	29.95
9	CROCODILE DUNDEE	COMEDY	PG13	HARRIS	2	69.95

Hint: **The fact that the movie NBR exists on the I_M Table in the M_NBR column means that the movie has been rented.**

9.9 TABLE JOINS — INTERPRETING SQL

1. SELECT MOV.* FROM MOV, I_M WHERE STAR BETWEEN 'HA' AND 'HZ' AND NBR = M_NBR;
2. SELECT * FROM CUST, INV, I_M, MOV
   WHERE CUSTID = ID AND
   INV.NBR = I_NBR AND
   MOV.NBR = M_NBR AND
   RATING = 'R';

9.10 ANSWER KEY

There may be more than one answer to an exercise. However, the one given emphasizes the topic discussed.

EXERCISE 9-1	SELECT TITLE, NBR, I_NBR, M_NBR FROM MOV, I_M WHERE NBR = M_NBR;
EXERCISE 9-2	SELECT * FROM INV, I_M WHERE I_NBR = NBR;
EXERCISE 9-3	SELECT LNAME, ID, NBR, CUSTID FROM INV, CUST WHERE ID = CUSTID ORDER BY LNAME;
EXERCISE 9-4	SELECT INV.*, ID, LNAME, FNAME FROM INV, CUST WHERE ID = CUSTID AND LNAME = 'DANGERFIELD';
EXERCISE 9-5	SELECT LNAME FROM INV, CUST WHERE ID = CUSTID AND RETURN IS NULL;
EXERCISE 9-6	SELECT DISTINCT MOV.* FROM MOV, I_M WHERE NBR = M_NBR AND TYPE = 'COMEDY' AND RATING = 'PG13';

Chapter 10
DATA DEFINITION STATEMENTS: CREATE, DROP, ALTER

PROCESSING CAPABILITIES

We recall that the only way to communicate with the DB2 or SQL/DS kernel is through the SQL language. Because of this, the SQL language contains commands for the complete spectrum of data processing activities. These include:

- Accessing (retrieving), and manipulating data: We've seen that SQL provides SELECT statements to accomplish this.
- Updating data: This includes maintaining the data in our database, and manipulating the stored values to keep the information current.
- Creating (defining), deleting, and modifying tables: The database we use must be kept current for our requirements.
- Controlling access to our data: System security and authorization must be controlled and maintained.

We've finished looking at SQL's SELECTion features for awhile. Now you will learn about data definition statements (CREATE, DROP, and ALTER).

The following topics will be discussed.

10.1 SQL Data Definition Statements
10.2 Creating Tables
10.3 SQL Data Names
10.4 SQL Character Data Types
10.5 SQL Numeric Data Types
10.6 Table Definition Exercise — CREATE TABLE
10.7 Table Definition Exercise — ALTER TABLE
10.8 Table Definition Exercise — DROP TABLE
10.9 Date/Time/Timestamp Datatypes

10.1 SQL DATA DEFINITION STATEMENTS

There are three SQL statements, called data definition statements, that allow you to define, destroy, and modify the tables in a database.

1. CREATE — Defines a new table to the database.
2. DROP — Removes an existing table from the database.
3. ALTER — Changes the structure of an existing table.

The CREATE and DROP verbs are used to define and delete other data objects besides tables (see Figure 10.1), including INDEXes, VIEWs, DATABASEs, and other more "physical" system objects like DBSPACES, STORPOOLS, TABLESPACEs and STORAGE GROUPs. ALTER is used mainly to modify table structures. We will only be learning about how to define, delete, and modify tables, indexes and views in this book.

10.1.1 SQL Data Access Controls

Remember that relational systems such as DB2 or SQL/DS are capable of executing all of the data definition statements, in fact, all SQL statements AT ANY TIME THE SYSTEM IS RUNNING. What this means is that at the same moment you might be querying a table:

- A technical support or database administrator could be reorganizing the table;
- an online program could be accessing or updating the table;
- another system user could be querying the same table;
- another system user could be adding a column to the table;
- another database administrator could be loading the table with new data;
- and so on!

This capability of concurrent processing could lead to system anarchy were it not for a sophisticated set of internal and external controls, the system security and authorization features. DB2 and SQL/DS support a complete arsenal of such mechanisms; they are covered in Chapter 17.

A final note before we continue: If you are working on a mainframe system, you will need to get what can be loosely termed "Resource Authority" from your database administrator or information-center personnel to be able to execute the data definition statements in this chapter. Contact them at this point if you are not sure that your ID currently has the authority to define, delete, and modify tables.

There are just a few considerations for using the SQL data definition statements. These are shown in Figure 10.1.

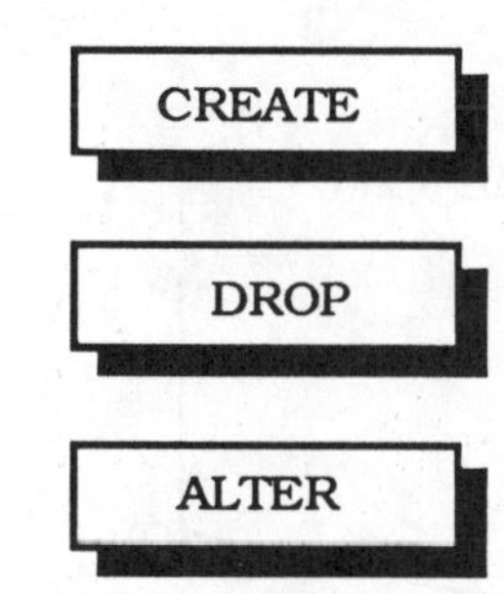

- You can CREATE and DROP base TABLES
- You can ALTER TABLES
- CREATE, DROP, and ALTER can be executed at any time
- This process, known as Dynamic Data Definition, allows for a flexible design process
- In order to use SQL data definition statements a user must have Resource Authority
- As the creator of a table, you have implicit authority for all processing privileges against that table.
- When you create a table, your user-id (logon-id) becomes a high-level Qualifier

Figure 10.1. Data definition statement considerations.

10.2 CREATING TABLES

Like SELECT, the CREATE TABLE statement is logical and easy to read and understand. Look at Figure 10.2. As you can see, it is composed of four major parts:

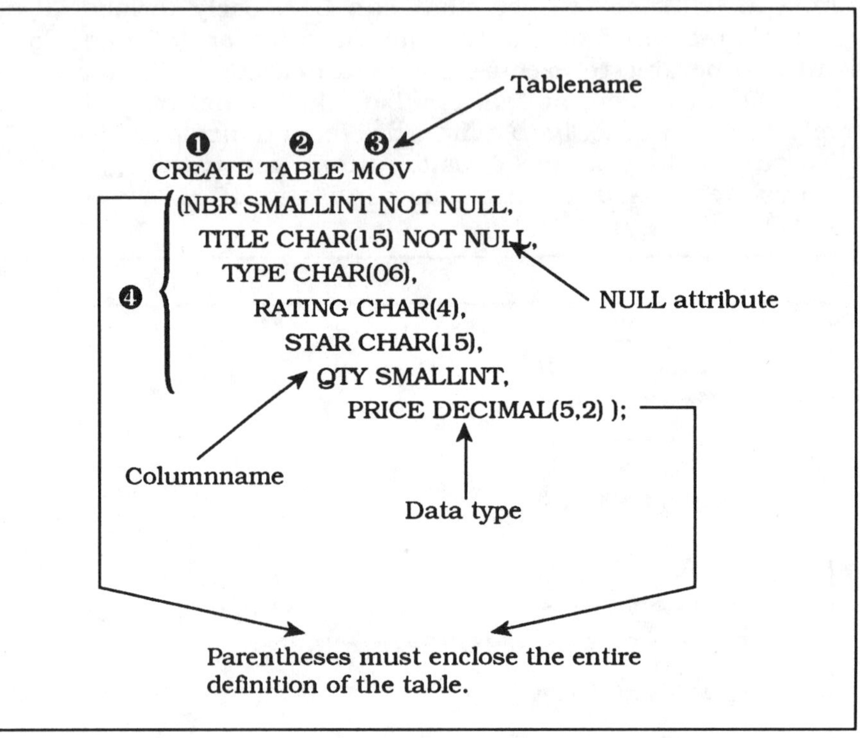

Figure 10.2. CREATE table (statement example).

❶ CREATE: This tells SQL to create a structure.

❷ TABLE: This tells what type of structure to create, in this case, a TABLE.

❸ <tablename>: Your table name must be unique within the tables previously defined by you, and must be a "legal" table name. ("Legal" names are discussed in section 10.3.)

❹ COLUMNS: For each column you create, you will need to state the following:

a. <columnname>: Your column name must be unique within the table, and must be a "legal" column name.

b. The column data type — Remember these from the chapter on WHERE clauses? We will be going into a little more depth about these in the next few chapters.

c. The NULL specifier — This consists of whether or not to permit NULL values in the column. The default is to permit them. If they are not wanted, you must code NOT NULL next to the column's data type.

There is a maximum limit of 255 columns in an SQL/DS table and 300 columns in a DB2 table.

When you execute a CREATE TABLE statement, a table description is placed in the system catalog defining your table and its columns to DB2 or SQL/DS. At this point, you've created an empty table structure. It has no data in it. Rows must be loaded into the table using SQL INSERT statements. There are other ways of loading data into tables, but we will not be covering them in this book.

Let's take a closer look at SQL table and column names.

10.3 SQL DATA NAMES

The rules for SQL naming conventions apply to almost all the logical data objects in the system — Tables, Columns, Views, and Indexes (as described in earlier). See Figure 10.3.

Naming convention rules are as follows:

1. Data names can be up to and including 18 characters in length.

2. Data names must start with an alphabetic or NAT.CHAR (national character).

   (Alphabetic characters are all the letters of the alphabet. NAT.CHAR are the special symbols @, #, $.)

3. The remaining characters in your data name can be composed of:

   a. Alphabetic characters (A–Z);

   b. Numbers (0 thru 9);

   c. National characters (@, #, $);

   d. The underscore (_). (Remember our I_M table, with its I_NBR and M_NBR columns?)

4. The last rule is that SQL data names may not contain DB2 or SQL/DS keywords (words reserved for SQL's use). A list of these words is shown in the Appendix.

- Valid names for tables, indexes, views, and columns:

  - Can be up to 18 characters long
  - Can be qualified (with USERID)

  - Must start with either an
  - ALPHABETIC character or a
  - NATIONAL character ($#@)

  - May be completed with:
  - ALPHABETIC character
  - NUMBER
  - NATIONAL character ($#@)
  - UNDERSCORE (_)

  - MAY NOT contain DB2 or SQL/DS reserved words (See Appendix 4)

Figure 10.3. SQL data names.

10.4 SQL CHARACTER DATA TYPES

As we've already learned, there are basically two categories of data types in a relational system — character and numeric. Which one you use to define your columns depends on:

1. The type of information you intend to store in the column;

2. Whether or not you will use the column arithmetically. If you anticipate either calculating results with it, or doing some kind of mathematical statistics on it, like averaging or summing the values in the columns, then it must be defined as numeric.

If you will be storing any nonnumeric data in a column (any other characters besides the numbers 0 thru 9, the plus or minus sign (+,-) and the decimal point), you must define the column with a character data type. You define a character data type with the word CHAR or VARCHAR (for variable characters) and a number enclosed by parentheses. There are two types of character data definitions:

Definition	*Meaning*
CHAR(n)	A column with a *fixed length of n* is defined. Example: CHAR(24) — The system sets aside 24 characters of storage for each field in your column. This column can hold a field up to and including 24 characters in length.
VARCHAR(n)	A column with a *maximum length of n* is defined. Example: VARCHAR(24) — The system sets aside up to and including 24 characters of storage for each field in your column. However, if the field is less than 24 characters long, DB2 trims the column to fit the data and saves storage space.

The maximum size for a CHAR or VARCHAR column is 254 characters. If VARCHAR columns save space, why not use them all the time? The reason is that at run-time the system must do some extra processing to calculate the actual size of each field, so unless the size of your data varies widely, the added expense of VARCHAR processing can outweigh the advantages of the space savings. Now let's look at the numeric data types.

10.5 SQL NUMERIC DATA TYPES

You define numeric columns with one of three SQL keywords: SMALLINT, INTEGER, or DECIMAL. Which one you use will depend on the kind and size of numeric data you intend to store in the column.

Definition	*Meaning*
SMALLINT	A column that will store small positive and negative integers — maximum *value* approximately ± 32,760.

INTEGER — A column that will store much larger positive and negative integers — maximum *value* approximately ±2,140,000,000.

DECIMAL (p,s) — A column that will store positive and negative decimal values. The *p, s* stands for precision and scale. *Precision* refers to the total number of digits in the number. *Scale* refers to the number of digits to the right of the decimal place. Decimal fields have a maximum precision of 15 digits (e.g., 123.456789012345 or .123456789012345).

If you are only storing whole numbers (integers), you have your choice of SMALLINT and INTEGER. If you need to store values with decimal numbers, you will have to use the DECIMAL data type. See Figure 10.4.

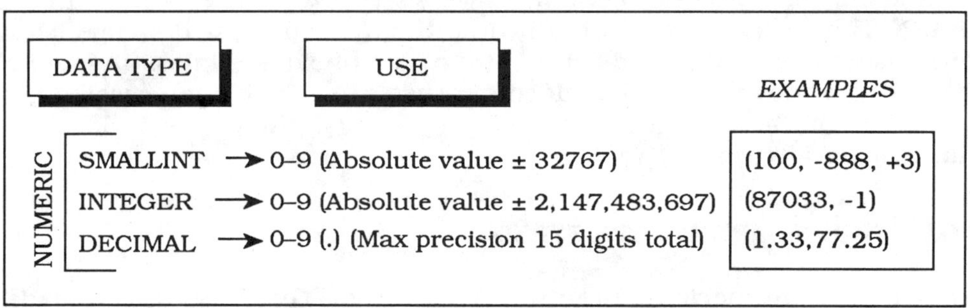

Figure 10.4. Numeric data types (most frequently used).

If you do not need to store decimal values, use SMALLINT or INTEGER. They are the most efficient numeric column data types.

To review all data types, see Figure 10.5.

	Length or Value	Description
CHAR	total length n <= 254	All fields the same or about the same length.
VARCHAR	total length n <=254 *	Fields differ in length or exceed 254 characters.
SMALLINT	total value n <=32767	Whole numbers
INTEGER	total value 2.1 billion	Whole numbers
DECIMAL	total length up to 15 digits precision	Decimal numbers

* There is a way to define columns with lengths greater than 254 characters. It is beyond the scope of this book and is generally considered to be very inefficient.

Figure 10.5. Data Types — Summary.

10.6 TABLE DEFINITION EXERCISE—CREATE TABLE

Let's look at a simple example of CREATE TABLE. We are going to create a table to hold customer savings account information. A description of the table is shown in Figure 10.6.

Table Name => CUST_SAV_ACCOUNT

	Columnname	Type/Size	NULL/Not NULL
1.	ACC_NO	INTEGER	Do Not Allow NULLS
2.	CUST_LNAME	VARCHAR(25)	Do not allow NULLS
3.	CUST_FNAME	CHAR(12)	Allow NULLS
4.	ACC_BALANCE	DECIMAL 10 chars total price 2 chars to right of decimal place.	Allow NULLS

Figure 10.6. CREATE TABLE example.

You may create this table by typing this statement and executing it.

```
CREATE TABLE  CUST_SAV_ACCOUNT

(ACC_NO  INTEGER  NOT  NULL,

CUST_LNAME  VARCHAR (25)  NOT NULL,

CUST_FNAME  CHAR(12),  ACC_BALANCE  DECIMAL (10,2));
```

Now try writing your own CREATE TABLE statement in the following exercise.

Our table definition task requires us to CREATE a table that will store information regarding our personal record collection. There are three separate items we want contained on the table:

1. REC_NBR: This column will contain the catalog identification number (or code) for the record.
2. TITLE This column will contain the record's title.
3. PERFORMER This column will contain the last name of the major artist or performing group on the record.

Code the SQL statement to CREATE our Record table, type it in, and execute it. If you are using DB2 or SQL/DS and find that your statement fails because you don't have the proper authority to CREATE tables, you can still see a correct statement by looking at the solutions presented at the Answer Key.

EXERCISE 10-1

❶ TOPIC: Data Definition Examples (CREATE TABLE)

❷ EXERCISE 10-1. Use the CREATE TABLE statement to create a record collection table, using the following specifications:

❸ INPUT TABLE:

COLUMNS	LENGTHS	ATTRIBUTES
REC_NBR	8	CHAR NO NULLS
TITLE	20	CHAR NO NULLS
PERFORMER	30	CHAR NULLS Allowed

CREATE TABLE FORMAT

❹

```
CREATE TABLE <tablename>

(<columnname>     <data type>     <NULL attribute>,
 <columnname>     <data type>     <NULL attribute>
 <columnname>     <data type>     <NULL attribute>) ;
```

❺ STEP 1. ______________________________

Write your answer.

____

STEP 2. Enter your answer into the computer.

❻ RESULTS TABLE:

TABLE CREATED

10.7 TABLE DEFINITION EXERCISE — ALTER TABLE

ALTER TABLE is used to modify the structure of an existing table. The table can be either empty or it can already contain data. In either case, when you execute an ALTER statement, SQL extends your table by adding a single column to the right of your current columns, and fills it with NULL values.*

This extension of a table only creates a NULL-filled column, it does not physically update any records. You have to do this explicitly using SQL UPDATE statements. ALTER TABLE is easy to code. As you can see from the format in Figure 10.7, it is made up of 4 parts:

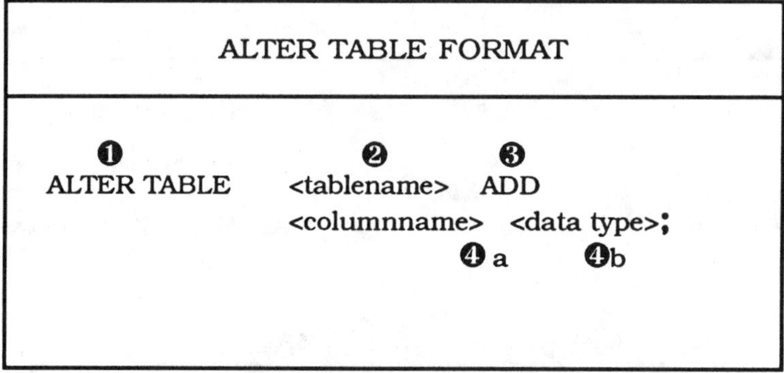

Figure 10.7. ALTER TABLE format.

* Note to SQL/DS Users. You may not alter tables and add columns that don't allow nulls (NOT NULL). This can be done with DB2.

❶ The SQL keywords ALTER TABLE.

❷ The table name of the table you wish to ALTER. (If you did not create the TABLE you will have to get authority to ALTER it from the table creator. You already have the authority for your own tables.)

❸ The SQL keyword ADD.

❹ The new column, which in turn is composed of 2 parts,

a. the column name

b. the column data type

For the DB2 system, the following table shows what will be inserted into new columns, if you code NOT NULL WITH DEFAULT:

Value	**Column data type**
zeros	SMALLINT, INTEGER, DECIMAL
blanks	CHAR, VARCHAR

The ALTER Example shown below would add a TIME RENTED column to our movie table. The column would be a numeric small integer data type.

Example:

```
ALTER TABLE MOV
     ADD TIME_RENTED SMALLINT;
```

EXERCISE 10-2

❶ TOPIC: Data Definition Examples (ALTER TABLE)

❷ EXERCISE 10-2. Use the ALTER table statement to add the following two columns to your Record table:

❸ INPUT TABLE:

COLUMNS	LENGTHS	ATTRIBUTES
CONDITION	4	CHAR
PRICE	5,2	DECIMAL

FORMAT

❹
```
ALTER TABLE <tablename>
ADD <columnname>  <datatype>;
```

❺ STEP 1. ______________________________

Write your answer.

____

STEP 2. Enter your answer into the computer.

❻ RESULTS TABLE:

COLUMN(S) ADDED

***Hint:* How do you add two columns with a single statement (or can you)?**

10.8 TABLE DEFINITION EXERCISE — DROP TABLE

DROP is the SQL verb used to delete tables from the system. It is easy to understand, and almost too easy to use. All that is necessary to delete entire tables is to execute the DROP TABLE statement. Once executed, the table is gone — all rows, any INDEXes associated with the table, and any VIEWs associated with the table. The system does NOT give a "last chance to bail out" prompt. Your data is gone for good! See Figure 10.8.

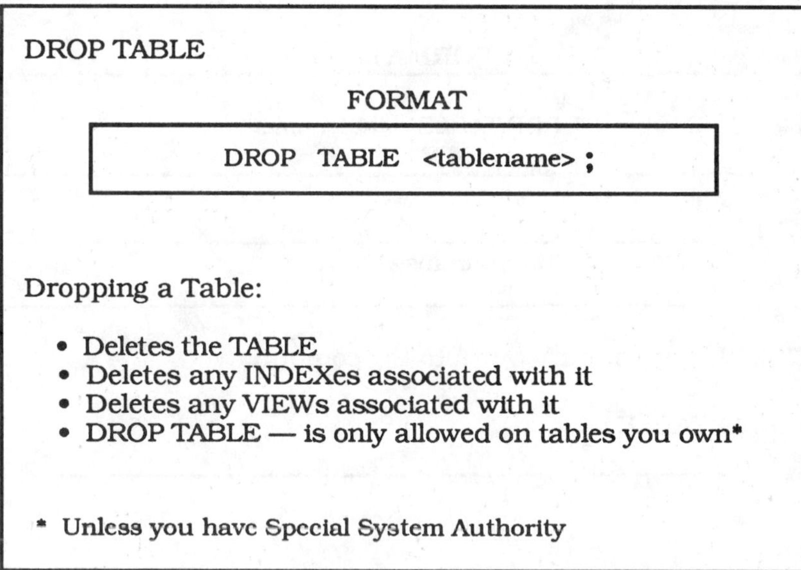

Figure 10.8. DROP TABLE statement.

SQL does ensure that you have the authority to DROP tables (usually only ones that you've created), before it executes the statement. However this is a dangerously powerful verb. It is particularly important to check and recheck the spelling of the table name to be deleted. You might inadvertently delete the wrong one! Now go ahead, and DROP the records table you just created.

EXERCISE 10-3

❶ TOPIC: DROP TABLE.
❷ EXERCISE 10-3. Drop the RECORD table
❸ INPUT TABLE:

We did not add any data into the table. Therefore, it is empty.

FORMAT

❹

DROP TABLE <tablename>;

❺ STEP 1. ______________________________
Write your answer.
____

STEP 2. Enter your answer into the computer.

❻ RESULTS TABLE:

TABLE DROPPED

❼ If your screen does not look like this, try again or look at the answer key and execute the solution given there.

10.9 DATE/TIME/TIMESTAMP DATATYPES

If you are working on a mainframe with SQL/DS 2.2 or later, DB2 Version 1.3, 2.1 or 2.2 (as of this writing, most installations are using DB2 1.3 and 2.1), you will be able to define columns that store valid date, time or timestamp values. These datatypes can be extremely useful in applications that need to do date or time dependent processing.

10.9.1 Date/Time Representation

DATE/TIME/TIMESTAMP values are represented internally by DB2 and SQL/DS in a format that is meant to be transparent to DB2 and SQL/DS system users. Externally, these datatypes are represented as character strings in one of four valid presentation formats illustrated in Figure 10.9. Which one you see on your screen depends on which one of the four your installation chose as a standard. We will be using ISO as the presentation format for this course. You will know which representation you are using when you complete the exercises.

Date Representation

FORMAT NAME	ABBREV.	FORMAT	EXAMPLE
INTERNATIONAL	ISO	yyyy-mm-dd	1986-12-25
STANDARDS ORG. IBM/USA	USA	mm/dd/yyyy	12/25/1986
IBM EUROPEAN	EUR	dd.mm.yyyy	25.12.1986
JAPANESE INDUSTRIAL STANDARDS	JIS	yyyy-mm-dd	1986-12-25

Time Representation

FORMAT NAME	ABBREV.	FORMAT	EXAMPLE
INTERNATIONAL	ISO	HH.MM.SS	13.01.05
STANDARDS ORG. IBM/USA	USA	HH.MM (AM or PM)	1:01 PM
IBM EUROPEAN	EUR	HH.MM.SS	13.01.06
JAPANESE INDUSTRIAL STANDARDS	JIS	HH:MM:SS	13:01:05

Figure 10.9. Date/Time/Timestamp datatypes.

10.9.2 Date Character Strings

A character string representation of a valid date is dependent on the type of external representation your system uses. In ISO format, a date will be defined as:

'YYYY-MM-DD'

Four digits to represent the year, a dash, two digits for the month, a dash and two digits to represent the day. Example:

'1989-09-23' - September 23, 1989.

When you define a table column as a DATE datatype, you simply use the reserved word DATE in the column attribute position. Columns with a date attribute can also allow NULLs and can be defined as NOT NULL WITH DEFAULT. The default NULL value for date datatypes is the current date.

```
CREATE TABLE DATE_TAB
(DATE_1  DATE,                            —   note, allow NULL
                                              values
 DATE _2 DATE NOT NULL,                   —   NULL value not
                                              allowed
 DATE_3  DATE NOT NULL WITH DEFAULT);     —   current date inserted
                                              in column for NULLs
```

SQL/DS does not allow default fill characters as of this writing. It is assumed IBM will add this feature in a subsequent release.

10.9.3 Time Character Strings

A character string representation of a valid date is also dependent on the type of external representation your system uses. In ISO format, a time will be defined as:

'HH.MM.SS'

Two digits to represent the hour, a period, two digits for the minute, a period and two digits to represent the second. Example:

'13.01.01' — One minute and one second after one o'clock PM.

When you define a table column as a TIME datatype, you simply use the reserved word TIME in the column attribute position. Columns with a time attribute can also allow NULLs and can be defined as NOT NULL WITH DEFAULT. The default NULL value for time datatypes is the current time.

```
CREATE TABLE TIME_TAB
(TIME_1  TIME,                            —   note, allow NULL
                                              values
 TIME_2  TIME, NOT NULL,                  —   NULL value not
                                              allowed
 TIME_3  TIME NOT NULL WITH DEFAULT;      —   current time in
                                              serted in column for
                                              NULLs
```

10.9.4 Timestamp Datatypes

A timestamp is a seven part value — a concatenation of:

YEAR–MONTH–DAY–HOUR.MINUTE.SECOND.MICROSECOND

Example: '1991–01–01–01.01.01.000001' — The first day, first hour, first minute, first second and first microsecond of 1991.

Timestamp columns are mainly used for data entry applications where operator transactions are timestamped for uniqueness, and scientific or manufacturing real-time applications.

EXERCISE 10-4 — Using DATE/TIME Datatypes

Create the DAYTIMER table as specified in the figure below.

COLUMNS	LENGTH	ATTRIBUTES
THE_DATE_IS	N/A	DATE — NO NULLS
THE_TIME_IS	N/A	TIME — NO NULLS
I_SHOULD_BE_DOING	50	VARIABLE CHARACTER NO NULLS

If you are using DB2 or SQL/DS on the mainframe, enter the statements necessary to define the table and execute them.

After your table definition is successful, insert some rows with the following statement format (We will cover this statement in depth in the next lesson).

```
INSERT INTO DAYTIMER VALUES
('1989-10-01','10.42.01','GET HAIRCUT — NO BEARD TRIM')
```

You may alter the values within the single quotes to make up your own row data. After you have successfully inserted several rows, execute SELECT * on the DAYTIMER table to retrieve your data.

10.10 CREATING TABLES IN SPECIFIC DBSPACES AND DATABASES

When you create tables using the statements we've covered so far in this chapter, the table is placed in a default DB2 database (DSNDB04), or your private DBSPACE in SQL/DS. This DBSPACE or DATABASE can grow too large, so many installations will create separate ones for each application. To put your table in a specific DBSPACE or DATABASE you must code an extension to the CREATE TABLE statement telling the user where to put the TABLE (see the examples, below).*

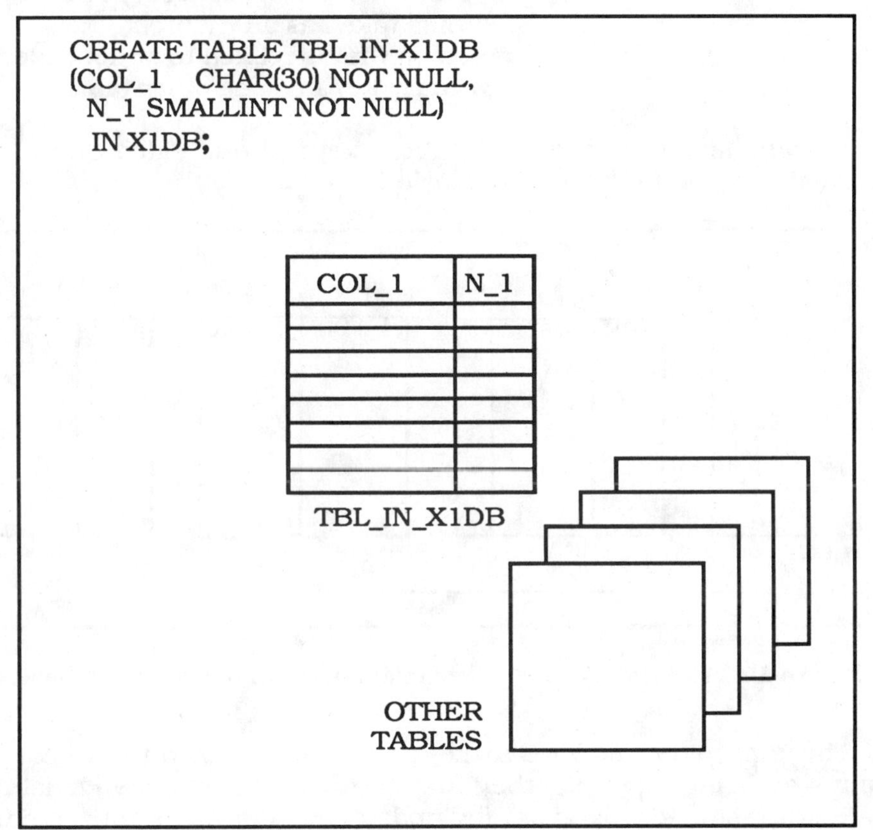

Figure 10.10. Example for SQL/DS-DBSPACE-X1DB.

* An example to create the same table in a specific DB2 database is:

```
CREATE TABLE TBL_IN_X1DB
    (COL_1 CHAR(30) NOT NULL, N_1 SMALLINT NOT NULL)
    IN DATABASE X1DB;
```

10.11 PRIMARY AND FOREIGN KEYS

Part of the table design process, is to identify key columns in each of our tables. These key columns establish row identity, and relationships among tables in a database. Recall from what we learned about table joins that there are two types of keys we are interested in identifying in our design:

Primary Key Columns — the column or combination of columns that uniquely identifies each row (sometimes a single column is not enough to guarantee uniqueness, but several columns taken as a unit are).

Foreign Key Columns — a column that is related to or dependent on a primary key in some other table.

Examples of primary and foreign keys in our Video Store Database are illustrated in Figure 10.11 below.

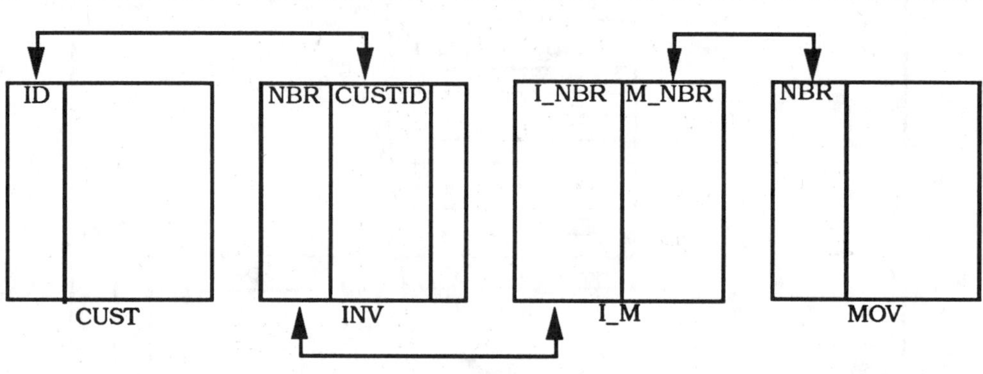

Figure 10.11. Primary/foreign key relationship in the video database.

These primary and foreign keys actually do more than designate columns as being "special," they maintain the logical dependencies for update processing in a database (by update here I mean all three modes of update—INSERT, DELETE and UPDATE processing).

10.12 DEFINING LOGICAL DEPENDENCIES FOR UPDATE PROCESSING

In the process of identifying primary and foreign keys for a table, it usually starts to become apparent that for data integrity to be guaranteed, update events should follow some logical sequence. This sequence is generally hierarchical in nature and conforms to the implicit parent/child relationships established through primary and foreign keys. This is illustrated below in Figure 10.12.

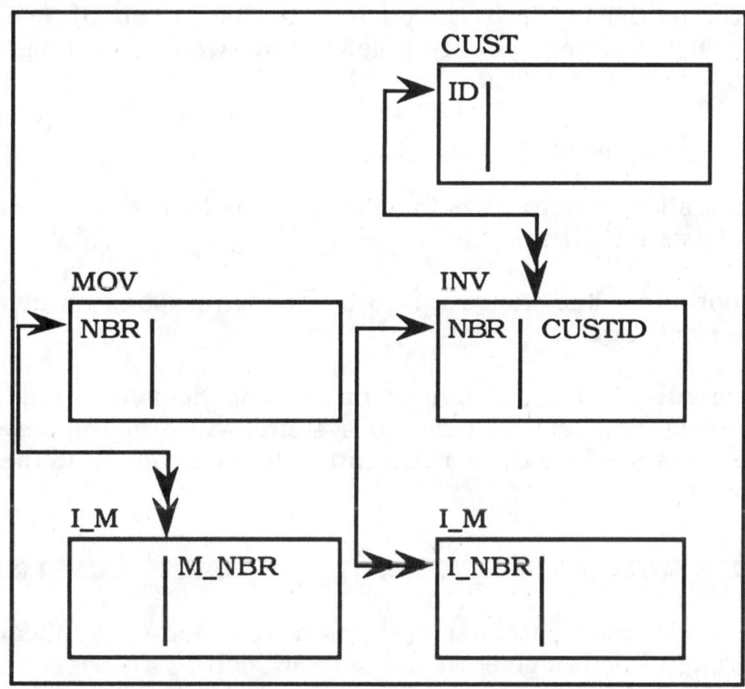

Figure 10.12. Logical dependencies for video database.

The illustration points out the dependent nature of the information in the tables of the database. Certain processing rules become clear from this illustration:

APPLICATION RULES FOR INSERT:

A row must exist for the customer in the CUST table before an Invoice is created: (ID<——————>> CUSTID)

APPLICATION RULES FOR INSERT

An invoice must exist in the INV table before line items are created in the I_M table: (NBR<————>> I_NBR)

The row inserted into the I_M table must also specify a valid movie in the MOV table: (M_NBR<<————> MOV.NBR)

APPLICATION RULES FOR DELETE:

If a customer is no longer a member, then that customer's row should be deleted from the customer table. Also, all invoices for that customer should be deleted, as well as all line item rows for those invoices.

Because of this strategy:

Do not allow invoice rows to be deleted, as long as a customer row exists for them.

Do not allow line item rows to be deleted as long as invoice rows exist for them.

These rules reflect business processing decisions made by the database designers. Your particular systems will probably have similar strategies. Let's see how these rules can be implemented into the physical table definition.

10.13 DB2 AND SQL/DS REFERENTIAL INTEGRITY CONSTRAINTS

Up to now, relational database systems have relied on application logic and a thorough knowledge of the table relationships and dependencies to maintain data integrity rules (called referential integrity). This reliance on programmer and/or end-user skills and up-to-date information on the state of the database caused the data processing industry to clamor for DBMS maintained referential integrity. Finally, IBM complied and defined extensions to the CREATE TABLE statement shown below:

PRIMARY KEY (column_names)

FOREIGN KEY (column_names) REFERENCES <primarykeytable> ON DELETE <option>

10.13.1 Specifying Primary Keys

The PRIMARY KEY specification defines a primary key composed of the identified column or columns. Considerations include:

- You can only specify one primary key per table
- Each column name can be named only once
- Each column must be declared as NOT NULL or NOT NULL WITH DEFAULT
- You are limited to sixteen columns per primary key
- The primary key must be indexed, and the table is unavailable for all intents and purposes until the index is defined
- SQL/DS automatically creates a unique index as the primary key. DB2 does not.*

Example:

```
CREATE TABLE CUST
      (ID        CHAR(2) NOT NULL
       LNAME     CHAR(8),
       FNAME     CHAR(8),
       CITY      CHAR(10),
       ST        CHAR(2)
PRIMARY KEY (ID));
```

10.13.2 Specifying Foreign Key Dependent Relationships

Each FOREIGN KEY specification defines a foreign key composed of the identified column or columns. Considerations include:

- FOREIGN KEY defines a referential integrity constraint — a parent/child relationship between the FOREIGN KEY table, and a PRIMARY KEY table
- You can specify many foreign keys per table
- Each column name can be named only once
- The primary key table named must exist and contain a defined primary key

---

* To create a unique index for a primary key in DB2 use the following type of statement:

```
CREATE UNIQUE INDEX <indexname>
      ON <TSOID.tablename (<columnlist>);
CREATE UNIQUE INDEX CUST_IDX
      ON QED.CUST (ID);
```

- The specified foreign key must have the same number, position, size and type of columns as the primary key of the specified primary key table
- You can optionally name the foreign key with a constraint name which is coded before the column names, and must be unique within the table
- SQL/DS will create a non-unique index on the foreign key. DB2 will not.*
- The columns that compose a FOREIGN KEY must match the PRIMARY KEY columns from the specified primary key table in position, datatype attribute, and NULL attribute

The delete rule determined by the ON DELETE clause has several options:

CASCADE — will delete all rows of the FOREIGN KEY table where FOREIGN KEY column values are equal to the specified PRIMARY KEY column values (basically what we want to happen in our CUST/INV/I_M relationship when a customer row is deleted). SQL/DS cascading updates and deletes are limited to one level below the parent table.

RESTRICT — will not allow the row of the PRIMARY KEY table to be deleted if a row with the same value exists as a FOREIGN KEY (this option would allow you to tell SQL not to permit deletions from the CUST table while rows associated with the keys existed on the INV/I_M tables)

SET NULL — will set all associated values of the FOREIGN KEY table to NULLs if a row with the same value is deleted from the PRIMARY KEY table.

---

* To create a non-unique index for a foreign key in DB2 use the following statement:

```
CREATE INDEX <indexname> ON <TSOID.tablename>
(<columnlist>);
```

Example:

```
CREATE INDEX CUST_INV_IDX ON QED.INV
(CUSTID);
```

The establishment of referential integrity constraints through PRIMARY KEY/FOREIGN KEY definitions will enforce all INSERT rules described in 10.9.

Example:

```
CREATE TABLE INV
      (NBR         SMALLINT NOT NULL,
       CUSTID      CHAR(2) NOT NULL,
       TOT         DEC(5,2),
       RENT        INTEGER,
       RETURN      INTEGER,
PRIMARY KEY (NBR),
FOREIGN KEY (CUSTID) REFERENCES CUST ON DELETE
CASCADE);
```

10.14 ANSWER KEY

```
EXERCISE 10-1  CREATE TABLE RECORD
               (REC_NBR CHAR(8) NOT NULL,
               TITLE CHAR(20) NOT NULL,
               PERFORMER CHAR(30));
EXERCISE 10-2  ALTER TABLE RECORD ADD CONDITION CHAR(4);
               PRICE DECIMAL 5,2);
EXERCISE 10-3  DROP TABLE RECORD;
```

10.15 ADDITIONAL EXERCISE

Here is an example of a CREATE TABLE statement. Look it over, then fill in the description information requested in the form.

```
CREATE TABLE             TRUCK_RENTAL
(MANUFAC_CODE            CHAR(25) NOT NULL,
SIZE                     DECIMAL (4,2),
WEIGHT                   SMALLINT,
COLOR                    CHAR (10),
MAKE                     VARCHAR (20),
MODEL                    VARCHAR (20),
STORAGE_CUBIC_FT         SMALLINT,
LAST_USED                INTEGER,
LAST_USED_BY             VARCHAR (50),
MILEAGE                  DECIMAL (7,1) NOT NULL,
RADIO                    CHAR (1),
CONDITION                VARCHAR (200));
```

Table Name ______________

Column Position	Column Name	DATATYPE	Max Size	NULLS Allowed (Y/N)
1				
2				
3				
4				
5				
6				
7				
8				
9				
10				
11				
12				

10.15.1 Additional exercise answer

Truck Rental

Column Position	Column Name	DATATYPE	Max Size	NULLS Allowed (Y/N)
1	MANUFAC_CODE	fixed char	25	N
2	SIZE	decimal number	99.99	Y
3	WEIGHT	integer number	32,767	Y
4	COLOR	fixed char	10	Y
5	MAKE	var. char	20	Y
6	MODEL	var. char	20	Y
7	STORAGE_CUBIC_FT	integer number	32,767	Y
8	LAST_USED	integer number	2,150,000,000	Y
9	LAST_USED_BY	var. char	50	Y
10	MILEAGE	decimal number	999999.9	N
11	RADIO	fixed char	1	Y
12	CONDITION	var. char	200	Y

Chapter 11
SQL UPDATE STATEMENTS: INSERT, DELETE, UPDATE

Here you will learn to manipulate data stored on the database using:

- INSERT — Adds a row (or rows) to a table;
- DELETE — Removes a row from a table;
- UPDATE — Changes a value stored in a row.

VERY IMPORTANT NOTICE:

You should make every effort to complete this section *without* logging off, exiting, or switching off the computer, because the examples and exercises build one upon the next. Your results may be unpredictable (different from ours), depending upon where you end your session.

The following topics are covered:

11.1 SQL UPDATE STATEMENTS

It seems that things keep coming in threes in SQL — SELECT, FROM, WHERE; CREATE, DROP, ALTER. So we are not surprised to find that there are three SQL statements that allow you to maintain your table information — INSERT, DELETE, and UPDATE. These statements are extremely simple to learn. Their structure and syntax are almost self-explanatory.

11.2 INSERT — FORMAT I—SINGLE ROW INSERTION

You use the INSERT statement to add new rows to your table. INSERT comes in two formats. Format I allows us to enter one new row at a time to an existing table. Format II allows us to SELECT a set of rows from an existing table, and INSERT them into our table. Let's begin by looking at Format I INSERT.

Format I INSERT statements add a single row made up of separate values to an existing table. You code the statements as shown in Figure 11.1.

- Single row data insertion

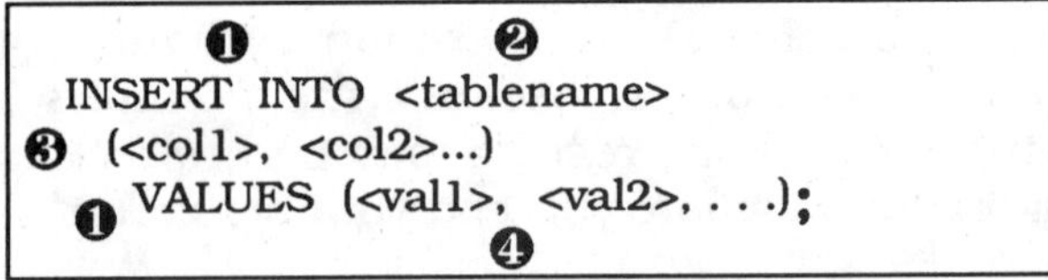

- Data values are inserted positionally as follows:

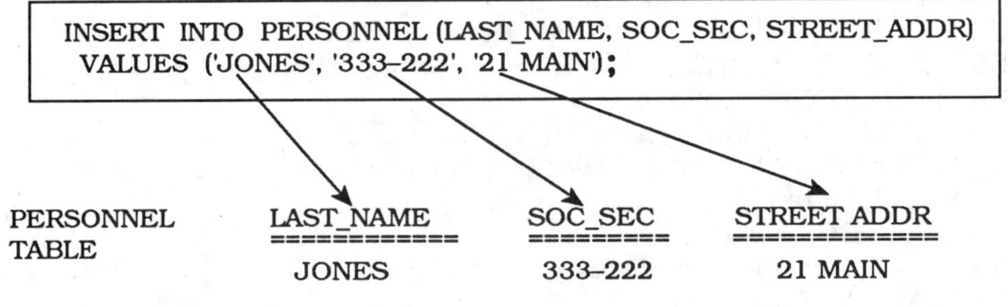

Figure 11.1. Single row insertion (Format I INSERT).

INSERT's syntax requires:

❶ The SQL keywords INSERT INTO, and VALUES entered as shown;

❷ The variable <tablename> — which must refer to a single existing table in the database, more specifically, to the table you intend to INSERT a new row INTO;

❸ The variables <columnname, columnname...> which must refer to existing columns in the table;

❹ A list of values enclosed in parentheses which are assigned by SQL to their positional columns in the <columnname> clause.

Your inserted values must match their column counterparts in three ways:

1. They must be the same datatype;
2. They must stay within the limits of their particular column's size;
3. There must be a one-to-one positional correspondence between values and columns (the first value is assigned to the first column, the second value to the second, etc.).

If you don't know a specific value for a field when you want to INSERT a row, you should still code a value for its column or insert a NULL field. If the column has not been defined with a NOT NULL data attribute, you can code the keyword NULL in place of a value, and the field will be NULL filled, when your statement is executed.

Try out the Format I INSERT statement by inserting this new row into the INV table. The statement is:

```
INSERT INTO INV (NBR, CUSTID, TOT, RENT, RETURN) VALUES
        (11, '11', 7.55, 87056, NULL);
```

The table will now look like:

NBR	CUSTID	TOT	RENT	RETURN
1	01	9.55	87055	
2	02	13.55	87030	
3	02	21.01	87041	87042
4	01	11.25	87042	87072
5	03	12.55	87045	87051
6	02	9.75	87047	87051
7	04	10.35	86355	
8	04	12.55	87051	87053
9	05	10.35	87052	
10	02	7.55	87030	87060
11	11	7.55	87056	

SELECT * from the table, to verify that your INSERT statement worked. Once you've accomplished this, notice that the database is in an inconsistent state. There is no customer 11 on the CUST table to be charged with an invoice. This is an example of a referential integrity problem that could have been avoided by defining referential integrity constraints on our tables as described in Chapter 10.

EXERCISE 11-1

❶ TOPIC: INSERT Format I

❷ EXERCISE 11-1. Insert yourself into the CUST table, fill in all the columns.
Note: You only have 6 characters for your first name!

❸ INPUT TABLE:

Note: The column sizes are as follows:

NAME	COLTYPE	LENGTH
ID	CHAR	2
LNAME	CHAR	12
FNAME	CHAR	6
CITY	CHAR	15
ST	CHAR	2

FORMAT

❹
```
INSERT INTO <tablename>
              (<columnname, columnname . . .>)
        VALUES
              (<VALUE1, VALUE2, . . .>);
```

❺ STEP 1. ______________________________
Write your answer.
____

Hint: **Insert into CUST values ('11', 'yourLNAME', 'yourFNAME', 'city', 'state'). You must be sure that the data you insert into the column will fit! See Chapter 3 for a description of the column sizes.**

STEP 2. Enter your answer into the computer.

After you've finished successfully inserting yourself, type in:

```
SELECT * FROM CUST;
```

You should have a new row in your CUST table.

❻ RESULTS TABLE: (including the Author's inserted Row)

ID	LNAME	FNAME	CITY	ST
01	DANGERFIELD	RODNEY	HARTFORD	CT
02	FIELD	SALLY	FRANKLIN	NY
03	NICHOLSON	JACK	HARTFORD	CT
04	MURRAY	FRED	BOZRAH	CT
05	MADDEN	JOHN	BRATTLEBORO	VT
06	WEST	MAE	PARK PLACE	CA
07	WOODWARD	JOANNE	GETTYSBURG	PA
08	ROGERS	ROY	HAPPY TRAILS	TX
09	RINGWALD	MOLLY	PINKSVILLE	CA
10	ROGERS	FRED	PARK PLACE	CA
11	**SAYLES**	**JON**	**COVENTRY**	**CT**

❼ If your screen does not look like this, try again or look at the answer key and execute the solution given there.

EXERCISE 11-2

❶ TOPIC: INSERT Format I.

❷ EXERCISE 11-2. Insert a new invoice into the INV table. Fill in all of the column values.

❸ INPUT TABLE:

Note: The column sizes are as follows:

NAME	COLTYPE	LENGTH
NBR	SMALLINT	99999
CUSTID	CHAR	2
TOT	DECIMAL	999.99
RENT	INTEGER	99999
RETURN	INTEGER	99999

FORMAT

❹
```
INSERT INTO <tablename>
            (<columnname, columnname . . .>)
      VALUES
            (<VALUE1, VALUE2, . . .>);
```

❺ STEP 1. ______________________________

Write your answer.

____

STEP 2. Enter your answer into the computer.

❻ RESULTS TABLE: (including the Author's inserted row)

NBR	CUSTID	TOT	RENT	RETURN
1	01	9.55	87055	
2	02	13.55	87030	
3	02	21.01	87041	87042
4	01	11.25	87042	87072
5	03	12.55	87045	87051
6	02	9.75	87047	87051
7	04	10.35	86355	
8	04	12.55	87051	87053
9	05	10.35	87052	
10	02	7.55	87030	87060
33	**11**	**0**	**87001**	**86001**

NOTICE THAT THE system does not do any checking for data accuracy or correctness. Look at the TOT, RENT and RETURN fields on the last entry on the results table above.

11.3 INSERT — FORMAT II—MULTIPLE ROW INSERTION

It would be tedious and time-consuming to have to load up a 30,000 row table one row at a time with Format I INSERT statements. SQL provides us with an alternative — multiple row insertion.

Format II INSERT statements add multiple rows to an existing table, made up of SELECTed values FROM an existing table. You code the statements as shown in Figure 11.2.

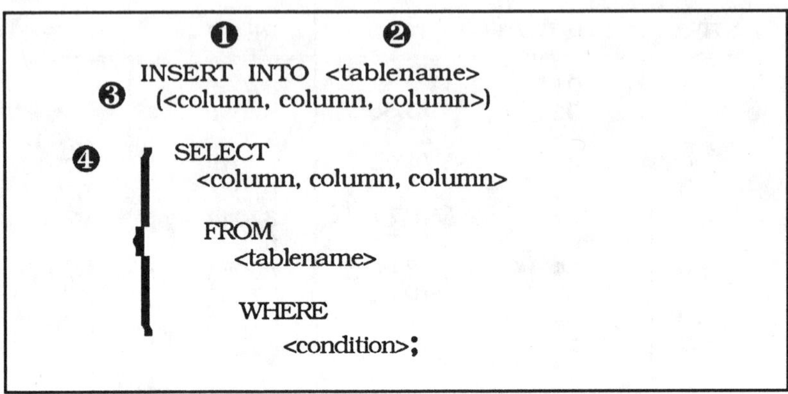

Figure 11.2. Format II INSERT.

INSERT's syntax requires:

❶ The SQL keywords INSERT and INTO — entered as shown;

❷ The variable <tablename>—which must refer to a single existing table in the database. Think of this table as the target table for the INSERT;

❸ The variables <columnname, columnname ...> — which must refer to existing columns in the table;

❹ A SELECT statement — which must SELECT columns from an existing table. The SELECT clause is like any ordinary SELECT. Think of the results table as being copied into the target table (INSERTED INTO Table), instead of merely being displayed.

Your SELECTed values must match their target column counterparts in three ways:

1. They must be the same datatype;

2. They must stay within the limits of their particular column's size;

3. There must be a one-to-one correspondence between the SELECTed columns and the target table columns (the first is assigned to the first column, the second to the second, etc.).

EXERCISE 11-3 Part 1

In order to explore the Format II INSERT statement we will need a new table. Create the table called "CHEAP_MOV" as shown below:

```
CREATE TABLE CHEAP_MOV
    (NBR SMALLINT,
    TITLE CHAR(25),
    PRICE DECIMAL (5,2));
```

RESULTS:

```
TABLE CREATED
```

If the table has already been created by another student, use the DROP command to remove it.

Type:

DROP TABLE CHEAP_MOV;

Now CREATE the table again.

EXERCISE 11-3 Part 2

❶ TOPIC: INSERT

❷ EXERCISE 11-3.2. Insert data into the CHEAP_MOV table.

Insert a set of rows into the new table from the current movie table. Insert only movies that have a QTY less than four, and a PRICE under $30. See the MOV Table.

❸ INPUT TABLE:

CHEAP_MOV

EMPTY

MOV

NBR	TITLE	TYPE	RATING	STAR	QTY	PRICE
1	GONE WITH THE WIND	DRAMA		GABLE	4	39.95
2	FRIDAY THE 13TH	HORROR	R	JASON	2	69.95
3	TOP GUN	DRAMA	PG	CRUISE	7	49.95
4	SPLASH	COMEDY	PG13	HANKS	3	29.95
5	101 DALMATIONS	COMEDY	G		3	59.95
6	BODY HEAT	DRAMA	R	TURNER	3	19.95
7	RISKY BUSINESS	COMEDY	R	CRUISE	2	44.55
8	COCOON	SCIFI	PG	AMECHE	2	
9	CROCODILE DUNDEE	COMEDY	PG13	HARRIS	2	69.95
10	TOOTSIE	COMEDY	PG	HOFFMAN	1	29.95

FORMAT

❹
```
INSERT INTO <tablename>
             (<columnname, columnname . . .>)
  SELECT (<columnname, columnname . . .>) FROM <tablename>
             WHERE <condition>;
```

❺ STEP 1. ______________________________

Write your answer.

____

STEP 2. Enter your answer into the computer.

After you've finished successfully inserting the rows, type in:

```
SELECT * FROM CHEAP_MOV;
```

You should get this results table:

❻ RESULTS TABLE:

NBR	TITLE	PRICE
4	SPLASH	29.95
6	BODY HEAT	19.95
10	TOOTSIE	29.95

Now that you're done with the exercise, type:

DROP TABLE CHEAP_MOV;

11.4 DELETE — REMOVING TABLE ROWS

Removing rows from tables in SQL is actually simpler than inserting them. For instance, all you need to do to remove each and every row from a table is:

```
DELETE FROM <tablename>;
```

There is no system prompt or second chance! Your table is emptied! **Be careful!**

DELETE is more or less permanent. You might be able to ask your data base administration group to recover your table, if those mechanisms are in place, but there is *no* UNDELETE command in SQL. This means that you should be careful when using this statement!

The important aspect of DELETE thus becomes the WHERE clause. It restricts the deletion to specified rows. You must code your search condition so that it pinpoints precisely those rows you want removed — AND NO OTHERS. This will probably mean more use of compound conditionals (using AND or OR) in your SELECT statement (see Figure 11.3).

```
DELETE FROM
        <tablename>

    WHERE
        <condition1> AND/OR
        <condition2>;
```

Figure 11.3. DELETE format.

To summarize some of the considerations for using DELETE:

1. You can DELETE:
   - single rows
   - multiple rows
   - all rows
   - no rows

   depending on the scope and accuracy of your WHERE condition.
2. You can only DELETE from one table at a time.
3. DELETE is a global operation. It will remove all rows that satisfy your WHERE condition.
4. DELETE does not remove your table (use DROP to do that).
5. DELETE permanently removes your rows, depending on the type of COMMIT processing you are using. (This will be explained in Section 11.8).
6. We suggest you "prototype" your DELETE operation by coding the statement first as a SELECT. This will allow you to see which rows will get deleted and verify the logic in your WHERE clause.
7. Like INSERT, DELETE can cause referential integrity problems.

EXERCISE 11-4

This exercise asks you to remove Fred Rogers from the CUST table. Notice that if you are successful in doing this, you will have made the database inconsistent again, as Fred's invoices on the INV table no longer relate to any CUSTomer. Once you've successfully DELETEd Fred's row from CUST, be sure to verify that your statements worked by displaying the CUST table.

❶ TOPIC: DELETE
❷ EXERCISE 11-4. Delete Fred Rogers from the customer table.
❸ INPUT TABLE:

CUST

ID	LNAME	FNAME	CITY	ST
01	DANGERFIELD	RODNEY	HARTFORD	CT
02	FIELD	SALLY	FRANKLIN	NY
03	NICHOLSON	JACK	HARTFORD	CT
04	MURRAY	FRED	BOZRAH	CT
05	MADDEN	JOHN	BRATTLEBORO	VT
06	WEST	MAE	PARK PLACE	CA
07	WOODWARD	JOANNE	GETTYSBURG	PA
08	ROGERS	ROY	HAPPY TRAILS	TX
09	RINGWALD	MOLLY	PINKSVILLE	CA
10	ROGERS	FRED	PARK PLACE	CA

FORMAT

❹
```
DELETE FROM <tablename>
  WHERE <condition>;
```

❺ STEP 1. ______________________________

Write your answer.

____

STEP 2. Enter your answer into the computer.

❻ RESULTS TABLE (From SELECT * FROM CUST;):

ID	LNAME	FNAME	CITY	ST
01	DANGERFIELD	RODNEY	HARTFORD	CT
02	FIELD	SALLY	FRANKLIN	NY
03	NICHOLSON	JACK	HARTFORD	CT
04	MURRAY	FRED	BOZRAH	CT
05	MADDEN	JOHN	BRATTLEBORO	VT
06	WEST	MAE	PARK PLACE	CA
07	WOODWARD	JOANNE	GETTYSBURG	PA
08	ROGERS	ROY	HAPPY TRAILS	TX
09	RINGWALD	MOLLY	PINKSVILLE	CA

EXERCISE 11-5

This exercise asks you to delete all of the January 1987 (87001 thru 87031) invoices that have been returned where RETURN contains a non-NULL value (use rent for the dates; and RETURN for overdue portion of the statement).

❶ TOPIC: DELETE.

❷ EXERCISE 11-5. Delete invoices with rent dates in January 1987, that are not overdue.

❸ INPUT TABLE:

INV

NBR	CUSTID	TOT	RENT	RETURN
1	01	9.55	87055	
2	02	13.55	87030	
3	02	21.01	87041	87042
4	01	11.25	87042	87072
5	03	12.55	87045	87051
6	02	9.75	87047	87051
7	04	10.35	86355	
8	04	12.55	87051	87053
9	05	10.35	87052	
10	02	7.55	87030	87060
11	11	7.55	87056	
33	11	0	87001	86001

❹ RESULTS TABLE:

NBR	CUSTID	TOT	RENT	RETURN
1	01	9.55	87055	
3	02	21.01	87041	87042
4	01	11.25	87042	87072
5	03	12.55	87045	87051
6	02	9.75	87047	87051
7	04	10.35	86355	
8	04	12.55	87051	87053
9	05	10.35	87052	

11.5 UPDATE — MODIFYING COLUMN VALUES

UPDATE is the SQL verb that changes or modifies data values in a table. There are actually four parts to the UPDATE command as shown in Figure 11.4. UPDATE's syntax requires:

❶ The SQL keywords UPDATE and SET — entered as shown;

❷ The variable <tablename> which must refer to a single existing table in the database;

❸ The variables <columnname> = <value>, <columnname> = <value>... which assign the value specified to the column;

❹ A WHERE condition that functions as it does in a SELECT statement. That is, it restricts the data update to certain rows.

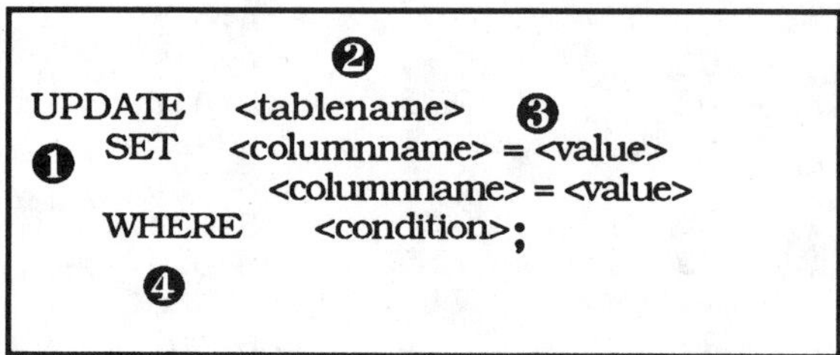

Figure 11.4. UPDATE format.

1. The UPDATE command has the following considerations:
   - single columns
   - several columns
   - single rows
   - multiple rows
   - all rows
   - no rows

   depending on the scope and accuracy of your WHERE condition.

2. You can only UPDATE one table at a time.
3. UPDATE is another "Global" operation. It will change the values on *all* table rows that satisfy your WHERE condition.
4. With UPDATE you can set a column equal to:
   - an explicit data value
   - an arithmetic expression
   - another column from the same row in the table
   - a NULL value
5. If you are not the creator of a table, you must be GRANTed UPDATE authority on every table you want to UPDATE.
6. UPDATE can cause referential integrity problems. If you update values on Table A that exist on Table B for the same business processing entity (like a person's SSN, Department, etc.), be sure to UPDATE Table B and those columns relating to the one you just UPDATEd on Table A.

Most of the time, you will be updating columns with literal data values. The values you code in your SET clause must match their column counterparts in two ways:

1. These values must be the same datatype as the column;
2. They must stay within the limits of their particular column's size.

EXERCISE 11-6

This exercise asks you to change the "Gone With the Wind" rating from NULL to 'G'. Once you've successfully UPDATEd the row, be sure to SELECT * from the table, to verify that your statement worked.

❶ TOPIC: (UPDATE a single row)

❷ EXERCISE 11-6. SET the "Gone With the Wind" rating to a 'G'.

❸ INPUT TABLE:

MOV

NBR	TITLE	TYPE	RATING	STAR	QTY	PRICE
1	GONE WITH THE WIND	DRAMA		GABLE	4	39.95
2	FRIDAY THE 13TH	HORROR	R	JASON	2	69.95
3	TOP GUN	DRAMA	PG	CRUISE	7	49.95
4	SPLASH	COMEDY	PG13	HANKS	3	29.95
5	101 DALMATIONS	COMEDY	G		3	59.95
6	BODY HEAT	DRAMA	R	TURNER	3	19.95
7	RISKY BUSINESS	COMEDY	R	CRUISE	2	44.55
8	COCOON	SCIFI	PG	AMECHE	2	
9	CROCODILE DUNDEE	COMEDY	PG13	HARRIS	2	69.95
10	TOOTSIE	COMEDY	PG	HOFFMAN	1	29.95

FORMAT

❹
```
UPDATE    <tablename>
   SET    <columnname = value, columnname = value . . .>
      WHERE <condition>;
```

❺ RESULTS TABLE:

NBR	TITLE	TYPE	RATING	STAR	QTY	PRICE
1	GONE WITH THE WIND	DRAMA	G	GABLE	4	39.95
2	FRIDAY THE 13TH	HORROR	R	JASON	2	69.95
3	TOP GUN	DRAMA	PG	CRUISE	7	49.95
4	SPLASH	COMEDY	PG13	HANKS	3	29.95
5	101 DALMATIONS	COMEDY	G		3	59.95
6	BODY HEAT	DRAMA	R	TURNER	3	19.95
7	RISKY BUSINESS	COMEDY	R	CRUISE	2	44.55
8	COCOON	SCIFI	PG	AMECHE	2	
9	CROCODILE DUNDEE	COMEDY	PG13	HARRIS	2	69.95
10	TOOTSIE	COMEDY	PG	HOFFMAN	1	29.95

Hint: **Your statement should look like this:**

```
UPDATE MOV SET RATING = 'G' WHERE <condition>;
```

Note: **You fill in the correct condition.**

EXERCISE 11-7

This exercise asks you to change the NULL values of the RETURN column on the INV table.

Once you've successfully UPDATEd the row, be sure to SELECT * from the table to verify that your statement worked.

❶ TOPIC: UPDATE Examples (UPDATE multiple rows).

❷ EXERCISE 11-7. Set all of the NULL values for the RETURN column in the invoice table to 99999.

❸ INPUT TABLE:

INV

NBR	CUSTID	TOT	RENT	RETURN
1	01	9.55	87055	
2	02	13.55	87030	
3	02	21.01	87041	87042
4	01	11.25	87042	87072
5	03	12.55	87045	87051
6	02	9.75	87047	87051
7	04	10.35	86355	
8	04	12.55	87051	87053
9	05	10.35	87052	
11	11	7.55	87056	

FORMAT

❹
```
UPDATE    <tablename>
   SET   <columnname = value, columnname = value . . .>
      WHERE  <condition>;
```

❺ RESULTS TABLE:

NBR	CUSTID	TOT	RENT	RETURN
1	01	9.55	87055	99999
2	02	13.55	87030	99999
3	02	21.01	87041	87042
4	01	11.25	87042	87072
5	03	12.55	87045	87051
6	02	9.75	87047	87051
7	04	10.35	86355	99999
8	04	12.55	87051	87053
9	05	10.35	87052	99999

11.6 UPDATE — USING CALCULATED VALUES

You can use arithmetic expressions in your SET statements to calculate new column values based on their current ones (see Figure 11.5). This is a very practical feature, and again is easy to understand and use. The arithmetic expressions must follow all the rules we covered so far.

```
UPDATE <tablename>
      SET <column> = <column * values>,
        <column> = <value * value>;
```

Figure 11.5. UPDATE with calculated value format.

There are two UPDATE considerations involved in the calculated values:

1. *Truncation.* The data type for the column dictates the impact of the UPDATE statement. What this means is that if, for example, you multiply an INTEGER column by 1.8, any fractional part will be lost or truncated — not rounded up. This could have a significant impact on your results.
2. *Overflow.* you can have an arithmetic overflow error if your UPDATE statement calculates values beyond the range of the column.

Example:

```
UPDATE <table> SET SMALLINT_COL = 1000 * 1000;
```

EXERCISE 11-8

This exercise asks you to multiply all of the drama movie prices by 1.075 (perhaps to calculate the replacement cost for our inventory).

Once you've successfully UPDATEd the rows, be sure to SELECT * from the table to verify that your statement worked.

❶ TOPIC: UPDATE Examples (UPDATE Using Calculated Values).

❷ EXERCISE 11-8. Adjust all drama TYPE movie PRICEs up by 7.5% (multiply PRICE * 1.075).

❸ INPUT TABLE:

MOV

NBR	TITLE	TYPE	RATING	STAR	QTY	PRICE
1	GONE WITH THE WIND	DRAMA	G	GABLE	4	39.95
2	FRIDAY THE 13TH	HORROR	R	JASON	2	69.95
3	TOP GUN	DRAMA	PG	CRUISE	7	49.95
4	SPLASH	COMEDY	PG13	HANKS	3	29.95
5	101 DALMATIONS	COMEDY	G		3	59.95
6	BODY HEAT	DRAMA	R	TURNER	3	19.95
7	RISKY BUSINESS	COMEDY	R	CRUISE	2	44.55
8	COCOON	SCIFI	PG	AMECHE	2	
9	CROCODILE DUNDEE	COMEDY	PG13	HARRIS	2	69.95
10	TOOTSIE	COMEDY	PG	HOFFMAN	1	29.95

FORMAT

❹

```
UPDATE   <tablename>
  SET   <columnname = value, columnname = value . . .>
    WHERE  <condition>;
```

❺ STEP 1. ______________________________________

Write your answer.

____

STEP 2. Enter your answer into the computer.

❻ RESULTS TABLE:

NBR	TITLE	TYPE	RATING	STAR	QTY	PRICE
1	GONE WITH THE WIND	DRAMA	G	GABLE	4	42.94
2	FRIDAY THE 13TH	HORROR	R	JASON	2	69.95
3	TOP GUN	DRAMA	PG	CRUISE	7	53.69
4	SPLASH	COMEDY	PG13	HANKS	3	29.95
5	101 DALMATIONS	COMEDY	G		3	59.95
6	BODY HEAT	DRAMA	R	TURNER	3	21.44
7	RISKY BUSINESS	COMEDY	R	CRUISE	2	44.55
8	COCOON	SCIFI	PG	AMECHE	2	
9	CROCODILE DUNDEE	COMEDY	PG13	HARRIS	2	69.95
10	TOOTSIE	COMEDY	PG	HOFFMAN	1	29.95

EXERCISE 11-9

This exercise asks you to change the value of the prices of movies 1, 5, and 9.

Once you've successfully UPDATEd the rows, be sure to SELECT * from the table to verify that your statement worked.

❶ TOPIC: UPDATE Examples (UPDATEing Using Calculated Values).

❷ EXERCISE 11-9. Make the PRICE column the product of the PRICE * QTY * 10, for movies numbered 1, 5, and 9.

❸ INPUT TABLE:

MOV

NBR	TITLE	TYPE	RATING	STAR	QTY	PRICE
1	GONE WITH THE WIND	DRAMA	G	GABLE	4	42.94
2	FRIDAY THE 13TH	HORROR	R	JASON	2	69.95
3	TOP GUN	DRAMA	PG	CRUISE	7	53.69
4	SPLASH	COMEDY	PG13	HANKS	3	29.95
5	101 DALMATIONS	COMEDY	G		3	59.95
6	BODY HEAT	DRAMA	R	TURNER	3	21.44
7	RISKY BUSINESS	COMEDY	R	CRUISE	2	44.55
8	COCOON	SCIFI	PG	AMECHE	2	
9	CROCODILE DUNDEE	COMEDY	PG13	HARRIS	2	69.95
10	TOOTSIE	COMEDY	PG	HOFFMAN	1	29.95

FORMAT

❹

```
UPDATE   <tablename>
   SET   <columnname = value, columnname = value . . .>
     WHERE  <condition>;
```

❺ STEP 1. ______________________________

Write your answer.

____

STEP 2. Enter your answer into the computer.

❻ RESULTS TABLE:

NBR	TITLE	TYPE	RATING	STAR	QTY	PRICE
1	GONE WITH THE WIND	DRAMA	G	GABLE	4	717.6
2	FRIDAY THE 13TH	HORROR	R	JASON	2	69.95
3	TOP GUN	DRAMA	PG	CRUISE	7	53.69
4	SPLASH	COMEDY	PG13	HANKS	3	29.95
5	101 DALMATIONS	COMEDY	G		3	798.5
6	BODY HEAT	DRAMA	R	TURNER	3	21.44
7	RISKY BUSINESS	COMEDY	R	CRUISE	2	44.55
8	COCOON	SCIFI	PG	AMECHE	2	
9	CROCODILE DUNDEE	COMEDY	PG13	HARRIS	2	399
10	TOOTSIE	COMEDY	PG	HOFFMAN	1	29.95

Note: **There is displacement of our decimal fields. To pretty this report up, we would need to use the Formatting Ability of QMF or ISQL.**

11.7 SQL DATA INTEGRITY

As detailed in Figure 11.6, SQL allows concurrent access to tables for multiple system users doing multiple data processing tasks. (Concurrent access means access of a specific item of information, stored on a database, by multiple system users at approximately the same time.) This can cause such undesirable effects as two users reading the same row from a table and both calculating new values for its columns, based on what they read. If user A updates the row with a value first, then user B updates the row, user A's updates are lost. This is one of several problems inherent in any system that allows concurrent processing.

- SQL allows CONCURRENT ACCESS to tables for:
  - Interactive processing (QMF/DB2I)
  - Online application programs
  - Batch application programs
  - Utility programs

- SQL ensures data consistency and integrity through the use of:
  - LOCKING Feature
  - COMMIT/ROLLBACK Processing
  - LOGICAL UNIT OF WORK (Synchronization point)

- SQL's Primary Data Integrity Consideration

No changes are made to the database until COMMIT processing is complete

- For interactive SQL, COMMIT processing executes in 2 modes:

AUTOCOMMIT — on AUTOCOMMIT — off

Figure 11.6. SQL data integrity.

SQL provides concurrency control, and prevents these situations by means of locks on table data. These locks basically associate SQL data with the first user who got to the data. The user "owns" or "holds" the data, until certain processing events occur. These events are called *commit points.*

Commit points are processing events that signal SQL that "all is well" with the database processing. Once SQL determines this, two things happen:

1. All updates are made permanent to the database;
2. SQL releases any locks held by the user on the updated data.

For interactive SQL, COMMIT processing executes in two modes. These are AUTOCOMMIT ON and AUTOCOMMIT OFF.

11.8 AUTOCOMMIT

AUTOCOMMIT is a feature of interactive DB2 processing that generates a COMMIT command to DB2 upon the successful execution of each SQL statement. (You may specify this on the ISQL main panel, the default is AUTOCOMMIT ON.)

DB2 also allows you to issue a COMMIT command:

```
COMMIT WORK;
```

11.9 COMMIT and ROLLBACK

COMMIT and ROLLBACK are manual commands to apply (or not apply) your INSERTs, DELETEs, and UPDATEs to the table. They establish a commit point for you, and signal SQL to make your changes permanent.

COMMIT format is:

```
COMMIT WORK;
```

This causes all table updates registered since the last commit point to be made permanent on the database.

ROLLBACK format is:

```
ROLLBACK WORK;
```

This causes all table updates registered since the last commit point to be *cancelled*, and a new commit point to be established.

Figure 11.7 summarizes this for you.

SET AUTOCOMMIT ON or OFF:

- If you are using DB2I, you get to set autocommit processing on the SPUFI Main Panel.
- For QMF and SQL/DS, users must contact their technical support personnel.

SET AUTOCOMMIT ON

- Updates against the database are made permanent after the successful execution of each SQL statement.

SET AUTOCOMMIT OFF

- Updates against the database are postponed until a COMMIT or ROLLBACK command is explicitly issued by the User.

- -

COMMIT or ROLLBACK:

- Are only used when AUTOCOMMIT is set off.
- Specify the end of a Logical Unit of Work (synchronization point).

COMMIT WORK

- Causes all table updates registered since the last COMMIT or ROLLBACK command to be made permanent on the database.

ROLLBACK WORK

- Causes all table updates registered since the last COMMIT or ROLLBACK command to be cancelled.

Figure 11.7. AUTOCOMMIT.

We suggest you do a ROLLBACK here to decommit your work. Enter:

ROLLBACK WORK;

Your tables are now the way they were.

11.10 ANSWER KEY

There may be more than one answer to an exercise. However, the one given emphasizes the topic discussed.

EXERCISE 11-1	INSERT INTO CUST VALUES ('11', 'SAYLES', 'JON', 'COVENTRY', 'CT');
EXERCISE 11-2	INSERT INTO INV VALUES (33, '11', 0, 87001, 86001);
EXERCISE 11-3	INSERT INTO CHEAP_MOV (NBR, TITLE, PRICE) SELECT NBR, TITLE, PRICE FROM MOV WHERE PRICE < 30 AND QTY < 4;
EXERCISE 11-4	DELETE FROM CUST WHERE LNAME = 'SAYLES';
EXERCISE 11-5	DELETE FROM INV WHERE RENT BETWEEN 87001 AND 87031 AND RETURN IS NOT NULL;
EXERCISE 11-6	UPDATE MOV SET RATING = 'G' WHERE TITLE LIKE 'GONE%';
EXERCISE 11-7	UPDATE INV SET RETURN = 99999 WHERE RETURN IS NULL;
EXERCISE 11-8	UPDATE MOV SET PRICE = PRICE *1.075 WHERE TYPE = 'DRAMA';
EXERCISE 11-9	UPDATE MOV SET PRICE = PRICE * QTY * 10 WHERE NBR IN (1,5,9);

Note: **In Exercise 11-4 your answer will have your name in place of SAYLES.**

Chapter 12
COLUMN FUNCTIONS

In this chapter you will learn how to:

1. SELECT the highest value in a column;
2. SELECT the lowest value in a column;
3. SELECT the average value in a column;
4. SELECT the sum of all the values in a column;
5. SELECT the sum of all the different values in a column;
6. SELECT a count of the number of rows in a results table;
7. SELECT a count of the number of different values in a column.

The following topics are covered:

12.1 Introduction to Functions
12.2 The Employee Database
12.3 Column Functions
12.4 Using Column Functions
12.5 Average — AVG
12.6 NULL VALUE ELIMINATION
12.7 AVG PRECISION AND TRUNCATION ERRORS
12.8 Maximum — MAX
12.9 Minimum — MIN
12.10 Totals — SUM
12.11 Count — COUNT
12.12 Answer Key

For many of the remaining exercises we will use a new database called the Employee database. It consists of three tables which are shown in Figures 12.1 through 12.3. These are:

1. The Employee Table (Figure 12.1)
2. The Pay Table (Figure 12.2)
3. The Project Table (Figure 12.3)

If you are going to do the exercises, you will have to enter the tables. You should be able to do this now because you did it earlier for the other database, Please review Chapter 10 and then proceed. There is a diskette available from QED, if you don't want to enter the tables.

Col	Information	Length	Data Type
NBR	Unique employee identification number	2	character
LNAME	Employee last name	10	character
FNAME	Employee first name	6	character
DOB	Employee Date-of-birth	99999	integer numeric
HIREDTE	Employee Date-of-hire	99999	integer numeric
PERF	Employee Performance Rating	2	character
JOB	Employee Job Type	4	character
DEPT	Employee Department	3	character
PROJ	Current Employee Project	2	character

NBR	LNAME	FNAME	DOB	HIREDTE	PERF	JOB	DEPT	PROJ
01	LOWE	ROB	53012	85012	4	PROG	FIN	01
02	SHIELD	BROOKE	59131	87001	3	MAN	MKT	01
03	MOORE	ROGER	48111	86002	1	DIR	MKT	04
04	EASTWOOD	CLINT	41091	60120	3	PROG	FIN	03
05	MOSTEL	ZERO	21365	84211		PRES		
06	BURNS	GEORGE	11178	49001	2	SYS	FIN	01
07	O'NEIL	RYAN	42189	60121	3	DIR	ACC	05
08	MARVIN	LEE	32187	51876	2	VP	ACC	02
09	LANCASTER	BURT	41091	79092	1	MAN	R&D	02
10	BLAIR	LINDA	54013	85012	1	PROG	MKT	

Figure 12.1. The EMPL Table.

Col	Information	Length	Data Type
NBR	Unique employee identification number	2	character
HOURS	Hours worked in 1 week	99.99	Decimal Numeric
RATE	Pay Rate	99.99	Decimal Numeric
DED	Payroll Deductions	999.99	Decimal Numeric
YTD	Year-To-Date Pay	99999.99	Decimal Numeric

NBR	HOURS	RATE	DED	YTD
01	8.89	43	128.78	11890.66
02	13.23	40	204.45	15840.78
03	6.11	49	94.76	11890.66
04	26.75	45	132.58	17605.66
05	67.82	37	394.69	79990.99
06	32.45	32	121.99	53421.23
07	26.75	49	101.56	32758.11
08	15.99	52	327.98	67870.01
09	43.59	24	0	28090.91
10	32.41	52	112.78	27000.01

Figure 12.2. The PAY Table.

Col	Information	Length	Data Type
NBR	Unique employee identification number	2	character
NAME	Project Title	10	character
DEPT	Sponsoring Department	3	character
MAJPROJ	If part of a major project, major project identification number	2	character

```
NBR      NAME          DEPT      MAJPROJ
=====    ========      ======    =========
 01      PHASERS       MKT
 05      SYSTEM X      MKT          03
 03      SYSTEM R      FIN
 04      LASER         ACC          01
 02      R*            FIN          03
 06      NEW PROJ      R&D          02
```

Figure 12.3. The PROJ Table.

If you need help, follow these directions:

Type:

```
CREATE TABLE EMPL (NBR CHAR(2),
LNAME VARCHAR(10),
FNAME VARCHAR(8),
DOB INTEGER,
HIREDTE INTEGER,
PERF SMALLINT,
JOB CHAR(4)
DEPT CHAR(3),
PROJ CHAR(2));
```

Type:

```
INSERT INTO EMPL,
  VALUES ('01', 'LOWE', 'ROB', 53012, 85012, 4, 'PROG', 'FIN', '01';
INSERT INTO EMPL
  VALUES ('02', 'SHIELD','BROOKE', 59131, 87001, 3, 'MAN',
  'MKT', '01');
INSERT INTO EMPL,
  VALUES ('03', 'MOORE', 'ROGER', 48111, 86002, 1, 'DIR', 'MKT',
'04');
INSERT INTO EMPL
  VALUES ('04', 'EASTWOOD', 'CLINT', 41091, 60120, 3, 'PROG',
  'FIN', '03');
INSERT INTO EMPL
  VALUES ('05', 'MOSTEL', ' ZERO' , 21365 , 84211 , NULL , 'PRES',
  NULL, NULL);
INSERT INTO EMPL
  VALUES ('06', 'BURNS', 'GEORGE', 11178, 49001, 2, 'SYS', 'FIN',
  '01');
INSERT INTO EMPL
  VALUES ('07', 'O'NEIL', 'RYAN', 42189, 60121, 3, 'DIR', 'ACC',
  '05');
INSERT INTO EMPL
  VALUES ('08', 'MARVIN', 'LEE', 32187, 51876, 2, 'VP', 'ACC', '02');
INSERT INTO EMPL
  VALUES ('09', 'LANCASTER', 'BURT', 41091, 79092, 1, 'MAN',
  'R&D', '02');
INSERT INTO EMPL
  VALUES ('10', 'BLAIR', 'LINDA', 54013, 85012, 1, 'PROG', 'MKT',
  NULL);
```

Type:

```
CREATE TABLE
  PROJ (NBR CHAR(2), NAME VARCHAR(10), DEPT VARCHAR(3),
  MAJPROJ VARCHAR(2));
INSERT INTO PROJ VALUES
  ('01', 'PHASERS', 'MKT', NULL);
INSERT INTO PROJ VALUES
  ('05', 'SYSTEM X', 'MKT', '03');
INSERT INTO PROJ VALUES
  ('03', 'SYSTEM R', 'FIN', NULL);
INSERT INTO PROJ VALUES
  ('04', 'LASERS', 'ACC', '01');
INSERT INTO PROJ VALUES
  ('02', 'R*', 'FIN', '03');
```

```
INSERT INTO PROJ VALUES
  ('06', 'NEW PROJ', 'R&D', '02');
```

Type:

```
CREATE TABLE
  PAY (NBR CHAR(2), HOURS DECIMAL (5,2), RATE DECIMAL (5,2),
  DED DECIMAL (5,2), YTD DECIMAL (7,2));
INSERT INTO PAY VALUES
  ('01', 8.89, 43,128.78,11890.66);
INSERT INTO PAY VALUES
  ('02', 13.23, 40, 204.45, 15840.78);
INSERT INTO PAY VALUES
  ('03', 6.11, 49, 94.76, 11890.66);
INSERT INTO PAY VALUES
  ('04', 26.75, 45, 132.58, 17605.66);
INSERT INTO PAY VALUES
  ('05', 67.82, 37, 394.69, 79990.99);
INSERT INTO PAY VALUES
  ('06', 32.45, 32, 121.99, 53421.23);
INSERT INTO PAY VALUES
  ('07', 26.75, 49, 101.56, 32758.11);
INSERT INTO PAY VALUES
  ('08', 15.99, 52, 327.98, 67870.01);
INSERT INTO PAY VALUES
  ('09', 43.59, 24, 0, 28090.91);
INSERT INTO PAY VALUES
  ('10', 32.41, 52, 112.78, 27000.01);
```

You should check out each table with a SELECT command. If they are all correct, proceed.

12.1 INTRODUCTION TO FUNCTIONS

SQL is a non-procedural data access language. There is no IF-THEN-ELSE construct, nor is there a facility to directly store and manipulate temporary values within statements. With the standard SELECT-FROM-WHERE operation, it is impossible to perform many common types of data analysis such as finding average, maximum or minimum value in a column. The founders of SQL realized this and, in their wisdom, they provided us with five built-in column functions:

AVG — which calculates a column average
SUM — which calculates a column total
MIN — which displays the lowest column value
MAX — which displays the highest column value
COUNT — which counts the number of rows in a results table

These column functions are easy to learn and use.

But before we begin studying functions, let's take a brief look at an Employee Database.

12.2 THE EMPLOYEE DATABASE

The tables in the Employee Database shown in Figures 12.1 through 12.3 will be used as our standard tables . Take a minute to acquaint yourself with the standard four items of interest about each table:

Table name
Column names
Column Datatypes
Table-Linking-Values for Joining tables

Please note that as in the INV table, dates are stored in Julian format on the EMPL table. Also note that NULL values are presented as blanks in our tables (they will be displayed as dashes in DB2 and question marks in SQL/DS).

12.3 COLUMN FUNCTIONS

The five column functions shown in Figure 12.4 are used in data analysis reports to summarize and present data in a more informational context.

AVG — which calculates a column average
SUM — which calculates a column total
MIN — which displays the lowest column value
MAX — which displays the highest column value
COUNT — which counts the number of rows in a results table

Figure 12.4 The five column function.

As illustrated in Figure 12.5, column functions process "vertically" through a table, and the resultant execution of a column function on a set of fields is always a single value that pertains to the column.

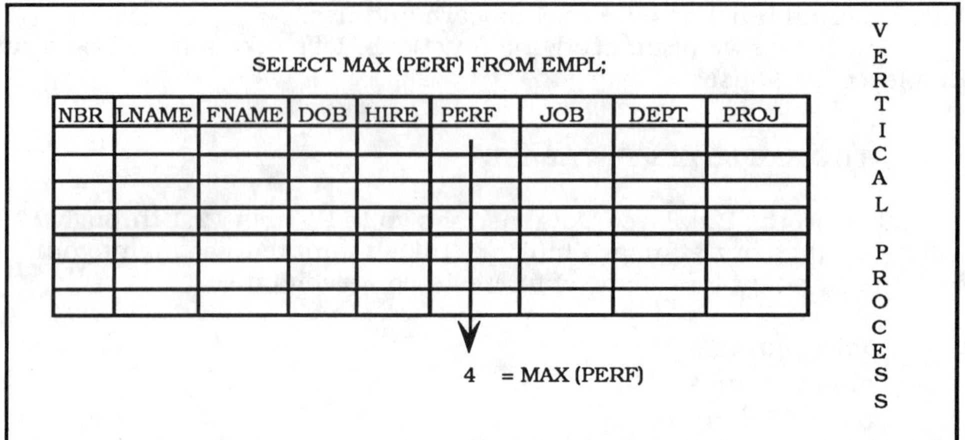

Figure 12. 5 Column functions process vertically.

12.4 USING COLUMN FUNCTIONS

Column functions are easy to understand, code and use. Just keep these things in mind when you use them:

1. The Column name you are processing is enclosed in parentheses

   Ex. SELECT AVG(PERF) FROM EMPL;

2. If you use a function, all columns in the SELECT clause must also be functions (unless you GROUP BY — Chapter 14)

   Ex. SELECT AVG (PERF), MIN(PERF), MAX(DOB) FROM EMPL;

3. Functions of functions are not possible in SQL (they are possible in QMF)

   Incorrect: SELECT MAX(AVG(PERF)) FROM EMPL;

4. You can apply function to arithmetic expressions

   Ex. SELECT AVG (PERF * 1.0) FROM EMPL;

   Note — you often use an arithmetic expression to change or upgrade the final precision of your result value. This is necessary when you are taking an average of INTEGER or SMALLINT columns as DB2 accepts the column datatype as the default precision for the result value. A truncated (not rounded) average will be your result. To boost the precision of this value to one decimal position, just multiply the column name WITHIN THE PARENTHESES by 1.0 as shown in the above example.

5. When NULL Values are found within a column they are ignored by most of the built-in functions.

6. You can mix and match the different column functions on the same or different columns in one query — AS LONG AS YOU DON'T SELECT A COLUMN WITHOUT A FUNCTION

   Ex. SELECT MAX(YTD), MIN(YTD), SUM(HOURS) FROM PAY;

   We'll start our exercises by looking at the average function.

12.5 AVG FUNCTION

Obtaining a column average is a standard type of processing request against a table. To satisfy the request we will use the AVG function in our SELECT clause. AVG displays the average value of all non-NULL fields in the column (the sum of the fields, divided by a count of the non-NULL rows).

AVG Format

```
SELECT AVG   (<column_name>)
             (DISTINCT  <column_name>)

       FROM  <table_name)

       WHERE <condition>;
```

The considerations for average include:

- AVG must only be used on numeric columns.
- AVG disregards rows with NULL values in the column you are averaging.
- AVG ignores duplicate values in a column if you code the keyword DISTINCT before the column name in the parentheses.
- AVG will give you the same precision in the result value as defined in the column datatype — UNLESS YOU OVERRIDE THIS WITH AN ARITHMETIC EXPRESSION.

Of these considerations the only important ones are the two regarding NULL value elimination and default precision.

12.6 NULL VALUE ELIMINATION

The classical definition of an average is the sum or total of all column values divided by the number of rows in the table. In cases where the AVG encounters NULL values in a column you end up with a quite different formula. AVG will calculate the result based on the total of all non-NULL values in a column divided by the number of rows that DID NOT HAVE null values in the column being averaged. This is illustrated in Figure 12.6.

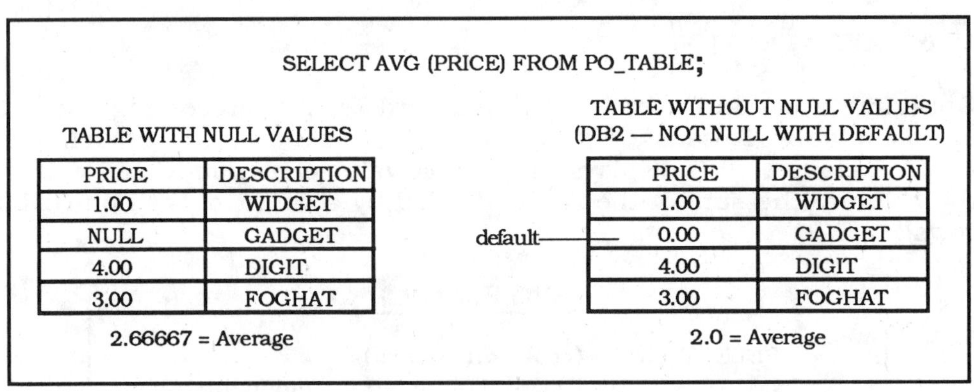

Figure 12.6. AVG calculation.

Most of the time you will want to obtain an average eliminating NULL values and their corresponding row count; however, this can vary from application to application.

12.7 AVG PRECISION AND TRUNCATION ERRORS

In DB2 and SQL/DS you will usually have to account for an integer or SMALLINT precision in a column you are averaging by multiplying the column by 1.0 within the parentheses. If you do not, SQL will give you a truncated average of the values in the column. This will not happen in SQLBase, but will on the mainframe.

EXERCISE 12-1

❶ TOPIC: Column Functions: Average.
❷ EXERCISE 12-1. Display the average YTD employee salary.
❸ INPUT TABLE:

PAY

NBR	HOURS	RATE	DED	YTD
01	8.89	43	128.78	11890.66
02	13.23	40	204.45	15840.78
03	6.11	49	94.76	11890.66
04	26.75	45	132.58	17605.66
05	67.82	37	394.69	79990.99
06	32.45	32	121.99	53421.23
07	26.75	49	101.56	32758.11
08	15.99	52	327.98	67870.01
09	43.59	24	0	28090.91
10	32.41	52	112.78	27000.01

FORMAT

❹
```
SELECT function (<column>)
        FROM table;
```

❺ STEP 1. ______________________________
Write your answer here
____

STEP 2. Enter your answer into the computer.

STEP 3. Check your results.

❻ RESULTS TABLE:

AVG (YTD)
34635.00

EXERCISE 12-2

❶ TOPIC: Column Functions: Average.

❷ EXERCISE 12-2. Calculate and print the average employee paycheck (RATE * HOURS – DED).:

❸ INPUT TABLE:

PAY

NBR	HOURS	RATE	DED	YTD

❻ RESULTS TABLE:

COL1
921.649

Hint: **The function can act on a single column, as in the previous exercise, or on an expression consisting of one or more columns, or literal numbers. Here, the function will act on an expression: Rate times Hours minus DED.**

Note: **If you are working on a mainframe you may see five or six extraneous trailing zeros (921.64900000). In the next chapter you will learn how to remove these with the Decimal Function.**

EXERCISE 12-3

❶ TOPIC: Column Functions: Average.

❷ EXERCISE 12-3. List the average performance evaluation for programmers.

❸ INPUT TABLE:

EMPL

NBR	LNAME	FNAME	DOB	HIREDTE	PERF	JOB	DEPT	PROJ
01	LOWE	ROB	53012	85012	4	PROG	FIN	01
02	SHIELD	BROOKE	59131	87001	3	MAN	MKT	01
03	MOORE	ROGER	48111	86002	1	DIR	MKT	04
04	EASTWOOD	CLINT	41091	60120	3	PROG	FIN	03
05	MOSTEL	ZERO	21365	84211		PRES		
06	BURNS	GEORGE	11178	49001	2	SYS	FIN	01
07	O'NEIL	RYAN	42189	60121	3	DIR	ACC	05
08	MARVIN	LEE	32187	51876	2	VP	ACC	02
09	LANCASTER	BURT	41091	79092	1	MAN	R&D	02
10	BLAIR	LINDA	54013	85012	1	PROG	MKT	

FORMAT

❹
```
SELECT  function  (<expression>)
            FROM table
                 WHERE col_a = 'value';
```

❺ STEP 1. ______________________________
Write your answer here.
____

STEP 2. Enter your answer into the computer.

❻ RESULTS TABLE:

COL1
2.66667

Hint: **Similar to the first exercise, but a Where condition is added, to select only programmers.**

Note: **If you do not get decimal precision, see Section 12.7.**

EXERCISE 12-4

❶ TOPIC: Column Functions: Average.

❷ EXERCISE 12-4. List the average hours worked for programmers.

❸ INPUT TABLE:

EMPL

NBR	LNAME	FNAME	DOB	HIREDTE	PERF	JOB	DEPT	PROJ
01	LOWE	ROB	53012	85012	4	PROG	FIN	01
02	SHIELD	BROOKE	59131	87001	3	MAN	MKT	01
03	MOORE	ROGER	48111	86002	1	DIR	MKT	04
04	EASTWOOD	CLINT	41091	60120	3	PROG	FIN	03
05	MOSTEL	ZERO	21365	84211		PRES		
06	BURNS	GEORGE	11178	49001	2	SYS	FIN	01
07	O'NEIL	RYAN	42189	60121	3	DIR	ACC	05
08	MARVIN	LEE	32187	51876	2	VP	ACC	02
09	LANCASTER	BURT	41091	79092	1	MAN	R&D	02
10	BLAIR	LINDA	54013	85012	1	PROG	MKT	

PAY

NBR	HOURS	RATE	DED	YTD

❹

FORMAT

```
SELECT function (<expression>)
       FROM table_a, table_b
            WHERE <Primary Key = Foreign Key>
            AND <column = 'value'>;
```

❺ STEP 1. ______________________________

Write your answer here.

____

STEP 2. Enter your answer into the computer.

❻ RESULTS TABLE:

COL1
22.683333

Hint **Joining two tables, with a Join Condition as we have learned in a previous chapter, and a Where condition to select only programmers.**

12.8 MAX FUNCTION

To display the largest value of all non-NULL fields in a column you use the MAX function.

MAX Format

```
SELECT MAX (<column_name>)

               FROM <table_name>

               WHERE <condition>;
```

The considerations for finding column maximums include:

- MAX can be applied to date, time, timestamp and character as well as numeric columns. When used with character columns, MAX returns the highest EBCDIC collating value found in the column. When used with numeric columns, MAX returns the highest ALGEBRAIC value found in the column. When used with date, time or timestamp columns, MAX returns the most recent value in the column.
- MAX disregards rows with NULL values in the column you are processing.

You should find little difficulty using the MAX function. Try the four exercises that follow.

EXERCISE 12-5

❶ TOPIC: Column Functions: Maximum.
❷ EXERCISE 12-5. Display the maximum YTD employee salary.
❸ INPUT TABLE:

PAY

NBR	HOURS	RATE	DED	YTD
01	8.89	43	128.78	11890.66
02	13.23	40	204.45	15840.78
03	6.11	49	94.76	11890.66
04	26.75	45	132.58	17605.66
05	67.82	37	394.69	79990.99
06	32.45	32	121.99	53421.23
07	26.75	49	101.56	32758.11
08	15.99	52	327.98	67870.01
09	43.59	24	0	28090.91
10	32.41	52	112.78	27000.01

FORMAT

❹
```
SELECT function (<column>)
            FROM table;
```

❺ STEP 1. ______________________________
Write your answer here.
____

STEP 2. Enter your answer into the computer.

❻ RESULTS TABLE:

COL1
79990.99

Hint: **Similar to the previous problem.**

EXERCISE 12-6

❶ TOPIC: Column Functions: Maximum.

❷ EXERCISE 12-6. Calculate and print the average employee paycheck (RATE * HOURS – DED).

❸ INPUT TABLE:

PAY

NBR	HOURS	RATE	DED	YTD
01	8.89	43	128.78	11890.66
02	13.23	40	204.45	15840.78
03	6.11	49	94.76	11890.66
04	26.75	45	132.58	17605.66
05	67.82	37	394.69	79990.99
06	32.45	32	121.99	53421.23
07	26.75	49	101.56	32758.11
08	15.99	52	327.98	67870.01
09	43.59	24	0	28090.91
10	32.41	52	112.78	27000.01

FORMAT

❹
```
SELECT function (<expression>)
          FROM table;
```

❺ STEP 1. ______________________________

Write your answer here.

____

STEP 2. Enter your answer into the computer.

❻ RESULTS TABLE:

COL1
2114.65

Hint: **We are just listing the largest paycheck, calculating it with the expression that we used in Exercise 12-2.**

EXERCISE 12-7

❶ TOPIC: Column Functions: Maximum.

❷ EXERCISE 12-7. Print the employee LNAME closest to the end of the alphabet.

❸ INPUT TABLE:

EMPL

NBR	LNAME	FNAME	DOB	HIREDTE	PERF	JOB	DEPT	PROJ
01	LOWE	ROB	53012	85012	4	PROG	FIN	01
02	SHIELD	BROOKE	59131	87001	3	MAN	MKT	01
03	MOORE	ROGER	48111	86002	1	DIR	MKT	04
04	EASTWOOD	CLINT	41091	60120	3	PROG	FIN	03
05	MOSTEL	ZERO	21365	84211		PRES		
06	BURNS	GEORGE	11178	49001	2	SYS	FIN	01
07	O'NEIL	RYAN	42189	60121	3	DIR	ACC	05
08	MARVIN	LEE	32187	51876	2	VP	ACC	02
09	LANCASTER	BURT	41091	79092	1	MAN	R&D	02
10	BLAIR	LINDA	54013	85012	1	PROG	MKT	

FORMAT

❹
```
SELECT  function  (<col>)
               FROM table;
```

❺ STEP 1. ______________________________

Write your answer here.

____

STEP 2. Enter your answer into the computer.

❻ RESULTS TABLE:

COL1
SHIELD

Hint: **The largest, or maximum, name is the one that is closest to the end of the alphabet.**

EXERCISE 12-8

❶ TOPIC: Column Functions: Maximum.

❷ EXERCISE 12-8. List the highest YTD programmer salary.

❸ INPUT TABLE:

EMPL

NBR	LNAME	FNAME	DOB	HIREDTE	PERF	JOB	DEPT	PROJ
01	LOWE	ROB	53012	85012	4	PROG	FIN	01
02	SHIELD	BROOKE	59131	87001	3	MAN	MKT	01
03	MOORE	ROGER	48111	86002	1	DIR	MKT	04
04	EASTWOOD	CLINT	41091	60120	3	PROG	FIN	03
05	MOSTEL	ZERO	21365	84211		PRES		
06	BURNS	GEORGE	11178	49001	2	SYS	FIN	01
07	O'NEIL	RYAN	42189	60121	3	DIR	ACC	05
08	MARVIN	LEE	32187	51876	2	VP	ACC	02
09	LANCASTER	BURT	41091	79092	1	MAN	R&D	02
10	BLAIR	LINDA	54013	85012	1	PROG	MKT	

PAY

NBR	HOURS	RATE	DED	YTD
01	8.89	43	128.78	11890.66
02	13.23	40	204.45	15840.78
03	6.11	49	94.76	11890.66
04	26.75	45	132.58	17605.66
05	67.82	37	394.69	79990.99
06	32.45	32	121.99	53421.23
07	26.75	49	101.56	32758.11
08	15.99	52	327.98	67870.01
09	43.59	24	0	28090.91
10	32.41	52	112.78	27000.01

FORMAT

❹

```
SELECT  function  (<table_b col_a>)
            FROM table_a,  table_b
                  WHERE <Primary Key = Foreign Key>
                  AND <column = 'value'>;
```

❺ STEP 1. ____

Write your answer here.

____

STEP 2. Enter your answer into the computer.

❻ RESULTS TABLE:

COL1
27000.01

Hint: **A very simple use of a function on a column with a table join.**

12.9 MIN FUNCTION

To display the smallest value of all non-NULL fields in a column, you use the MIN function.

MIN Format

```
SELECT MIN (<column_name>)
            FROM <table_name>
            WHERE <condition>;
```

The considerations for finding column maximums include:

- MIN can be applied to date, time, timestamp and character as well as numeric columns. When used with character columns, MIN returns the lowest EBCDIC collating value found in the column. When used with numeric columns, MIN returns the lowest ALGEBRAIC value found in the column. When used with date, time or timestamp columns, MIN returns the least current (OLDEST) value in the column.
- MIN disregards rows with NULL values in the column you are processing.

You should find little difficulty using the MIN function. They work exactly as the MAX function. Do exercises 12-5 through 12-8, substituting MIN for MAX.

12.10 SUM FUNCTION

SUM calculates and displays the total or sum value of all non-NULL fields in the column. It is similar in nature to the AVG function.

SUM Format

```
SELECT SUM  (<column_name>)
       SUM  (DISTINCT <column_name>)

            FROM <table_name>

            WHERE <condition>;
```

The considerations for column totals include:

- SUM must only be used on numeric columns.
- SUM disregards rows with NULL values in the column you are totalling.
- SUM ignores duplicate values in a column if you code the keyword DISTINCT before the column name in the parentheses.
- SUM will give you the same precision in the result value as defined in the column datatype — UNLESS YOU OVERRIDE THIS WITH AN ARITHMETIC EXPRESSION.

Because we are usually interested only in a precision equal to the column default, we typically will not need to override it with an arithmetic expression. And since adding the default of zero to the sum will have the exact same effect as skipping a NULL valued field, we will not need to account for NULLs in columns as we did with AVG.

EXERCISE 12-9

❶ TOPIC: Column Functions: Sum.
❷ EXERCISE 12-9. Display the total of all salaries.
❸ INPUT TABLE:

PAY

NBR	HOURS	RATE	DED	YTD
01	8.89	43	128.78	11890.66
02	13.23	40	204.45	15840.78
03	6.11	49	94.76	11890.66
04	26.75	45	132.58	17605.66
05	67.82	37	394.69	79990.99
06	32.45	32	121.99	53421.23
07	26.75	49	101.56	32758.11
08	15.99	52	327.98	67870.01
09	43.59	24	0	28090.91
10	32.41	52	112.78	27000.01

FORMAT

❹
```
SELECT   function   (<column>)
                 FROM table;
```

❺ STEP 1. ______________________________

Write your answer here.

____

STEP 2. Enter your answer into the computer.

❻ RESULTS TABLE:

COL1
346359.02

EXERCISE 12-10

❶ TOPIC: Column Functions: Sum.
❷ EXERCISE 12-10. Display the total of all weekly paychecks.
❸ INPUT TABLE:

PAY

NBR	HOURS	RATE	DED	YTD
01	8.89	43	128.78	11890.66
02	13.23	40	204.45	15840.78
03	6.11	49	94.76	11890.66
04	26.75	45	132.58	17605.66
05	67.82	37	394.69	79990.99
06	32.45	32	121.99	53421.23
07	26.75	49	101.56	32758.11
08	15.99	52	327.98	67870.01
09	43.59	24	0	28090.91
10	32.41	52	112.78	27000.01

FORMAT

❹
```
SELECT   function   (<expression>)
              FROM table;
```

❺ STEP 1. ______________________________________

Write your answer here.

____

STEP 2. Enter your answer into the computer.

❻ RESULTS TABLE:

COL1
9216.49

***Hint:* A function acting on an expression.**

EXERCISE 12-11

❶ TOPIC: Column Functions: Sum with DISTINCT modifier.
❷ EXERCISE 12-11. Display the total of all the different salaries.
❸ INPUT TABLE:

PAY

NBR	HOURS	RATE	DED	YTD
01	8.89	43	128.78	11890.66
02	13.23	40	204.45	15840.78
03	6.11	49	94.76	11890.66
04	26.75	45	132.58	17605.66
05	67.82	37	394.69	79990.99
06	32.45	32	121.99	53421.23
07	26.75	49	101.56	32758.11
08	15.99	52	327.98	67870.01
09	43.59	24	0	28090.91
10	32.41	52	112.78	27000.01

FORMAT

❹
```
SELECT   function   (<DISTINCT column>)
              FROM table;
```

❺ STEP 1. ______________________________
Write your answer here.
____

STEP 2. Enter your answer into the computer.

❻ RESULTS TABLE:

COL1
334468.36

Hint: **This will not count any salary that is the same as any other salary.**

EXERCISE 12-12

❶ TOPIC: Column Functions: Sum.

❷ EXERCISE 12-12. Display the total of all programmer YTD salaries.

❸ INPUT TABLE:

EMPL

NBR	LNAME	FNAME	DOB	HIREDTE	PERF	JOB	DEPT	PROJ
01	LOWE	ROB	53012	85012	4	PROG	FIN	01
02	SHIELD	BROOKE	59131	87001	3	MAN	MKT	01
03	MOORE	ROGER	48111	86002	1	DIR	MKT	04
04	EASTWOOD	CLINT	41091	60120	3	PROG	FIN	03
05	MOSTEL	ZERO	21365	84211		PRES		
06	BURNS	GEORGE	11178	49001	2	SYS	FIN	01
07	O'NEIL	RYAN	42189	60121	3	DIR	ACC	05
08	MARVIN	LEE	32187	51876	2	VP	ACC	02
09	LANCASTER	BURT	41091	79092	1	MAN	R&D	02
10	BLAIR	LINDA	54013	85012	1	PROG	MKT	

PAY

NBR	HOURS	RATE	DED	YTD
01	8.89	43	128.78	11890.66
02	13.23	40	204.45	15840.78
03	6.11	49	94.76	11890.66
04	26.75	45	132.58	17605.66
05	67.82	37	394.69	79990.99
06	32.45	32	121.99	53421.23
07	26.75	49	101.56	32758.11
08	15.99	52	327.98	67870.01
09	43.59	24	0	28090.91
10	32.41	52	112.78	27000.01

FORMAT

❹

```
SELECT   function   (<table_b col_a>)
              FROM table_a,  table_b
                   WHERE <Primary Key = Foreign Key>
                   AND <column = 'value'>;
```

❺ STEP 1. ____________________

Write your answer here.

____

STEP 2. Enter your answer into the computer.

❻ RESULTS TABLE:

COL1
56496.33

Hint: **We're joining two tables, so we need a Join Condition, and we need to select only programmers, so we need an additional condition.**

12.11 COUNT FUNCTION

COUNT is a dual-purpose function that either:

1. Returns an integer equal to the number of rows in a results table

   SELECT COUNT (*)

2. Returns an integer equal to the number of distinct values in a column

   SELECT COUNT(DISTINCT <column_name>)

COUNT Format

```
SELECT  COUNT(*)
             (DISTINCT  <column_name>)

        FROM  <table_name>

        WHERE  <condition>;
```

The considerations for column totals include:

- COUNT(*) is interpreted as "Count the number of rows SELECTed into this results table." Because COUNT(*) works at the row level, it will count rows with NULL fields in the columns.
- COUNT(DISTINCT <column_name>) is interpreted as "Count the number of different values found in this column." Because COUNT(DISTINCT <column_name>) works at the column level, it does not count NULL fields.

Reasons for using COUNT(*) are fairly obvious. You would use it to answer questions like:

- How many employees are on the payroll table?

```
SELECT COUNT(*)  FROM PAY;
```

- How many invoices have not been returned?

```
SELECT COUNT(*)  FROM INV WHERE RETURN IS NULL;
```

- How many customers live out of state?

```
SELECT COUNT(*)  FROM CUST WHERE NOT ST = 'CT';
```

A few ways to use COUNT(DISTINCT <column_name> are presented below:

- How many different salary levels are on the payroll table:

```
SELECT COUNT(DISTINCT RATE)  FROM PAY;
```

- How many different projects are my programmers involved in?

```
SELECT COUNT(DISTINCT PROJ) FROM EMPL
    WHERE JOB = 'PROG';
```

EXERCISE 12-13

❶ TOPIC: Column Functions: Count.
❷ EXERCISE 12-13. Display the total number of employees.
❸ INPUT TABLE:

EMPL

NBR	LNAME	FNAME	DOB	HIREDTE	PERF	JOB	DEPT	PROJ
01	LOWE	ROB	53012	85012	4	PROG	FIN	01
02	SHIELD	BROOKE	59131	87001	3	MAN	MKT	01
03	MOORE	ROGER	48111	86002	1	DIR	MKT	04
04	EASTWOOD	CLINT	41091	60120	3	PROG	FIN	03
05	MOSTEL	ZERO	21365	84211		PRES		
06	BURNS	GEORGE	11178	49001	2	SYS	FIN	01
07	O'NEIL	RYAN	42189	60121	3	DIR	ACC	05
08	MARVIN	LEE	32187	51876	2	VP	ACC	02
09	LANCASTER	BURT	41091	79092	1	MAN	R&D	02
10	BLAIR	LINDA	54013	85012	1	PROG	MKT	

FORMAT

❹
```
SELECT  function  (<*>)
             FROM table;
```

❺ STEP 1. ______________________________________
Write your answer here.
____

STEP 2. Enter your answer into the computer.

❻ RESULTS TABLE:

COL1
10

Hint: **The function COUNT(*) will just count the rows selected,**

EXERCISE 12-14

❶ TOPIC: Column Functions: Count.

❷ EXERCISE 12-14. Display the total number of programmers.

❸ INPUT TABLE:

EMPL

NBR	LNAME	FNAME	DOB	HIREDTE	PERF	JOB	DEPT	PROJ
01	LOWE	ROB	53012	85012	4	PROG	FIN	01
02	SHIELD	BROOKE	59131	87001	3	MAN	MKT	01
03	MOORE	ROGER	48111	86002	1	DIR	MKT	04
04	EASTWOOD	CLINT	41091	60120	3	PROG	FIN	03
05	MOSTEL	ZERO	21365	84211		PRES		
06	BURNS	GEORGE	11178	49001	2	SYS	FIN	01
07	O'NEIL	RYAN	42189	60121	3	DIR	ACC	05
08	MARVIN	LEE	32187	51876	2	VP	ACC	02
09	LANCASTER	BURT	41091	79092	1	MAN	R&D	02
10	BLAIR	LINDA	54013	85012	1	PROG	MKT	

FORMAT

❹
```
SELECT  function  (<*>)
              FROM table
              WHERE <col_a = 'value'>;
```

❺ STEP 1. ______________________________

Write your answer here.

____

STEP 2. Enter your answer into the computer.

❻ RESULTS TABLE:

COL1
3

Hint: **Like the previous exercise, but with an additional condition to select only programmers.**

EXERCISE 12-15

❶ TOPIC: Column Functions: Count with DISTINCT modifier.

❷ EXERCISE 12-15. Display the total number of different DEPTs.

❸ INPUT TABLE:

EMPL

NBR	LNAME	FNAME	DOB	HIREDTE	PERF	JOB	DEPT	PROJ
01	LOWE	ROB	53012	85012	4	PROG	FIN	01
02	SHIELD	BROOKE	59131	87001	3	MAN	MKT	01
03	MOORE	ROGER	48111	86002	1	DIR	MKT	04
04	EASTWOOD	CLINT	41091	60120	3	PROG	FIN	03
05	MOSTEL	ZERO	21365	84211		PRES		
06	BURNS	GEORGE	11178	49001	2	SYS	FIN	01
07	O'NEIL	RYAN	42189	60121	3	DIR	ACC	05
08	MARVIN	LEE	32187	51876	2	VP	ACC	02
09	LANCASTER	BURT	41091	79092	1	MAN	R&D	02
10	BLAIR	LINDA	54013	85012	1	PROG	MKT	

FORMAT

❹
```
SELECT   function   (DISTINCT <column>)
              FROM table;
```

❺ STEP 1. ______________________________

Write your answer here.

____

STEP 2. Enter your answer into the computer.

❻ RESULTS TABLE:

COL1
4

Hint: **Since we need the number of different departments, we need to put the DISTINCT qualifier before the column selected.**

EXERCISE 12-16

❶ TOPIC: Column Functions: Count with DISTINCT modifier.

❷ EXERCISE 12-16. Display the total number of different DEPTs that have programmers in them.

❸ INPUT TABLE:

EMPL

NBR	LNAME	FNAME	DOB	HIREDTE	PERF	JOB	DEPT	PROJ
=====	============	=========	=======	=========	======	======	======	=====
01	LOWE	ROB	53012	85012	4	PROG	FIN	01
02	SHIELD	BROOKE	59131	87001	3	MAN	MKT	01
03	MOORE	ROGER	48111	86002	1	DIR	MKT	04
04	EASTWOOD	CLINT	41091	60120	3	PROG	FIN	03
05	MOSTEL	ZERO	21365	84211		PRES		
06	BURNS	GEORGE	11178	49001	2	SYS	FIN	01
07	O'NEIL	RYAN	42189	60121	3	DIR	ACC	05
08	MARVIN	LEE	32187	51876	2	VP	ACC	02
09	LANCASTER	BURT	41091	79092	1	MAN	R&D	02
10	BLAIR	LINDA	54013	85012	1	PROG	MKT	

FORMAT

❹
```
SELECT  function  (<DISTINCT column>)
          FROM table
          WHERE <col_a = 'value'>;
```

❺ STEP 1. ______________________________

Write your answer here.

____

STEP 2. Enter your answer into the computer.

❻ RESULTS TABLE:

COL1
==========
2

Hint: **Like the previous exercise, but with an additional condition to select only the rows that have programmers in them.**

12.12 ANSWER KEY

There may be more than one answer to an exercise. However, the one given emphasizes the topic discussed.

EXERCISE	12-1	SELECT AVG (YTD) FROM PAY;
EXERCISE	12-2	SELECT AVG (RATE * HOURS – DED) FROM PAY;
EXERCISE	12-3	SELECT AVG (PERF * 1.0) FROM EMPL WHERE JOB = 'PROG';
EXERCISE	12-4	SELECT AVG(HOURS) FROM EMPL, PAY WHERE EMPL.NBR = PAY.NBR AND JOB = 'PROG';
EXERCISE	12-5	SELECT MAX(YTD) FROM PAY;
EXERCISE	12-6	SELECT MAX(RATE * HOURS – DED) FROM PAY;
EXERCISE	12-7	SELECT MAX(LNAME) FROM EMPL;
EXERCISE	12-8	SELECT MAX(YTD) FROM EMPL, PAY WHERE EMPL.NBR = PAY.NBR AND JOB = 'PROG';
EXERCISE	12-9	SELECT SUM(YTD) FROM PAY;
EXERCISE	12-10	SELECT SUM(RATE * HOURS – DED) FROM PAY;
EXERCISE	12-11	SELECT SUM (DISTINCT YTD) FROM PAY;
EXERCISE	12-12	SELECT SUM(YTD) FROM EMPL, PAY WHERE EMPL.NBR = PAY.NBR AND JOB = 'PROG';
EXERCISE	12-13	SELECT COUNT(*) FROM EMPL;
EXERCISE	12-14	SELECT COUNT(*) FROM EMPL WHERE JOB = 'PROG';
EXERCISE	12-15	SELECT COUNT(DISTINCT DEPT) FROM EMPL;
EXERCISE	12-16	SELECT COUNT(DISTINCT DEPT) FROM EMPL WHERE JOB = 'PROG';

Chapter 13
ROW FUNCTIONS

Here you will learn how to:

1. Manipulate character, date, time and timestamp values;
2. Create column data using the SQL concatenation operations;
3. Translate NULL values into non-NULL values;
4. Perform date, time and timestamp arithmetic.

The following topics are covered:

While Row Functions are not much harder conceptually than the column functions we learned previously, their syntax can be a little confusing.

Also, there are several fairly exotic Row functions which I could not find a good use for in my application. The ones I feel are important are covered in the usual fashion. Some are represented with examples which

---

Note: You must have SQL/DS 2.2 or 2.3, or DB2 1.3, 2.1 or 2.2 to use the ROW functions in this chapter.

may seem a bit arbitrary. And there are a few functions which I have left out entirely. These can be learned in the *IBM DB2 Application Programming Manual.* If after learning the how and why about any of the row functions you don't see any point in working on (say functions like Hex or Floating point translation), feel free to skip them when they appear in the text.

13.1 COLUMN VERSUS ROW FUNCTIONS

Column functions process tables vertically. The result is always a single value representing a cumulative fact about your data. In contrast, Row functions are applied to each field in a row, processing the values in a table horizontally. This is illustrated in Figure 13.1.

As you can see, the average salary (column function AVG) when applied to the Salary column in the table produces the single value 1773.94, the average of all the fields in the Salary column.

The Row function SUBSTR when applied to the LNAME column produces one value for each row in the table — the first single character in each field.

Column functions produce statistical summary information about the values in a table. Row functions are used to transform individual fields into a more readable or useable format.

13.2 ROW FUNCTIONS

As of this writing there are about twenty Row functions for SQL/DS and DB2. They can be logically grouped into two separate categories: data conversion functions which translate values from their stored format for reporting purposes: and date, time and timestamp functions which are used in date manipulation and date arithmetic routines.

13.2.1 Data Conversion Functions

CHAR	—	Translates dates into different character representations.
VALUE	—	Translates NULL values into non-NULL values.
SUBSTR	—	Returns a portion of a character string.
LENGTH	—	Returns the length of the field.
TRUNC	—	Truncates signification digits.*
DECIMAL	—	Displays decimal data with specified precision.*
DIGITS	—	Returns the non-decimal portion of a decimal number.*

---

* DB2 only.

COLUMN vs ROW FUNCTIONS

Row (Scalar) Functions

Column (Aggregate) Functions

LNAME	FNAME	SALARY	DOB
SMITH	EVAN	1200.00	1968-09-12
JONES	K.C.	3200.00	1934-12-25
AARDVARK	LAWRENCE	2499.55	1952-10-01
MOSTEL	HOSTEL	1388.34	1953-01-02
BURNS	ALLEN	2477.99	1900-07-04
DOHNANYI	ERNST	1388.54	1822-03-16
RICHTER	SVIATOSLAV	2466.81	1946-02-18
WATTS	ANDRE	1398.34	1951-08-17
BERNSTEIN	LEONARD	9817.27	1949-05-28

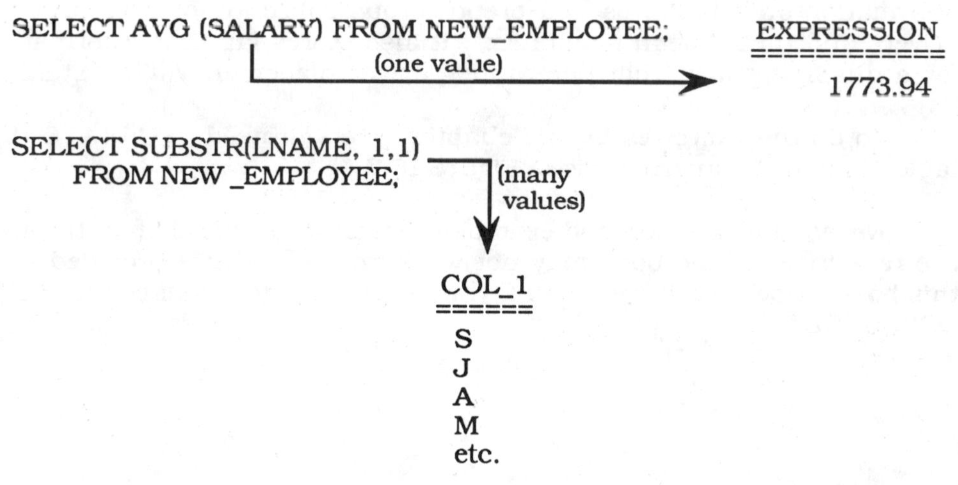

Figure 13.1. Column versus Row functions.

FLOAT	—	Translates numbers in scientific notation (Floating Point).
INTEGER	—	Returns the Integer portion of a number. (DB2 only.)
||	—	Concatenates separate character strings into a single column.

13.2.2 Date, Time, Timestamp Functions

DATE	—	Extracts the date portion of a timestamp.
DAY	—	Displays the day portion of a date or timestamp.
MONTH	—	Returns the month portion of a date or timestamp.
YEAR	—	Returns the year portion of a date or timestamp.
DAYS	—	Returns the number of days from 0001 AD.
HOUR	—	Extracts the hours portion of a time or timestamp.
MINUTE	—	Displays the minutes portion of a time or timestamp.
SECOND	—	Returns the seconds portion of a time or timestamp.
TIME	—	Extracts the time portion of a timestamp.
TIMESTAMP	—	Merges a date, time and unique microsecond.

13.3 THE CELEBRITY_RACE TABLE

Our final table for this course, the CELEBRITY_RACE table contains information on a fictitious and preposterous collection of amateur race drivers and their "mean machines." It also stores lap-time information for a qualifying heat that determines driver placement at the starting flag.

To do the exercises in this chapter, you will need to CREATE this table. Follow the instructions in Figure 13.2.

Note: The exercises and examples presented in this chapter through the remainder of the book may not work on the diskette provided with this book. However, it will work if you are using a terminal connected to DB2 or SQL/DS.

CELEBRITY_RACE TABLE

ENTRANT	CAR	DRIVER	PRICE	MILEAGE	PURCHASE
1	LINCOLN	MARIO ANDERSON	3400.88	5400	1987-01-01
2	SEAHAWK	MRS. PAUL	1940.25	2100	1984-05-29
3	300 ZX	PAUL NEWMAN	189.99	12788	1988-05-22
4	LAND ROVER	MAGGIE THATCHER	----------	15400	1985-05-22
5	------------	MILLARD FILLMORE	1999.99	----------	1983-10-31

```
CREATE TABLE CELEBRITY_RACE
 ( ENTRANT      NUMBER,
   CAR          CHAR(12),
   DRIVER       VARCHAR(18),
   PRICE        NUMBER,
   MILEAGE      NUMBER,
   PURCHASE     DATE);

INSERT INTO CELEBRITY_RACE VALUES
   (1, 'LINCOLN', 'MARIO ANDERSON', 3400.88,5400,'1987-01-01');
INSERT INTO CELEBRITY_RACE VALUES
   (2, 'SEAHAWK', 'MRS. PAUL',1940.25,2100, '1984-05-29');
INSERT INTO CELEBRITY_RACE VALUES
   (3, '300 ZX', 'PAUL NEWMAN', , 189.99,12788, '1988-05-22');
INSERT INTO CELEBRITY_RACE VALUES
   (4, 'LAND ROVER', 'MAGGIE THATCHER',NULL,15400, '1985-05-22');
INSERT INTO CELEBRITY RACE_VALUES
   (5, NULL, 'MILLARD FILLMORE', 1999.99,NULL, '1983-10-31');
```

Figure 13.2. SQL statements to build the Celebrity_Race table.

13.4 DATE, TIME AND TIMESTAMP DATATYPES

DB2 and SQL/DS store date, time and timestamp internally as encrypted numeric data. Externally they present date, time and timestamp information in one of four formats:

INTERNATIONAL (ISO) FORMAT
STANDARD USA FORMAT
EUROPEAN FORMAT
JAPANESE INDUSTRIAL FORMAT

These formats are depicted in Figures 13.3 and 13.4.

ISO is the default format for DB2 and SQL/DS date installation, however, your shop may use any of the other three representations. To find out what your installation date format is, type in the following statement and execute it:

SELECT PURCHASE FROM CELEBRITY_RACE

If your answer is not in ISO format you will have to account for this when doing the other exercises.

As you can see from Figure 13.3, dates are broken up into three portions divided by a dash in ISO format:

A valid year — four digits from 0001 AD to 9999 AD
A valid month — two digits from 01 to 12
A valid day — two digits from 01 to 31

DB2 and SQL/DS enforce legal dates. For example, you cannot store February 29, 1981 in a date column, etc.

Figure 13.4 shows that times are broken up into three portions divided by a period in ISO format:

A valid hour — two digits from 00 to 24 (military format)
Valid minutes — two digits from 00 to 59
Valid seconds — two digits from 00 to 59

DB2 and SQL/DS enforce legal times. That is, you cannot store seventy-one seconds as 00.00.71; you must use 00.01.11 (one minute, eleven seconds).

Timestamps are stored as the concatenation of a:

Valid date
Valid time
Microsecond — six digits computer generated

Date Representation

FORMAT NAME	ABBREV.	FORMAT	EXAMPLE
INTERNATIONAL	ISO	yyyy-mm-dd	1986-12-25
STANDARDS ORG. IBM/USA	USA	mm/dd/yyyy	12/25/1986
IBM EUROPEAN	EUR	dd.mm.yyyy	25.12.1986
JAPANESE INDUSTRIAL STANDARDS	JIS	yyyy-mm-dd	1986-12-25

Time Representation

FORMAT NAME	ABBREV.	FORMAT	EXAMPLE
INTERNATIONAL	ISO	HH.MM.SS	13.01.05
STANDARDS ORG. IBM/USA	USA	HH.MM (AM or PM)	1:01 PM
IBM EUROPEAN	EUR	HH.MM.SS	13.01.06
JAPANESE INDUSTRIAL STANDARDS	JIS	HH:MM:SS	13:01:05

Figure 13.3. Date/Time/Timestamp datatypes.

Example: Six microseconds after 11:00 am, May 30, 1989, would be represented:

1989–05–30–11.00.00.000006

13.5 DB2 AND SQL/DS CURRENT DATE, TIME, TIMESTAMP REGISTERS

DB2 and SQL/DS supports three hardware registers that contain the system date, time and timestamp.

CURRENT DATE — Returns the system maintained date
CURRENT TIME — Returns the system maintained time
CURRENT TIMESTAMP — Returns the system maintained timestamp

- **Date**
  - — Year 0001 to 9999
  - — Month 1 to 12
  - — Day (30) for months 4, 6, 9, 11
    (31) for months 1, 3, 5, 7, 8, 10
    (28) for month 2 — leap year 29
  - — 1988-09-12
- **Time**
  - — Hour 0 to 24
  - — Minute 0 to 59
  - — Second 0 to 59
  - — 11.04.18
- **Timestamp**
  - — Date (YEAR–MONTH–DAY.)
  - — Time (HOUR.MINUTE.SECOND.)
  - — Microsecond (NNNNNN)
  - — 1988–09–12-11.04.18.000000
- **Dates can be character strings in the format:**

  'YYYY–MM–DD' '1988–09–12'

- **DB2 and SQL/DS support three SPECIAL REGISTERS for Date/Time**

  CURRENT DATE
  CURRENT TIME
  CURRENT TIMESTAMP

Figure 13.4. Date/Time/Timestamp Examples.

These values may be used in SELECT statement either in the SELECT clause or the WHERE condition.

Example: SELECT CURRENT DATE, CURRENT TIME, CURRENT TIMESTAMP FROM EMPL;

If you are using the mainframe version of this course, try typing in the above example and executing it.

13.6 ROW FUNCTION FORMAT

As I mentioned in the introduction to this chapter, the Row function formats are not as straightforward as the column functions from Chapter 2.

The basic format is shown below in Figure 13.5.

```
SELECT ROW-FUNCTION(COLUMN_NAME, ARGUMENTS)  FROM <TABLE_NAME>;
```

Figure 13.5. Row function format.

As you can see, the Row functions are coded with the target column names and function argument(s) within the parentheses. The function arguments represent parameters passed to the function processor telling it exactly how you want the data in the column to be converted.

Example:

DECIMAL(PRICE,9,3) (DB2 only) — Display the fields in the PRICE column from their current Decimal precision to a total precision of nine digits, with three digits to the right of the decimal place.

CHAR(PURCHASE,USA) — Display the fields in the PURCHASE column from the current ISO Date representation to USA standard representation.

SUBSTR(CAR,2,1) — Return one character of each field in the CAR column, at position 2.

13.7 DATA CONVERSION FUNCTIONS

SQL supports nine Row functions which can be used to convert data stored internally in one format to another for external reporting or manipulating purposes. THE ACTUAL VALUES AND THEIR INTERNAL REPRESENTATION ARE NOT CHANGED. Only the resultant fields, having been modified by the conversion function, appear in their translated format. The nine data conversion functions are:

CHAR	—	Translates dates into different character representations.
VALUE	—	Translates NULL values into non-NULL values.
SUBSTR	—	Returns a portion of a character string.
LENGTH	—	Returns the length of the field.
TRUNC	—	Truncates significant digits. (DB2 only)
DECIMAL	—	Displays decimal data with specified precision. (DB2 only)
DIGITS	—	Returns the non-decimal portion of a decimal number. (DB2 only)
FLOAT	—	Translates numbers in scientific notation (Floating Point).
HEX	—	Translates data into hexadecimal format.
INTEGER	—	Returns the Integer portion of a number.
||	—	Concatenates separate character strings into a single column.

Of these nine functions, we will be concentrating on what I believe to be the more valuable ones: CHAR, SUBSTR, VALUE and || (Concatenation). The rest will be shown as examples without corresponding reinforcement exercises. If you would like to try the examples out on the computer, feel free to type them in and execute them.

13.8 CHAR — DATE TO CHARACTER STRING CONVERSION

CHAR is used to translate date information from whatever representation it is stored in internally to a different presentational format. Recall the four different DB2 date representations:

Let's look at an example:

Present the Purchase date (PURCHASE) information on the CELEBRITY_RACE table in European Industrial Format:

```
SELECT CHAR(PURCHASE,EUR) FROM CELEBRITY_RACE;
```

PURCHASE
01.01.1987
29.05.1984
22.05.1988
25.12.1985
31.10.1983

EXERCISE 13-1

❶ TOPIC: Row Functions: Conversion.

❷ EXERCISE 13-1. Print the cars and purchase dates out in USA standard date format.

❸ INPUT TABLE:

CELEBRITY_RACE TABLE

ENTRANT	CAR	DRIVER	PRICE	MILEAGE	PURCHASE
1	LINCOLN	MARIO ANDERSON	3400.88	5400	1987-01-01
2	SEAHAWK	MRS. PAUL	1940.25	2100	1984-05-29
3	300 ZX	PAUL NEWMAN	189.99	12788	1988-05-22
4	LAND ROVER	MAGGIE THATCHER	----------	15400	1985-05-22
5	------------	MILLARD FILLMORE	1999.99	----------	1983-10-31

FORMAT

❹
```
SELECT <column>, function (<column>, arguments)
        FROM table;
```

❺ RESULTS TABLE:

CAR	PURCHASE
LINCOLN	01/01/1987
SEAHAWK	05/29/1984
300 ZX	05/22/1988
LAND ROVER	12/25/1985
----------	10/31/1983

13.9 SUBSTR — EXTRACT A PORTION OF A CHARACTER STRING

SUBSTR is used to display a specific portion — substring — of a column value. The column must be defined as CHAR or VARCHAR in order to be used with SUBSTR. SUBSTR must have two arguments along with the column name within parentheses.

```
SUBSTR(<column_name>,<start_pos>, <number_of_chars>)
```

Example: SELECT SUBSTR(LNAME,2,4) FROM EMPL WHERE DEPT = 'MKT';

```
==========
HIEL
OORE
LAIR
```

EXERCISE 13-2

❶ TOPIC: Row Functions: Substring.
❷ EXERCISE 13-2. Print last names and first initial of employees.
❸ INPUT TABLE:

EMPL

NBR	LNAME	FNAME	DOB	HIREDTE	PERF	JOB	DEPT	PROJ

FORMAT

❹
```
SELECT <column>, function (<column, argument1,argument2>)
       FROM table;
```

❺ RESULTS TABLE:

LNAME	COL1
LOWE	R
SHIELD	B
MOORE	R
EASTWOOD	C
MOSTEL	Z
BURNS	G
O'NEIL	R
MARVIN	L
LANCASTER	B
BLAIR	L

13.10 VALUE — CONVERT A NULL VALUE TO A NON-NULL VALUE

NULL values present many problems to application developers. They can't be evaluated in arithmetic expressions (the result is always NULL), and they don't look too terribly attractive on reports. The VALUE function is used to give a default value to any NULL found for a field.

13.10.1. Process

As VALUE is applied to the column, each time it encounters a NULL field, it replaces the NULL with whatever is specified in the argument. If a NULL is not encountered, VALUE leaves the field untouched.

VALUE typically replaces numeric columns with zeros and non-numeric columns with spaces or some other character value. Here's an example:

Format: VALUE(<column_name>,<replacement_value>)

Example: SELECT VALUE (PROJ,'XX') FROM EMPL;

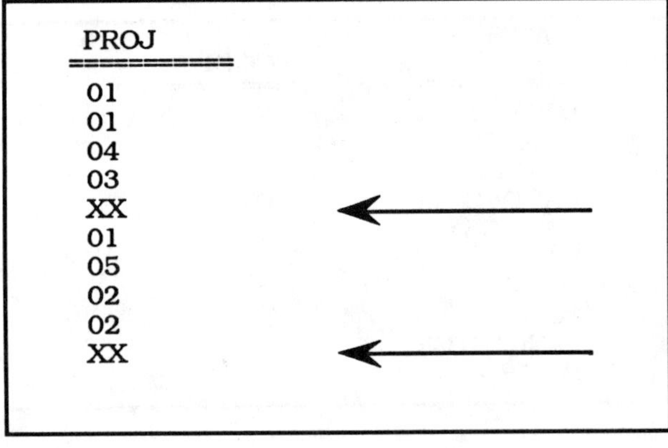

EXERCISE 13-3

❶ TOPIC: Row Functions: Value.

❷ EXERCISE 13-3. Create the report shown in the results table. This report is a replacement value insurance report for the cars on the CELEBRITY_RACE. A ten percent inflation factor is built into the replacement value.

❸ INPUT TABLE:

CELEBRITY_RACE TABLE

ENTRANT	CAR	DRIVER	PRICE	MILEAGE	PURCHASE
1	LINCOLN	MARIO ANDERSON	3400.88	5400	1987-01-01
2	SEAHAWK	MRS. PAUL	1940.25	2100	1984-05-29
3	300 ZX	PAUL NEWMAN	189.99	12788	1988-05-22
4	LAND ROVER	MAGGIE THATCHER	----------	15400	1985-05-22
5	------------	MILLARD FILLMORE	1999.99	----------	1983-10-31

FORMAT

❹
```
SELECT  FUNC  (<column_name>,  <non-null_value>)  *  1.10
            FROM table;
```

❺ STEP 1. ____

Write your answer here.

____

STEP 2. Enter your answer into the computer.

❻ RESULTS TABLE:

PRICE
3740.968
2134.275
208.989
0.000
2199.989

Hint: **The replacement value must match the column data type.**

13.11 || — CONCATENATE SEPARATE CHARACTER STRINGS INTO A SINGLE COLUMN

The long vertical bars (||) on top of the "1" key on most IBM 3270 terminals are used to combine or concatenate separate character strings into a single column. This can be useful for certain report formats and is very easy to use.

Format: <column_one> || <column_two>

Example:

```
SELECT JOB || ' ' || DEPT FROM EMPL;
```

```
        COL1
  ================
  PROG      FIN
  MAN       MKT
  DIR       MKT
  PROG      FIN
* - - - - - - - - - -
  PRES
  SYS       FIN
  VP        ACC
  MAN       R&D
  PROG      MKT
```

A few considerations for using Concatenation:

- The columns you are concatenating must be CHAR or VARCHAR.
- You may also concatenate Character translations of dates.
- You will often need to use literals for either spacing or readability in your reports.
- You may *not* concatenate any *non-character data*. This includes:

  Numeric columns
  Date, time, timestamp columns
  Functions that return numeric data
  Numeric literals

Now let's try doing an exercise that uses both concatenation and the VALUE function.

---

* Note: Dotted line represents NULL results row (DB2/SPUFI representation).

EXERCISE 13-4

❶ TOPIC: Row Functions: Concatenate character strings.

❷ EXERCISE 13-4. Create a report listing the cars and their drivers. If a car's name is not known, substitute the phrase 'HEAP'.

❸ INPUT TABLE:

CELEBRITY_RACE TABLE

ENTRANT	CAR	DRIVER	PRICE	MILEAGE	PURCHASE
1	LINCOLN	MARIO ANDERSON	3400.88	5400	1987-01-01
2	SEAHAWK	MRS. PAUL	1940.25	2100	1984-05-29
3	300 ZX	PAUL NEWMAN	189.99	12788	1988-05-22
4	LAND ROVER	MAGGIE THATCHER	----------	15400	1985-05-22
5	------------	MILLARD FILLMORE	1999.99	----------	1983-10-31

FORMAT

❹
```
SELECT  <column>  ||  'literal'  ||  function(<column>,'literal'))
          FROM table;
```

❺ STEP 1. ____

Write your answer here.

____

STEP 2. Enter your answer into the computer.

❻ RESULTS TABLE:

COL
MARIO ANDERSON DRIVES A LINCOLN
MRS. PAUL DRIVES A SEAHAWK
PAUL NEWMAN DRIVES A 300 ZX
MAGGIE THATCHER DRIVES A LAND ROVER
MILLARD FILLMORE DRIVES A HEAP

13.12 MISCELLANEOUS DATA CONVERSION FUNCTIONS (DB2 ONLY)

Here is an example using some of the other conversion row functions. All of the Conversion Functions are simple to code and useful if you need these types of processing in your application.

```
SELECT PRICE, DECIMAL(PRICE,7,0),  DIGITS(PRICE), FLOAT(PRICE)
          FROM CELEBRITY_RACE
               WHERE CAR = 'LINCOLN';
```

PRICE	COL1	COL2	COL3
3400.88	3400.	0340088	3.400880000000E+04

DECIMAL — Displays decimal data with specified precision

DIGITS — Returns the non-decimal portion of a decimal number

FLOAT — Translates numbers in scientific notation (Floating Point)

HEX — Translates data into hexadecimal format

EXERCISE 13-5

❶ TOPIC: Row Functions: Conversion.
❷ EXERCISE 13-5. Create the report shown in the RESULTS TABLE.
❸ INPUT TABLE:

CELEBRITY_RACE TABLE

ENTRANT	CAR	DRIVER	PRICE	MILEAGE	PURCHASE
1	LINCOLN	MARIO ANDERSON	3400.88	5400	1987-01-01
2	SEAHAWK	MRS. PAUL	1940.25	2100	1984-05-29
3	300 ZX	PAUL NEWMAN	189.99	12788	1988-05-22
4	LAND ROVER	MAGGIE THATCHER	---------	15400	1985-05-22
5	------------	MILLARD FILLMORE	1999.99	----------	1983-10-31

FORMAT

❹
```
SELECT   function  (<column>),   function(<column>)
              FROM table;
```

❺ STEP 1. ______________________________
Write your answer here.
____

STEP 2. Enter your answer into the computer.

❻ RESULTS TABLE:

COL1	PRICE
D3C9D5C3D6D3D5404040	0340088

13.13 DATE, TIME, TIMESTAMP FUNCTIONS

The Date, Time and Timestamp functions basically return a portion of a Date, Time or Timestamp value. They are coded the same as the Conversion Row functions with the target column names and function argument(s) within the parentheses. Remember, the function arguments represent parameters passed to the function processor, in this case, telling it exactly how you want the date/time/timestamp to be manipulated (most of the important functions do not need arguments as you shall see).

13.13.1 Date, Time, Timestamp Functions

DATE	—	Extracts the date portion of a timestamp
DAY	—	Displays the day portion of a date or timestamp
MONTH	—	Returns the month portion of a date or timestamp
YEAR	—	Returns the year portion of a date or timestamp
DAYS	—	Returns the number of days from 0001 AD
HOUR	—	Extracts the hours portion of a time or timestamp
MINUTE	—	Displays the minutes portion of a time or timestamp
SECOND	—	Returns the seconds portion of a time or timestamp
TIME	—	Extracts the time portion of a timestamp
TIMESTAMP	—	Is used to merge a date, time and unique microsecond

The resultant value of a Date, Time or Timestamp function is a date duration of some kind (5 months or 31 days or 11 seconds etc.), which is considered NUMERIC data by DB2. And just like the Conversion Row functions, Date, Time and Timestamp Functions can be coded anywhere in the SELECT or WHERE clause. Let's take a look at some examples.

```
SELECT DAY (PURCHASE) FROM CELEBRITY_RACE;
```

COL1
1
29
22
25
31

SELECT MONTH (PURCHASE) FROM CELEBRITY_RACE WHERE YEAR(PURCHASE) < 1988;

COL1
1
5
12
10

SELECT DATE (CURRENT TIMESTAMP) FROM CELEBRITY_RACE;

COL1
1990-02-05

(The above answer is the date portion of the current timestamp and of course will probably be different for you).

Ready to try a few yourself?

EXERCISE 13-6

❶ TOPIC: Row Functions: Date and time.

❷ EXERCISE 13-6. Create a report listing all cars purchased in the month of May and in the year 1988.

❸ INPUT TABLE:

CELEBRITY_RACE TABLE

ENTRANT	CAR	DRIVER	PRICE	MILEAGE	PURCHASE
1	LINCOLN	MARIO ANDERSON	3400.88	5400	1987-01-01
2	SEAHAWK	MRS. PAUL	1940.25	2100	1984-05-29
3	300 ZX	PAUL NEWMAN	189.99	12788	1988-05-22
4	LAND ROVER	MAGGIE THATCHER	----------	15400	1985-05-22
5	------------	MILLARD FILLMORE	1999.99	----------	1983-10-31

FORMAT

❹
```
SELECT *
FROM   table
       WHERE   <function1(column)  =  literal
       AND     <function1(column)  =  literal;
```

❺ STEP 1. ____________________

Write your answer here.

____

STEP 2. Enter your answer into the computer.

❻ RESULTS TABLE:

CELEBRITY_RACE TABLE

ENTRANT	CAR	DRIVER	PRICE	MILEAGE	PURCHASE
3	300 ZX	PAUL NEWMAN	189.99	12788	1988-05-22

Hint: **You compare date functions with numeric literals.**

13.14 NESTING FUNCTIONS

DB2 and SQL/DS permit the nesting of certain functions. This means you can apply a function on a function. This means it computes the result of a function and then does another function on the result.

Actually there are only three ways you can nest functions:

- You can nest Row functions with other Row functions.
- You can nest Row functions with Column functions.
- You can nest Column functions with Row functions.

Note that you cannot nest Column functions with other Column functions. This was mentioned earlier and I repeat it for emphasis so as to avoid confusion. Let's see a few examples:

```
SELECT SUBSTR ((CHAR(PURCHASE,USA)),6,5)  FROM CELEBRITY_
                        RACE;
```

```
COL1
==========
/1987
/1984
/1988
/1985
/1983
```

(This example extracts a substring of a converted date (from ISO to USA format).

```
SELECT MAX(YEAR(PURCHASE))  FROM CELEBRITY_RACE;
```

```
COL1
==========
1988
```

```
SELECT YEAR(MAX(PURCHASE))  FROM CELEBRITY_RACE;
```

```
COL1
==========
1988
```

Nesting Functions is not a hard concept to grasp. However, I do want to caution you about syntax snags. You must be very careful to use the proper number of parentheses and to place them correctly around your functions, columns and arguments. DB2 and SQL/DS are not very forgiving about this and will present you with an error message that is almost incomprehensible.

13.15 DATE, TIME AND TIMESTAMP ARITHMETIC

In addition to formatting date fields with the Date functions, you can also perform arithmetic operations on date fields. The following operations can be performed:

Date + Date duration
Date – Date duration
Date – Date

Time + Time duration
Time – Time duration
Time – Time

Timestamp + Timestamp duration
Timestamp – Timestamp duration
Timestamp – Timestamp

I have decided in the interest of time and expediency to work only with dates in this text. The time and timestamp arithmetic operations are identical in structure with the date operations and should pose no problem for you should you decide to pursue them independently using the IBM manuals.

13.15.1 Date Durations

As you can see, date arithmetic can be performed with either valid dates and date representations, or valid dates and what are called "date durations." What are date durations? Basically a date duration is one of two types of values that represents an interval of time:

A label duration — 25 YEARS, 1 MONTH, 12 DAYS etc.
A date duration — A three part value consisting of year month and day, represented as YYYYMMDD

Some examples of legal date arithmetic operations include:

SELECT PURCHASE + 1 YEAR FROM CELEBRITY_RACE WHERE CAR = 'LINCOLN'

COL1
1988-01-01

SELECT PURCHASE – 1 YEAR – 1 MONTH – 1 DAY FROM CELEBRITY_RACE WHERE CAR = 'LINCOLN'

COL1
1985-11-30

SELECT (DAYS(CURRENT DATE) – DAYS(PURCHASE)) / 7 FROM CELEBRITY_RACE WHERE CAR = 'LINCOLN'

COL1
64

(This query generates the age of the Lincoln in weeks. It uses CURRENT DATE so if you run it your results will differ.)

13.15.2 Date Arithmetic Peculiarities

Before you start on the date arithmetic exercises, a note of caution. Adding a month from a given date does not always result in the same day of the next month. This can be demonstrated easily using February as an example. If we were to add a row with an ARIES K for Lee Iacocca to the CELEBRITY_RACE table, and the purchase date of the car was January 30, 1989 (1989-01-30), and we then issued the following SELECT statement:

```
SELECT PURCHASE + 1 MONTH FROM CELEBRITY_RACE WHERE CAR
= 'ARIES K';
```

We would get back the following result:

COL1
1989-02-28

Why did the day portion of the date change? Well, we've run out of days in February, so DB2 decides to gives us the last day of the month by default. Basically DB2 says "If adding months to a date gives a valid date, that's okay. Otherwise, make the result the last day of the month." DB2 does issue a warning that an adjustment has been made to compensate for month differences.

An alternative to adding months might be to add a specific number of days. If we were to recode the query as follows:

```
SELECT PURCHASE + 30 DAYS FROM CELEBRITY_RACE WHERE CAR =
'ARIES K';
```

We would get back the following result:

COL1
1989-03-01

Let's try some exercises.

EXERCISE 13-7

❶ TOPIC: Row Functions: Date/time arithmetic.

❷ EXERCISE 13-9. Create a report listing the purchase date of the cars in the format shown in the results table. Add a year and 1 month to all purchase dates.

❸ INPUT TABLE:

CELEBRITY_RACE TABLE

ENTRANT	CAR	DRIVER	PRICE	MILEAGE	PURCHASE
1	LINCOLN	MARIO ANDERSON	3400.88	5400	1987-01-01
2	SEAHAWK	MRS. PAUL	1940.25	2100	1984-05-29
3	300 ZX	PAUL NEWMAN	189.99	12788	1988-05-22
4	LAND ROVER	MAGGIE THATCHER	----------	15400	1985-05-22
5	------------	MILLARD FILLMORE	1999.99	----------	1983-10-31

❹

FORMAT

```
SELECT column + numeric-literal YEAR + numeric-literal MONTH
       FROM table;
```

❺ STEP 1. ____

Write your answer here.

____

STEP 2. Enter your answer into the computer.

❻ RESULTS TABLE:

PURCHASE
1988-02-01
1985-06-29
1989-06-22
1987-01-25
1984-11-30

13.16 ANSWER KEY

There may be more than one answer to an exercise. However, the one given emphasizes the topic discussed.

EXERCISE 13-1	SELECT CAR, CHAR(PURCHASE,USA) FROM CELEBRITY_RACE;
EXERCISE 13-2	SELECT LNAME, SUBSTR(FNAME,1,1) FROM EMPL;
EXERCISE 13-3	SELECT VALUE (PRICE,0) * 1.10) FROM CELEBRITY_RACE;
EXERCISE 13-4	SELECT DRIVER || ' DRIVES A ' || VALUE(CAR,'HEAP') FROM CELEBRITY_RACE;
EXERCISE 13-5	SELECT HEX (CAR), DIGITS (PRICE) FROM CELEBRITY_RACE;
EXERCISE 13-6	SELECT * FROM CELEBRITY_RACE WHERE MONTH (PURCHASE) = 5 AND YEAR(PURCHASE) = 1988;
EXERCISE 13-7	SELECT PURCHASE + 1 YEAR + 1 MONTH, PURCHASE FROM CELEBRITY_RACE;

Chapter 14
SUMMARIZING DATA

Here you will learn to:

1. Summarize table values by groups;
2. Eliminate unwanted groups from your reports;
3. Merge rows from tables.

The following topics are discussed:

14.1 OVERVIEW OF CONTROL BREAK REPORTS

Up to now, we've been using SQL to process our tables in their entirety. That is, all the operators we have executed against our tables have dealt with ALL the rows in the table or all rows that met the conditions specified in our WHERE clause.

Sometimes however, we might like to operate on specific logical sub-groups of data in a table. For instance:

> Display the lowest, highest and average performance evaluation by DEPT;
> Display a count of the Invoices for customers by state;
> Display the lowest, highest and average salary by JOB;
> Display a count of the number of rentals by movie type;

This type of report is similar to the classic control breaks where done on application programs would:

1. Pre-sort the data on the field(s) the control breaks were done on (DEPT, ST, JOB, TYPE etc.);
2. Process each sub-group of records independently;
3. Produce a summary line for each sub-group;
4. Produce a grand total line for the report.

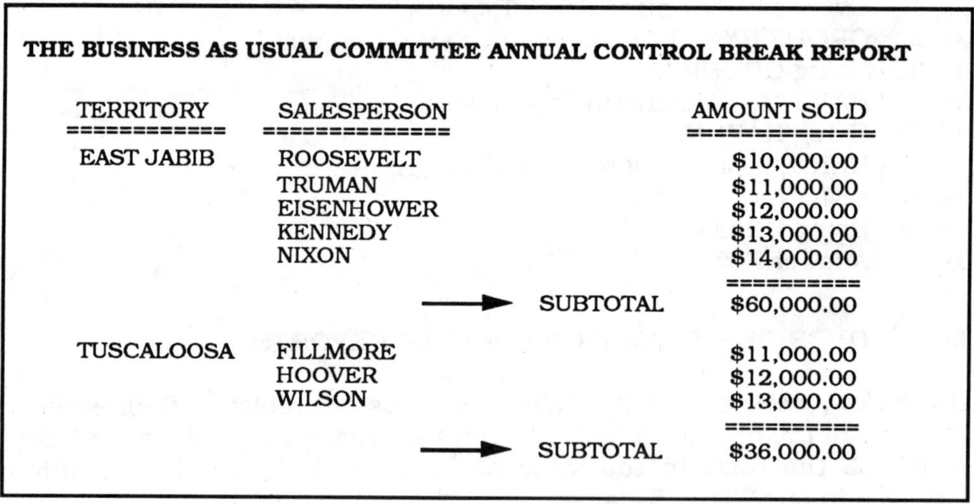

THE BUSINESS AS USUAL COMMITTEE ANNUAL CONTROL BREAK REPORT

TERRITORY	SALESPERSON		AMOUNT SOLD
EAST JABIB	ROOSEVELT		$10,000.00
	TRUMAN		$11,000.00
	EISENHOWER		$12,000.00
	KENNEDY		$13,000.00
	NIXON		$14,000.00
		→ SUBTOTAL	$60,000.00
TUSCALOOSA	FILLMORE		$11,000.00
	HOOVER		$12,000.00
	WILSON		$13,000.00
		→ SUBTOTAL	$36,000.00

Figure 14.1. Control break report.

This is, in essence, what the GROUP BY operator does, except that GROUP BY only produces the subtotal lines shown above, not the detail lines. However, as we are about to see, using GROUP BY is a whale of a lot easier than writing a Control Break Report program.

14.2 GROUP BY — PROCESSING TABLE SUB-GROUPINGS

GROUP BY is used in summary reports or whenever you need to apply the DB2 Column functions (MAX, MIN, AVG, SUM and COUNT) to table rows segregated by groups of the same data values. GROUP BY works in a two step process:

1. Rows are reordered (grouped) around each distinct value in a column.
2. The functions in the SELECT clause are applied to each of the sub-groups of data formed in step one.

This is illustrated in Figure 14.2.

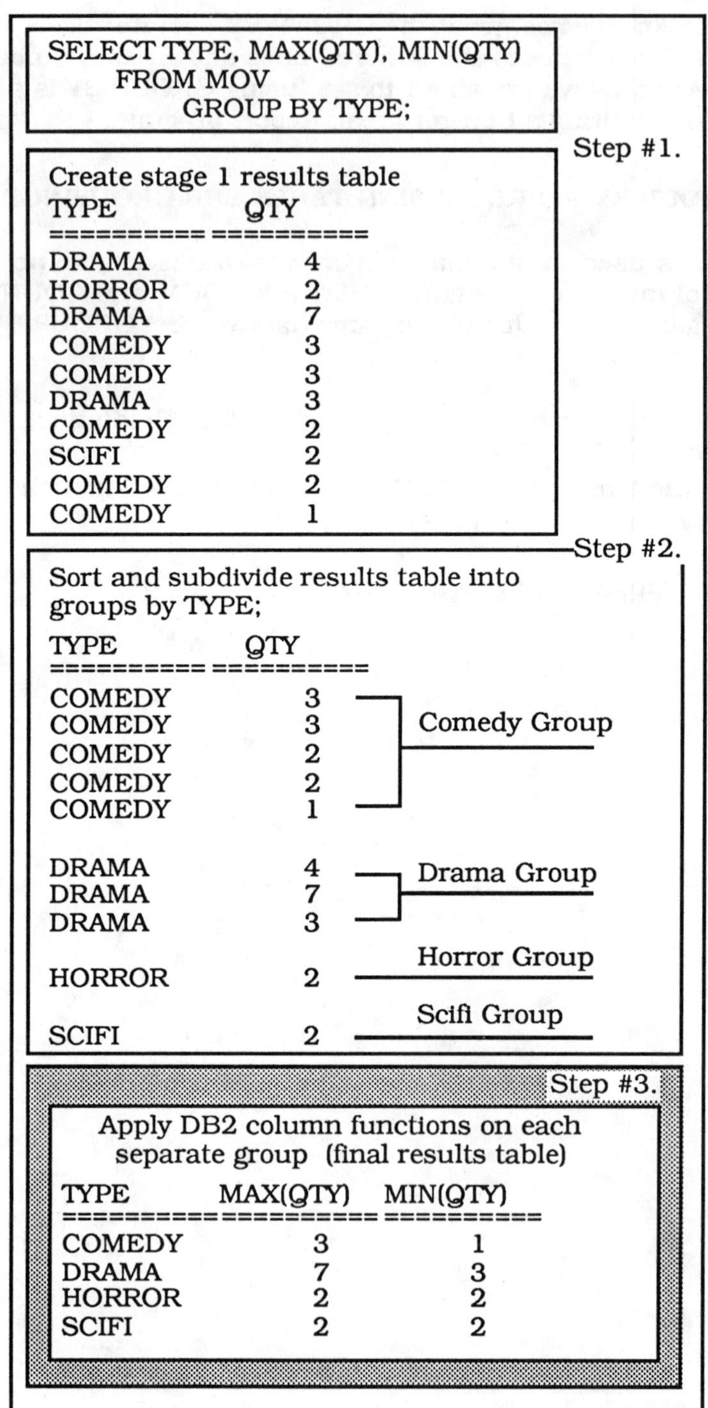

Figure 14.2. GROUP BY process.

14.3 USING GROUP BY

GROUP BY is easy to use. Here are the rules:

1. The GROUP BY clause comes after the:
   WHERE clause, or;
   FROM clause if there is no WHERE clause.
2. GROUP BY partitions your table rows, creating sub-groups of rows with the same values in the GROUPed BY column.
3. GROUP BY treats NULL values as a distinct group of values.
4. You may create sub-groups within sub-groups:
   GROUP BY RATING, TYPE — breaks the movie table out into sub-groups of all different RATING/TYPE values.
5. With GROUP BY you are able to avoid the SQL constraint against SELECTing Column functions and "non-functioned" table columns in the same SELECT clause.

Note that to accomplish its slight-of-hand, GROUP BY must first sort your table on the GROUPed BY column. This can be expensive if you are working with large tables.

EXERCISE 14-1

❶ TOPIC: Summarizing Data.

❷ EXERCISE 14-1. Display each department and the lowest, highest, and average performance evaluations within each department.

❸ INPUT TABLE:

EMPL

NBR	LNAME	FNAME	DOB	HIREDTE	PERF	JOB	DEPT	PROJ
01	LOWE	ROB	53012	85012	4	PROG	FIN	01
02	SHIELD	BROOKE	59131	87001	3	MAN	MKT	01
03	MOORE	ROGER	48111	86002	1	DIR	MKT	04
04	EASTWOOD	CLINT	41091	60120	3	PROG	FIN	03
05	MOSTEL	ZERO	21365	84211		PRES		
06	BURNS	GEORGE	11178	49001	2	SYS	FIN	01
07	O'NEIL	RYAN	42189	60121	3	DIR	ACC	05
08	MARVIN	LEE	32187	51876	2	VP	ACC	02
09	LANCASTER	BURT	41091	79092	1	MAN	R&D	02
10	BLAIR	LINDA	54013	85012	1	PROG	MKT	

FORMAT

```
SELECT <column>, function (<column>), function(<column>)
    FROM table
        GROUP BY <column>, <column>;
```

❹

❺ STEP 1. ______________________________

Write your answer. here

____

STEP 2. Enter your answer into the computer.

❻ RESULTS TABLE:

=========	===========	===========	=============
ACC	2	3	2.5
FIN	2	4	3
MKT	1	3	1.66666667
R&D	1	1	1

Hint: **You can select a column and a function in the same Select if you have a GROUP BY.**

Question: **How do you obtain decimal precision on the AVG(PERF)?**

EXERCISE 14-2

❶ TOPIC: Summarizing data.

❷ EXERCISE 14-2. Display each department, job, and the lowest, highest and average performance evaluations broken out by job within each department.

❸ INPUT TABLE:

EMPL

NBR	LNAME	FNAME	DOB	HIREDTE	PERF	JOB	DEPT	PROJ
=====	=======	=======	=====	=========	=====	====	=====	=====

FORMAT

❹ STEP 1. ______________________________

Write your answer here.

____

STEP 2. Enter your answer into the computer.

❺ RESULTS TABLE:

DEPT	JOB	MIN(PERF)	MAX(PERF)	AVG(PERF)
======	=======	===========	===========	=============
	PRES			
ACC	DIR	3	3	3
ACC	VP	2	2	2
FIN	PROG	3	4	3.5
FIN	SYS	2	2	2
MKT	DIR	1	1	1
MKT	MAN	3	3	3
MKT	PROG	1	1	1
R&D	MAN	1	1	1

Hint: **Similar to the previous, but with two columns GROUPed.**

Note: **Mainframe users will see rows in different sort sequence.**

EXERCISE 14-3

❶ TOPIC: Summarizing Data.

❷ EXERCISE 14-3. Create a report listing the average employee salary broken down by DEPT.

❸ INPUT TABLE:

EMPL

NBR	LNAME	FNAME	DOB	HIREDTE	PERF	JOB	DEPT	PROJ

PAY

NBR	HOURS	RATE	DED	YTD

FORMAT

❹
```
SELECT   <column>,   function(<column>),
         FROM  table, table
         GROUP BY  <column>;
```

❺ STEP 1. ______________________________

Write your answer here.

____

STEP 2. Enter your answer into the computer.

❻ RESULTS TABLE:

DEPT	AVG(YTD)
	79990.99
ACC	50314.06
FIN	27639.666
MKT	18243.8166
R&D	28090.91

Hint: **Contains a table join, a Select of a column and a function, and GROUP BY.**

14.4 HAVING — RESTRICTING GROUP SELECTION

You might not always want or need to see all of the various sub-groups in a table on a single report. In our miniature Employee Database with tables of ten rows and only a handful of distinct values per column, this is not an issue. However, in the real world with typical table sizes well over 50,000 rows and with perhaps hundreds or even thousands of distinct values in potential GROUPed BY columns, the number of summary lines produced may become an issue. You use HAVING to eliminate sub-groups of rows you do not want to see on your report.

Why not just use WHERE? We know that the WHERE clause filters rows out of the pool of values we apply GROUP BY to. This may not really help us because:

1. WHERE is applied before GROUP BY is executed.
2. WHERE can only restrict rows based on standard column/value comparisons.

HAVING restricts groups or rows not meeting a statistical criteria (HAVING AVG(YTD) <40000) etc. This is illustrated in Figure 14.3 where we are producing the same report as in Figure 14.2 except that we only want to see summary lines for movies that have more than two entries per type.

14.5 USING HAVING

HAVING is also easy to use. Here are the rules:

1. The HAVING clause comes after GROUP BY, as a matter of fact.
2. HAVING is only used in combination with GROUP BY — you never use HAVING without a GROUP BY clause.
3. HAVING usually references the five DB2 Column Functions:
   HAVING MAX
   HAVING MIN
   HAVING AVG
   HAVING COUNT
   HAVING SUM
4. You can specify multiple HAVING clauses connected by AND/OR:
5. HAVING MAX(PERF) = 3 OR AVG(PERF) > 1.0

You can combine HAVING and WHERE in the SELECT statement:

```
SELECT RATING, TYPE, AVG(PRICE), AVG(QTY)
    FROM MOV
        WHERE PRICE < 40
    GROUP BY RATING, TYPE
        HAVING COUNT(*) >2;
```

(Translation: Create a summary report of the average movie price and quantity for movies costing less than $40. Break the report down by TYPE within RATING for sub-groups of movies with more than two entries in each group.)

To see a GROUP BY/HAVING order of events look at Figure 14.3.

HAVING

```
SELECT TYPE, RATING, MIN(QTY), MAX(QTY), MIN(PRICE), MAX(PRICE)
     FROM MOV
     WHERE RATING LIKE 'PG%'
     GROUP BY RATING, TYPE
     HAVING MIN(QTY) > 2
```

1) Filter rows from MOV with WHERE condition

3	TOP GUN	DRAMA	PG	CRUISE	7	49.95
4	SPLASH	COMEDY	PG13	HANKS	3	29.95
8	COCOON	SCIFI	PG	AMECHE	2	
9	CROCODILE DUNDEE	COMEDY	PG13	HARRIS	2	69.95
10	TOOTSIE	COMEDY	PG	HOFFMAN	1	29.95

2) Sort and group rows on GROUP BY column

10	TOOTSIE	COMEDY	PG	HOFFMAN	1	29.95
3	TOP GUN	DRAMA	PG	CRUISE	7	49.95
8	COCOON	SCIFI	PG	AMECHE	2	
4	SPLASH	COMEDY	PG13	HANKS	3	29.95
9	CROCODILE DUNDEE	COMEDY	PG13	HARRIS	2	69.95

3) Apply Built-in functions

TYPE	RATING	MIN(QTY)	MAX(QTY)	MIN(PRICE)	MAX(PRICE)
COMEDY	PG	1	1	29.95	29.95
DRAMA	PG	7	7	49.95	49.95
SCIFI	PG	2	2		
COMEDY	PG13	2	3	29.95	69.95

4) Apply HAVING clause

TYPE	RATING	MIN(QTY)	MAX(QTY)	MIN(PRICE)	MAX(PRICE)
DRAMA	PG	7	7	49.95	49.95

Figure 14.3. GROUP BY/HAVING internal sequence of operations.

EXERCISE 14-4

❶ TOPIC: Summarizing Data. HAVING.

❷ EXERCISE 14-4. Display each department and the lowest, highest and average performance evaluations within each department for departments with MORE THAN ONE EMPLOYEE IN THEM.

❸ INPUT TABLE:

EMPL

NBR	LNAME	FNAME	DOB	HIREDTE	PERF	JOB	DEPT	PROJ
01	LOWE	ROB	53012	85012	4	PROG	FIN	01
02	SHIELD	BROOKE	59131	87001	3	MAN	MKT	01
03	MOORE	ROGER	48111	86002	1	DIR	MKT	04
04	EASTWOOD	CLINT	41091	60120	3	PROG	FIN	03
05	MOSTEL	ZERO	21365	84211		PRES		
06	BURNS	GEORGE	11178	49001	2	SYS	FIN	01
07	O'NEIL	RYAN	42189	60121	3	DIR	ACC	05
08	MARVIN	LEE	32187	51876	2	VP	ACC	02
09	LANCASTER	BURT	41091	79092	1	MAN	R&D	02
10	BLAIR	LINDA	54013	85012	1	PROG	MKT	

FORMAT

❹

```
SELECT  <column>, function(<column>), function(<column>),
        function(<column * numeric-literal>)
        FROM table
        GROUP BY <column>
        HAVING function(*) > numeric-literal;
```

❺ STEP 1. ______________________________

Write your answer here.

____

STEP 2. Enter your answer into the computer.

❻ RESULTS TABLE:

DEPT	MIN(PERF)	MAX(PERF)	AVG(PERF)
ACC	2	3	2.5
FIN	2	4	3
MKT	1	3	1.66666667

Hint: **Selecting a column, several functions, with a GROUP BY restricted by a HAVING.**

EXERCISE 14-5

❶ TOPIC: Summarizing Data. HAVING.

❷ EXERCISE 14-5. Display each department and the lowest, highest and average performance evaluations within each department for departments that have AN AVERAGE PERFORMANCE EVALUATION GREATER THAN 2.

❸ INPUT TABLE:

EMPL

NBR	LNAME	FNAME	DOB	HIREDTE	PERF	JOB	DEPT	PROJ
=====	=======	=======	=====	=========	=====	====	=====	=====

FORMAT

❹
```
SELECT <column>, function(<column>), function(<column>),
       function(<column * numeric-literal>)
       FROM table
       GROUP BY <column>
HAVING function(<column * numeric-literal>) > numeric-literal;
```

❺ STEP 1. ______________________________

Write your answer here.

____

STEP 2. Enter your answer into the computer.

❻ RESULTS TABLE:

DEPT	MIN(PERF)	MAX(PERF)	AVG(PERF)
=========	===========	===========	=============
ACC	2	3	2.5
FIN	2	4	3

Hint: **Similar to the previous , but the HAVING has the result of a function being compared to a numeric literal. What precision is the PERF column?**

EXERCISE 14-6

❶ TOPIC: Summarizing Data. HAVING.

❷ EXERCISE 14-6. Display each department and the lowest, highest and average performance evaluations within each department for departments that have AN AVERAGE PERFORMANCE EVALUATION GREATER THAN 2 OR HAVE MORE THAN ONE EMPLOYEE.

❸ INPUT TABLE:

EMPL

NBR	LNAME	FNAME	DOB	HIREDTE	PERF	JOB	DEPT	PROJ

FORMAT

❹
```
SELECT   <column>, function(<column>), function(<column>),
         function(<column * numeric-literal>)
         FROM table
         GROUP BY <column>
HAVING   function(<column * numeric-literal>) > numeric-literal
OR       function(<column * numeric-literal>) > numeric-literal;
```

❺ STEP 1. ______________________________________

Write your answer here.

____

STEP 2. Enter your answer into the computer.

❻ RESULTS TABLE:

DEPT	MIN(PERF)	MAX(PERF)	AVG(PERF)
ACC	2	3	2.5
FIN	2	4	3
MKT	1	3	1.66666667

Hint: **Similar to the previous, exercise but the HAVING has two conditions to be fulfilled — you can connect multiple HAVING clauses with AND/OR.**

14.6 UNION — MERGING ROWS FROM TABLES

UNION is an operation used to merge the results of two or more queries into a single results table. It is often used when combining lists of values retrieved from different base tables, but that is not a restriction.

Example:

Display all employees hired before 1953 and all employees making over 20,000:

```
SELECT NBR FROM EMPL WHERE HIREDTE < 53000
UNION
SELECT NBR FROM PAY WHERE YTD > 20000;
```

```
COL1
===========
05
06
07
08
09
10
```

This smells suspiciously like our old friend the table join. Let's see the differences between UNION and table joins.

14.7 UNION VERSUS JOIN

We have already learned how to combine data from separate tables by Joining tables on a common column value. This is similar to the Venn diagram shown in Figure 14.4. We see that the Join operation creates a subset of table rows where column values intersect between two or more tables.

There is another operation (UNION) that also combines data from different tables. UNION works a little differently. As shown below in Figure 14.4, UNION merges ALL rows form two or more tables. Actually, it sort of appends the rows from one SELECT execution below the rows from the next for all UNIONed queries.

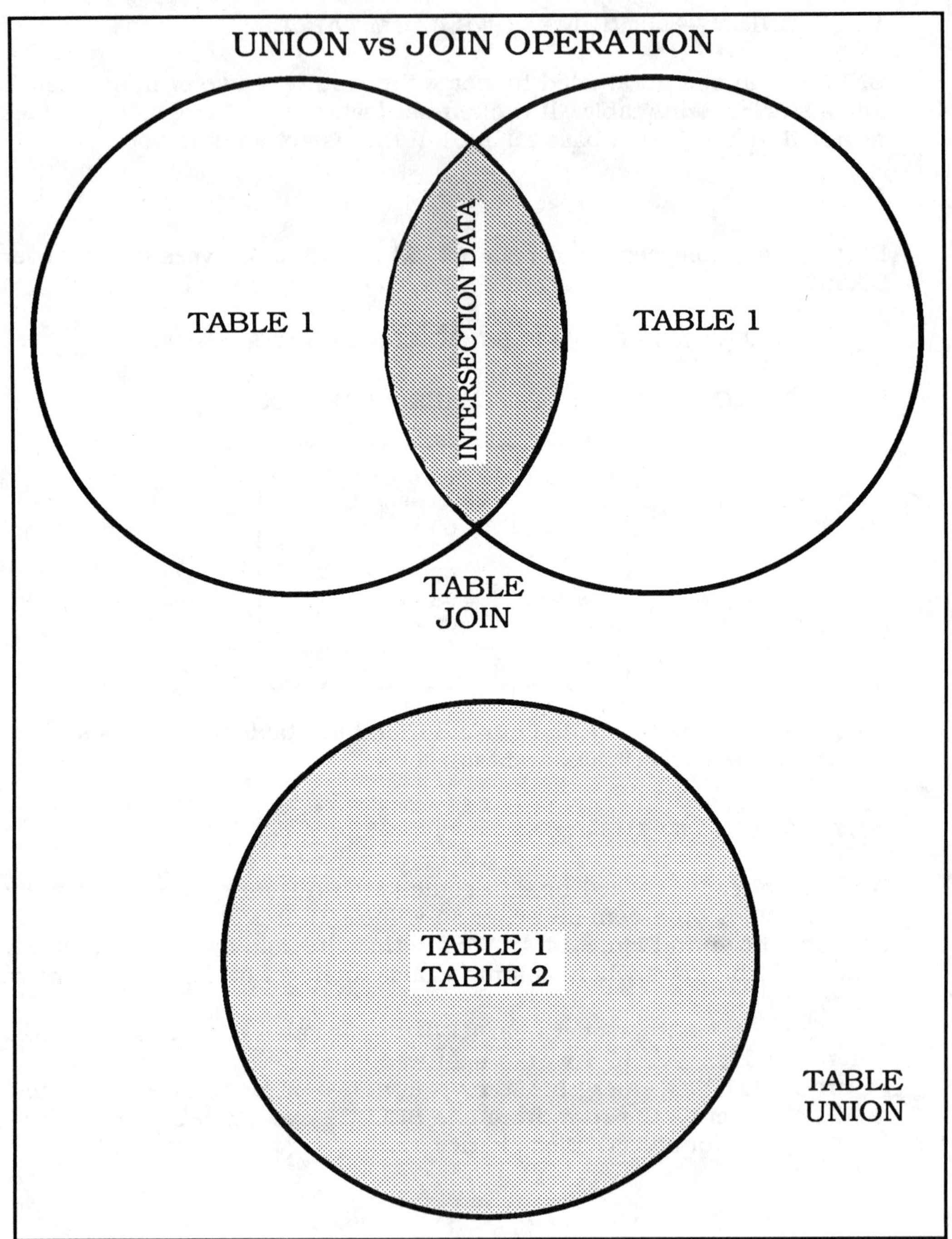

Figure 14.4. Venn diagram Union vs Join operator.

So we see one of the differences between UNION and Join:

> Join is used to combine columns from two or more base tables using common column values.
>
> UNION is used to merge results rows from two or more queries.

Let's look at some considerations for using UNION.

14.8 USING UNION

UNION is easy to use. Here are the rules:

1. The SELECT statements UNIONed together must SELECT the same number of columns.
2. The UNIONed columns SELECTed must be equivalent datatypes:

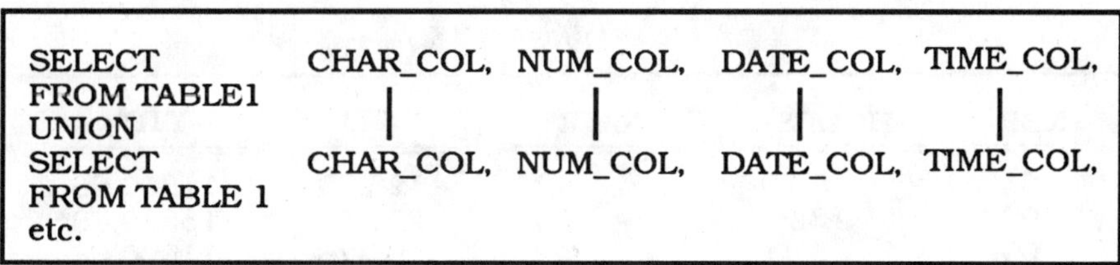

Figure 14.2. Equivalent column datatypes in UNION.

3. You may use the UNION operator between as many SELECTs as you need.
4. UNION may not appear in a SUBQUERY or VIEW definition (more on those topics in the next few chapters).
5. If you are going to use ORDER BY it must appear after the last UNIONed query, further;
6. ORDER BY must indicate the relative column position (ORDER BY 4) rather than the column name.
7. UNION will remove duplicate rows from your results table, unless you specify UNION ALL which will allow duplicates to remain in your results table.
8. In SQL/DS the SELECTed columns that you union together must match in datatype, size, specifications and null attributes.

EXERCISE 14-7

❶ TOPIC: UNION of two tables.

❷ EXERCISE 14-7. Display the NBRs for employees with a performance rating higher than 2 and for employees earning over $70,000 (YTD).

❸ INPUT TABLE:

EMPL

NBR	LNAME	FNAME	DOB	HIREDTE	PERF	JOB	DEPT	PROJ
01	LOWE	ROB	53012	85012	4	PROG	FIN	01
02	SHIELD	BROOKE	59131	87001	3	MAN	MKT	01
03	MOORE	ROGER	48111	86002	1	DIR	MKT	04
04	EASTWOOD	CLINT	41091	60120	3	PROG	FIN	03
05	MOSTEL	ZERO	21365	84211		PRES		
06	BURNS	GEORGE	11178	49001	2	SYS	FIN	01
07	O'NEIL	RYAN	42189	60121	3	DIR	ACC	05
08	MARVIN	LEE	32187	51876	2	VP	ACC	02
09	LANCASTER	BURT	41091	79092	1	MAN	R&D	02
10	BLAIR	LINDA	54013	85012	1	PROG	MKT	

PAY

NBR	HOURS	RATE	DED	YTD
01	8.89	43	128.78	11890.66
02	13.23	40	204.45	15840.78
03	6.11	49	94.76	11890.66
04	26.75	45	132.58	17605.66
05	67.82	37	394.69	79990.99
06	32.45	32	121.99	53421.23
07	26.75	49	101.56	32758.11
08	15.99	52	327.98	67870.01
09	43.59	24	0	28090.91
10	32.41	52	112.78	27000.01

FORMAT

❹
```
SELECT   (col_A FROM <table_A>
         WHERE <column> > numeric literal>
UNION
SELECT   <column_a> FROM <table_name>
         WHERE
         <column> > <numeric literal>);
```

❺ STEP 1. ______________________________________

Write your answer here.

____

STEP 2. Enter your answer into the computer.

❻ RESULTS TABLE:

```
    NBR
===========
    01
    02
    04
    05
    07
```

Hint: **This one is rather involved, but if you take it one part at a time, and follow the rules listed above, it should be no trouble.**

14.9 ANSWER KEY

There may be more than one answer to an exercise. However, the one given emphasizes the topic discussed.

EXERCISE 14.1 SELECT DEPT, MIN(PERF), MAX(PERF), AVG(PERF*1.0) FROM EMPL GROUP BY DEPT;

EXERCISE 14.2 SELECT DEPT, JOB, MIN(PERF), MAX(PERF), AVG(PERF*1.0) FROM EMPL GROUP BY DEPT, JOB;

EXERCISE 14.3 SELECT DEPT, AVG(YTD) FROM EMPL, PAY WHERE EMPL.NBR=PAY.NBR GROUP BY DEPT;

EXERCISE 14.4 SELECT DEPT, MIN(PERF), MAX(PERF), AVG(PERF * 1.0) FROM EMPL GROUP BY DEPT HAVING COUNT(*) > 1;

EXERCISE 14.5 SELECT DEPT, MIN(PERF), MAX(PERF), AVG(PERF * 1.0) FROM EMPL GROUP BY DEPT HAVING AVG(PERF * 1.0) >2;

EXERCISE 14.6 SELECT DEPT, MIN(PERF), MAX (PERF), AVG(PERF * 1.0) FROM EMPL GROUP BY DEPT HAVING AVG (PERF * 1.0) > 2 OR COUNT(*) > 1;

EXERCISE 14.7 SELECT NBR FROM EMPL WHERE PERF > 2 UNION SELECT NBR FROM PAY WHERE YTD > 70000;

Chapter 15
SUBQUERIES

Here you will learn how to code queries that contain nested SELECTs, and how to update tables using subqueries. You will also learn how to connect two or more subqueries with AND/OR, how to combine subqueries and the table joins, and how to handle subSELECTs which return multiple values. You will also learn how to write subqueries which generate a computed value for each row in the outer-level SELECT table.

Topics include:

INTRODUCTION TO SUBQUERIES

Some people find learning how and when to write subqueries a little less transparent or intuitive than writing standard SELECT statements. Because of this, I suggest that you read the material in this and the next two sections carefully, and don't hesitate to look up the answers for the first few exercises.

15.1 STANDARD QUERIES — ROW SELECTION BASED ON DEFINED SEARCH ARGUMENT VALUES

Up to now, the queries you've been using have contained WHERE clauses with explicitly defined, literal search argument values.

Example:

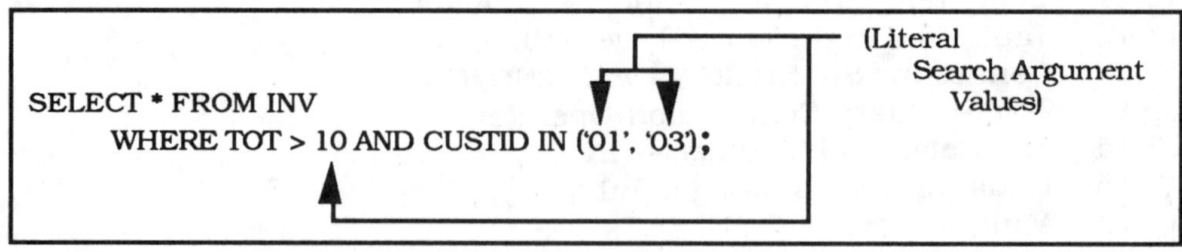

The types of business requests you've been satisfying with your SQL queries were requests where the search argument values have been clearly defined in the business processing specifications:

- List all PG DRAMA, and R COMEDY moves.
- Find all employees who were born before January 1, 1966 (66001).
- List all employees from the MKT, FIN and R&D departments who worked over 40 hours last week.

To satisfy these business specifications, you searched for rows that contained some known specific value or values. Further, you specified these values as literal search arguments in the WHERE clause. Most of your SQL database processing will involve these standard types of queries.

15.2 ROW SELECTION BASED ON UNKNOWN OR COMPUTED SEARCH ARGUMENT VALUES

Nevertheless, sooner or later you will need to find something out about the information in your tables based on search argument values you must compute — and therefore cannot specify as literals in the WHERE clause.

Example: Retrieve all employees whose performance evaluation is greater than the company average.

The only way we could satisfy the above request with our current SQL coding techniques would be to run two separate queries, the first to compute the company average:

```
SELECT AVG(PERF * 1.0) FROM EMPL;

        ==========
            2.2223
```

the second, to substitute the computed average as the search argument after the WHERE clause:

```
SELECT LNAME FROM EMPL WHERE PERF > computed_average ;
                    LNAME
                    ==========
                    LOWE
                    SHIELD
                    EASTWOOD
                    O'NEIL
```

There are several problems with this approach:

- It is time-consuming;
- It is error prone.

Let's look at a way to combine both SELECTs into one query.

15.3 SUBQUERIES — COMBINING SELECT STATEMENTS

Subqueries are a way of combining multiple SELECT statements into a single statement. What is a subquery? An SQL query containing independent SELECT statements, one SELECT nested inside the other.

```
select from where
        (select from where) ;
```

An example of a subquery used to solve the problem in 15-1 is shown below:

```
SELECT * FROM EMPL WHERE PERF >
        (SELECT AVG(PERF) FROM EMPL);
```

The way SQL would execute such query is shown in Figure 15.1.

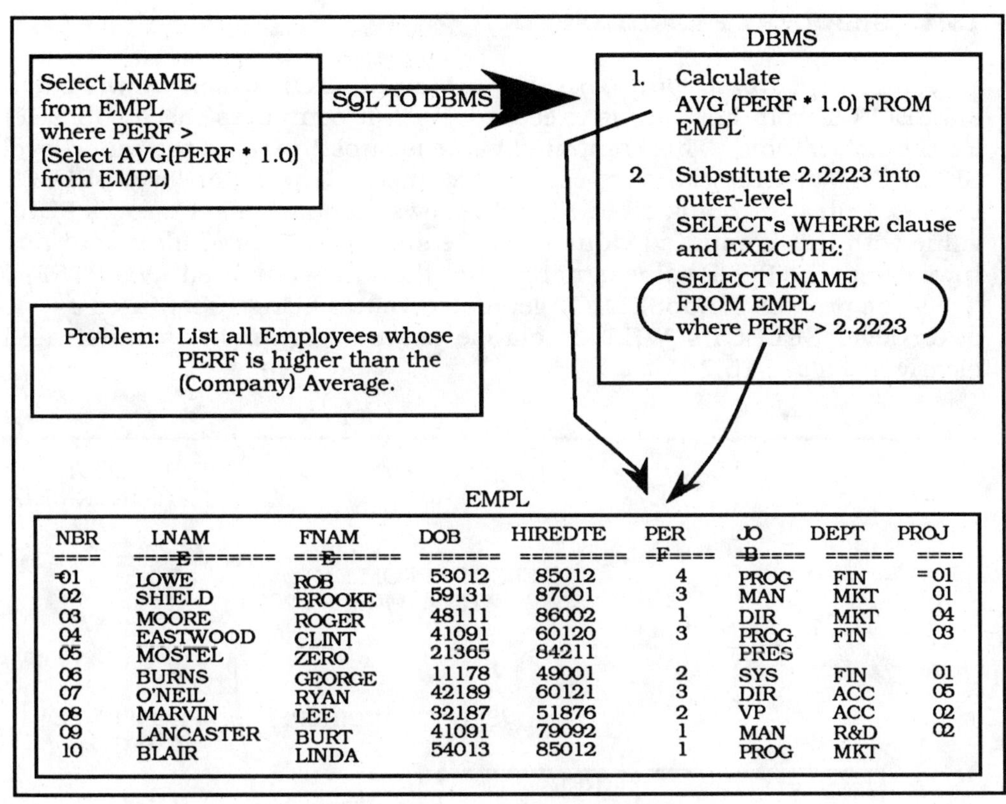

Figure 15.1. Subquery execution.

15.4 SUBQUERY EXECUTION

As you can see from Figure 15.1, the nested SELECT (which we will call a subSELECT from here on) is executed first. It computes the AVG(PERF) for the EMPL table. This computed value is substituted as the outer-level SELECT's, WHERE clause's, search argument. The outer-level SELECT then executes. It reads all EMPL table rows comparing each row's PERF value with this computed value from the subSELECT, and filters out any rows where the PERF value is not greater than the computed AVG(PERF).

The process of subSELECT generated values being substituted as the outer-level SELECT's WHERE clause search argument is illustrated clearly in Figure 15.2.

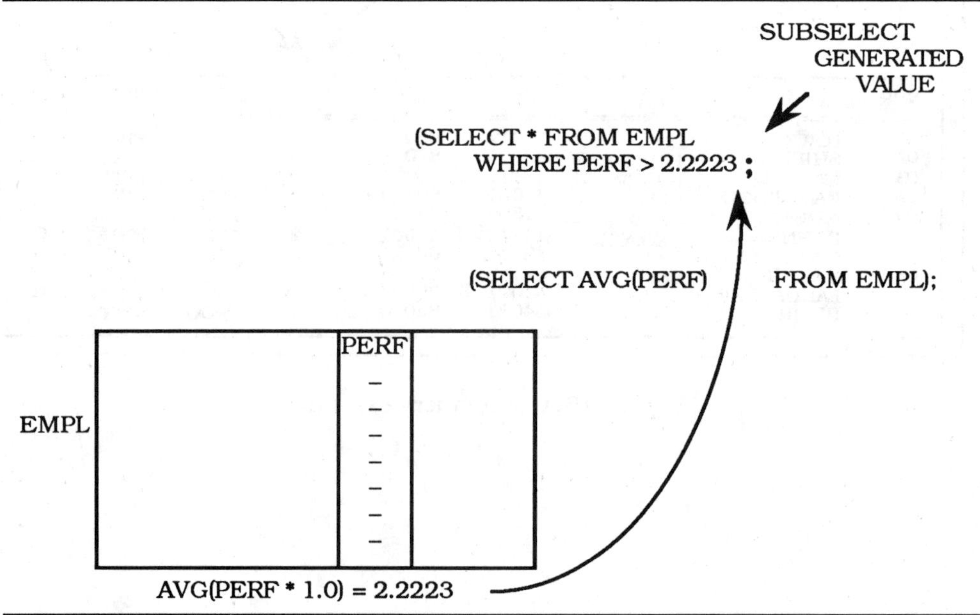

Figure 15.2. SubSELECT value substitution.

15.5 SubSELECT — NESTED SELECT USED TO GENERATE SEARCH VALUES

The sole purpose of the subSELECT is to compute or generate the search argument literals that you cannot manually specify. In the example we've been using, the literal value needed to complete our processing was average employee performance, which we obtained by SELECTing

AVG(PERF), a DB2 column function, but we may use subSELECTs to generate other types of values:

- A single column value
- A list of values from a single column
- An arithmetic expression
- A DB2 column function
- The existence of a value

The key to solving business requests using subqueries lies in understanding three things about writing subSELECTS:

- What information is missing, or needs to be computed by the subSELECT;
- How to code the proper subSELECT to generate the value or values;
- How to link the subSELECT with the outer-level SELECT query's WHERE clause.

The first two items relate directly to the specifications for the business request. Usually the subSELECT will be the first query you'd have had to execute *before* you could continue to solve the rest of the problem. The third item (How to link the subSELECT with the outer-level SELECT query's WHERE clause) will be covered shortly.

Let's back up for a minute, and reflect on what we've learned so far about subqueries.

15.6 SUBQUERY SUMMARY

A subquery allows you to SELECT rows from your table based on unknown search argument values that are generated or computed by a subSELECT instead of:

- Running a SELECT statement to generate the needed values;
- Writing down your results;
- Manually substituting your result in the WHERE clause and Running another SELECT statement to obtain your information

This allows you in a single query to ask the following types of questions about the data in our Employee database.

- What Departments contain employees who worked over the average hours last week?
- What is the name of the oldest employee in the company?
- What programmers are younger than managers?
- What projects have more than two programmers working on them?

15.7 CODING SUBQUERIES

There are just a few rules associated with coding subqueries:

1. The subSELECT must be enclosed in parenthesis.
2. You can have up to 15 levels of subSELECTs.
3. Each subSELECT must be enclosed in its own parenthesis.
4. Open and close parenthesis must be balanced.
5. The subSELECT must name a SINGLE COLUMN.
6. The type of comparison operator you use between the outer-level SELECT and the subSELECT depends on how many distinct values are returned by the subSELECT.

This last point deserves further explanation.

15.8 SUBQUERY COMPARISON OPERATORS

If your subselect will only return a single value, you can use the standard comparison operators shown below:

STANDARD COMPARISON OPERATORS

```
WHERE PERF    =    SELECT AVG(PERF)
WHERE PERF    >    SELECT AVG(PERF)
WHERE PERF    <    SELECT AVG(PERF)
WHERE PERF    >=   SELECT AVG(PERF)
WHERE PERF    <=   SELECT AVG(PERF)
WHERE PERF    ¬=   SELECT AVG(PERF)
```

But if the subselect returns more than a single distinct value as a result of executing, you must further qualify the standard comparison operators with one of the relational modifiers shown below:

QUANTIFIED RELATIONAL OPERATORS

```
WHERE EMPL.NBR <standard operator> ALL   (SELECT PAY.NBR FROM . . .
WHERE EMPL.NBR <standard operator> ANY   (SELECT PAY.NBR FROM . . .
WHERE EMPL.NBR <standard operator> IN    (SELECT PAY.NBR FROM . . .
```

Why is this?

Let's say you executed the following subquery to find employees who worked over 20 hours last week:

```
SELECT * FROM EMPL WHERE NBR = (SELECT NBR FROM PAY
                    WHERE HOURS > 20);
```

SQL executes the subSELECT and returns a list of six separate employee numbers (04, 05, 06, 07, 09, 10). This list is substituted back in the outer-level SELECT's WHERE clause as shown below:

```
SELECT * FROM EMPL WHERE NBR = '04', '05', '06', '07', '09', '10';
```

The resultant query doesn't really make any sense does it? Employee number '04' is equal to one value and not equal to the rest. Should the row with employee number '04' be SELECTed or not? SQL gets very confused in this case, and returns an error message at run-time.

We need to modify the comparison operator of subqueries where the subSELECT returns multiple values, telling SQL whether the row in the outer-level SELECT should be retrieved if the value is:

Equal/Less than/Greater than-ALL the values returned by the subSELECT;

Equal/Less than/Greater than-ANY the values returned by the subSELECT;

Equal to any one of the values in the list returned by the subSELECT.

How can you know ahead of time whether or not the subSELECT will return a single or multiple values? It depends on what you are selecting, the "scope" of the subSELECT's WHERE clause. If you SELECT a function, or the WHERE clause references a column with a unique index on it (like a primary key, for instance) and you compare with the equal sign (=), then you know you will be generating a single value. In all other subSELECTs you must use the relational modifiers shown above. These modifiers will be covered in the next two sections. In this section we will be writing subSELECTs that generate a single value. Let's try the first one together.

15.9 SUBQUERY EXAMPLE 1

REQUEST:

Retrieve the names of all the employees hired on the same date as BLAIR.

Before you start typing, let's analyze this a bit.

Question # 1 — What is the unknown value I need to generate with my subSELECT?
Answer #1 BLAIR's HIREDTE — The query to find this out is:

(SELECT HIREDTE FROM EMPL WHERE LNAME = 'BLAIR');

Question # 2 — What column will I use to link the outer-level query with my subSELECT?
Answer #2 The same column I SELECT in my subSELECT:

WHERE HIREDTE = (SELECT HIREDTE FROM EMPL WHERE LNAME = 'BLAIR');

One important point to make about the above example: If there were two or more employees named BLAIR, you would have had to use a relational modifier with the comparison operator:

```
SELECT LNAME FROM EMPL WHERE HIREDTE = ANY
    (SELECT HIREDTE FROM EMPL WHERE LNAME = 'BLAIR')
```

Question # 3 — What information do I want in my report?
Answer # 3 Last names

```
SELECT LNAME FROM EMPL WHERE HIREDTE =
        (SELECT HIREDTE FROM EMPL WHERE LNAME
        ='BLAIR');
```

LNAME
LOWE
BLAIR

Building our subqueries using the "bottom up" method described here makes it easier to visualize the process. Remember that this is also the way SQL executes subqueries. Let's do one more together.

REQUEST:

Find Rob Lowe's salary (YTD) using a subquery. Note that we will be looking at two tables for this request.

Question # 1 — What is the missing information?

Answer #1 Because I am linking the PAY and EMPL tables to satisfy this request, the missing information becomes: What is Rob Lowe's employee number. When I know that, I can use it as a search argument in the PAY table to extract YTD data.

```
(SELECT NBR FROM EMPL WHERE LNAME = 'LOWE' AND FNAME =
'ROB');
```

Question # 2 — What is the linking column used to connect the outer-level SELECT with the subSELECT?

Answer # 2 As mentioned above, employee number:

```
WHERE PAY.NBR = (SELECT NBR FROM EMPL
       WHERE LNAME = 'LOWE' AND FNAME = 'ROB');
```

Question # 3 — What information do I need in our report?

Answer # 3 The salary value (YTD):

```
SELECT YTD FROM PAY WHERE PAY.NBR = (SELECT NBR FROM EMPL
       WHERE LNAME = 'LOWE' AND FNAME = 'ROB');
```

```
       YTD
==========
  11890.66
```

If you feel ready to try some yourself, forge ahead. If not, try re-reading the first few sections again, then continue.

EXERCISE 15-1

❶ TOPIC: Subquery using a subSELECT which generates a single value.

❷ EXERCISE15-1. Find the employee LNAME and DOB of any employee who is the same age as Burt Lancaster.

❸ INPUT TABLE:

EMPL

NBR	LNAME	FNAME	DOB	HIREDTE	PERF	JOB	DEPT	PROJ
01	LOWE	ROB	53012	85012	4	PROG	FIN	01
02	SHIELD	BROOKE	59131	87001	3	MAN	MKT	01
03	MOORE	ROGER	48111	86002	1	DIR	MKT	04
04	EASTWOOD	CLINT	41091	60120	3	PROG	FIN	03
05	MOSTEL	ZERO	21365	84211		PRES		
06	BURNS	GEORGE	11178	49001	2	SYS	FIN	01
07	O'NEIL	RYAN	42189	60121	3	DIR	ACC	05
08	MARVIN	LEE	32187	51876	2	VP	ACC	02
09	LANCASTER	BURT	41091	79092	1	MAN	R&D	02
10	BLAIR	LINDA	54013	85012	1	PROG	MKT	

❹ FORMAT

```
SELECT   <column>
     FROM <table>
          WHERE <column> <operator>
      (SELECT <column> FROM <table> WHERE <condition>);
```

❺ STEP 1. ______________________________

Write your answer.

____

STEP 2. Enter your answer into the computer.

❻ RESULTS TABLE:

LNAME	DOB
LANCASTER	41091
EASTWOOD	41091

EXERCISE 15-2

❶ TOPIC: Subquery using a subSELECT which generates a single value.

❷ EXERCISE 15-2. Display the employee numbers of all employees who work on the same project as George Burns.

❸ INPUT TABLE:

EMPL

NBR	LNAME	FNAME	DOB	HIREDTE	PERF	JOB	DEPT	PROJ

❹ FORMAT

```
SELECT    <column>
     FROM <table>
          WHERE <column> <operator>
      (SELECT <column> FROM <table> WHERE <condition>);
```

❺ STEP 1. ______________________________

Write your answer.

____

STEP 2. Enter your answer into the computer.

If you are using a computer, complete the exercise. Otherwise check your answer by looking at the answer key.

❻ RESULTS TABLE:

NBR
01
02
06

EXERCISE 15-3

❶ TOPIC: Subquery using a subSELECT which generates a single value.

❷ EXERCISE 15-3. Find the names of all the programmers working on the SYSTEM R project.

❸ INPUT TABLES:

EMPL

NBR	LNAME	FNAME	DOB	HIREDTE	PERF	JOB	DEPT	PROJ

PROJ

NBR	NAME	DEPT	MAJPROJ
01	PHASERS	MKT	
05	SYSTEM X	MKT	03
03	SYSTEM R	FIN	
04	LASERS	ACC	01
02	R*	FIN	03
06	NEW PROJ	R&D	02

❹ FORMAT

```
SELECT   <column>
    FROM <table>
         WHERE <column> <operator>
       (SELECT <column> FROM <table> WHERE <condition>);
```

❺ STEP 1. ______________________________

Write your answer.

____

STEP 2. Enter your answer into the computer.

If you are using a computer, complete the exercise. Otherwise check your answer by looking at the answer key.

❻ RESULTS TABLE:

LNAME
EASTWOOD

EXERCISE 15-4

❶ TOPIC: Subquery using a subSELECT which computes a statistical function

❷ EXERCISE 15-4. Find the employee number of the employee with the highest salary (YTD).

❸ INPUT TABLE:

PAY

NBR	HOURS	RATE	DED	YTD
01	8.89	43	128.78	11890.66
02	13.23	40	204.45	15840.78
03	6.11	49	94.76	11890.66
04	26.75	45	132.58	17605.66
05	67.82	37	394.69	79990.99
06	32.45	32	121.99	53421.23
07	26.75	49	101.56	32758.11
08	15.99	52	327.98	67870.01
09	43.59	24	0	28090.91
10	32.41	52	112.78	27000.01

❹ FORMAT

```
SELECT   <column>
    FROM <table>
         WHERE  <column> <operator>
    (SELECT <func> FROM <table>);
```

❺ STEP 1. ______________________________
Write your answer.
____

STEP 2. Enter your answer into the computer.

❻ RESULTS TABLE:

NBR
05

EXERCISE 15-5

❶ TOPIC: Subquery using a subSELECT which computes a statistical function

❷ EXERCISE 15-5: Display the LNAME of any *programmer* with a higher PERF rating than *all of the directors.*

❸ INPUT TABLE:

EMPL

NBR	LNAME	FNAME	DOB	HIREDTE	PERF	JOB	DEPT	PROJ
01	LOWE	ROB	53012	85012	4	PROG	FIN	01
02	SHIELD	BROOKE	59131	87001	3	MAN	MKT	01
03	MOORE	ROGER	48111	86002	1	DIR	MKT	04
04	EASTWOOD	CLINT	41091	60120	3	PROG	FIN	03
05	MOSTEL	ZERO	21365	84211		PRES		
06	BURNS	GEORGE	11178	49001	2	SYS	FIN	01
07	O'NEIL	RYAN	42189	60121	3	DIR	ACC	05
08	MARVIN	LEE	32187	51876	2	VP	ACC	02
09	LANCASTER	BURT	41091	79092	1	MAN	R&D	02
10	BLAIR	LINDA	54013	85012	1	PROG	MKT	

❹ FORMAT

```
SELECT <column>
   FROM <table>
      WHERE <literal condition> AND <subquery condition>
   (SELECT <func> FROM <table> WHERE <condition>);
```

❺ STEP 1. ______________________________

Write your answer.

____

STEP 2. Enter your answer into the computer.

If you are using a computer, complete the exercise. Otherwise check your answer by looking at the answer key.

❻ RESULTS TABLE:

LNAME
LOWE

***Hint:* Your OUTER and inner level SELECT must contain extra conditions.**

15.10 INTERPRETING SQL

Translate the following subqueries into English:

1. SELECT LNAME, HOURS FROM EMPL, PAY WHERE EMPL.NBR = PAY.NBR AND HOURS = (SELECT MAX(HOURS) FROM PAY);
2. SELECT LNAME, HOURS FROM EMPL, PAY WHERE EMPL.NBR = PAY.NBR AND
   ((HOURS = (SELECT MIN (HOURS) FROM PAY)) OR
   (PERF = (SELECT MIN (PERF) FROM EMPL)));
3. SELECT LNAME, DOB, JOB FROM EMPL WHERE JOB = 'MAN' AND DOB > (SELECT MAX(DOB) FROM EMPL WHERE JOB = 'PROG');

INTERMEDIATE SUBQUERIES

In spite of this section's title, if you finished the exercises in the previous section successfully you shouldn't have much trouble with this chapter. The same concepts of subSELECT value generation and substitution apply. We're just going to add a bit more functionality and expand what you can do with subqueries.

15.11 COMBINING SUBQUERIES AND TABLE JOINS

If your outer-level SELECT or subSELECT needs to combine columns from more than one table in order to satisfy your reporting or search requirements, you may have to join tables. Joining tables in a subquery is no different than joining tables in a standard SQL query. You SELECT your columns, specify your tables in the FROM clause, and link the tables together in the WHERE clause.

Example:

List the employee last name and hours worked for the employee who worked the most hours:

```
SELECT   LNAME, HOURS FROM EMPL,PAY WHERE EMPL.NBR = PAY.NBR
     AND HOURS = (SELECT MAX(HOURS) FROM PAY);
```

Table joins are perfectly acceptable in either the subSELECT or outer-level SELECT statement. But the reasons for putting them in one or the other are very different.

Generally, you would join tables in the outer-level SELECT if you needed columns from more than one table or for your final report. This is illustrated in the above example.

You would join tables in the subSELECT if information from Table_A was needed in order to evaluate a search condition in Table_B — *IN ORDER TO GENERATE THE SUBSELECT'S VALUE(S).*

15.12 TABLE JOINS IN THE OUTER-LEVEL SELECT

If your report requires data in columns on more than one table, you would join tables in the outer-level SELECT statement.

Example:

Find the LNAME, YTD, and hours worked of the employee with the highest performance rating.

```
SELECT LNAME, YTD, HOURS FROM EMPL, PAY WHERE EMPL.NBR =
    PAY.NBR AND PERF = (SELECT MAX(PERF) FROM EMPL);
```

Notice that the join condition comes first in the WHERE clause. A rule of thumb is that the condition referencing the linking column between the outer-level SELECT and the subSELECT always goes last. Other conditions come before it.

EXERCISE 15-6

❶ TOPIC: Subquery with table join in outer-level SELECT, and function in the subSELECT.

❷ EXERCISE 15-6. Find the employee LNAME and salary (YTD) of the employee with the highest salary.

❸ INPUT TABLE:

EMPL

NBR	LNAME	FNAME	DOB	HIREDTE	PERF	JOB	DEPT	PROJ
01	LOWE	ROB	53012	85012	4	PROG	FIN	01
02	SHIELD	BROOKE	59131	87001	3	MAN	MKT	01
03	MOORE	ROGER	48111	86002	1	DIR	MKT	04
04	EASTWOOD	CLINT	41091	60120	3	PROG	FIN	03
05	MOSTEL	ZERO	21365	84211		PRES		
06	BURNS	GEORGE	11178	49001	2	SYS	FIN	01
07	O'NEIL	RYAN	42189	60121	3	DIR	ACC	05
08	MARVIN	LEE	32187	51876	2	VP	ACC	02
09	LANCASTER	BURT	41091	79092	1	MAN	R&D	02
10	BLAIR	LINDA	54013	85012	1	PROG	MKT	

PAY

NBR	HOURS	RATE	DED	YTD
01	8.89	43	128.78	11890.66
02	13.23	40	204.45	15840.78
03	6.11	49	94.76	11890.66
04	26.75	45	132.58	17605.66
05	67.82	37	394.69	79990.99
06	32.45	32	121.99	53421.23
07	26.75	49	101.56	32758.11
08	15.99	52	327.98	67870.01
09	43.59	24	0	28090.91
10	32.41	52	112.78	27000.01

❹ FORMAT

```
SELECT   <column, column>
    FROM <table, table>
        WHERE <join condition> and <subquery condition>
            (SELECT FUNCTION FROM <table>);
```

❺ STEP 1. ____
Write your answer.
____

STEP 2. Enter your answer into the computer.

❻ RESULTS TABLE:

LNAME	DOB
MOSTEL	79990.99

15.13 CONNECTING SUBSELECTS WITH AND/OR

Sometimes more than one subSELECT can be used to compute the necessary values for the request specifications. You can use the SQL connectors AND/OR to link up two or more subSELECTs with an outer-level SELECT.
Example:

Find employees with the lowest and highest performance evaluation in the company.

```
SELECT NBR, PERF FROM EMPL WHERE PERF =
(SELECT MAX (PERF) FROM EMPL) OR PERF =
(SELECT MIN  (PERF) FROM EMPL);
```

NBR	PERF
01	4
03	1
09	1
10	1

(Remember, you must repeat the entire OR condition between subSELECTS:

<column_name> <operator> <value>

when you use the OR connector.)

EXERCISE 15-7

❶ TOPIC: Find the LNAME, and DOB for the youngest and oldest employees in the company.

❷ EXERCISE 15-7. Find the LNAME, and DOB for the youngest and oldest employees in the company.

❸ INPUT TABLE:

EMPL

NBR	LNAME	FNAME	DOB	HIREDTE	PERF	JOB	DEPT	PROJ
01	LOWE	ROB	53012	85012	4	PROG	FIN	01
02	SHIELD	BROOKE	59131	87001	3	MAN	MKT	01
03	MOORE	ROGER	48111	86002	1	DIR	MKT	04
04	EASTWOOD	CLINT	41091	60120	3	PROG	FIN	03
05	MOSTEL	ZERO	21365	84211		PRES		
06	BURNS	GEORGE	11178	49001	2	SYS	FIN	01
07	O'NEIL	RYAN	42189	60121	3	DIR	ACC	05
08	MARVIN	LEE	32187	51876	2	VP	ACC	02
09	LANCASTER	BURT	41091	79092	1	MAN	R&D	02
10	BLAIR	LINDA	54013	85012	1	PROG	MKT	

❹ FORMAT

```
SELECT <column, column>
   FROM <table>
         WHERE <subquery condition>
   (SELECT FUNCTION FROM <table>)
         OR <subquery condition>
   (SELECT FUNCTION FROM <table>);
```

❺ STEP 1. ______________________________

Write your answer.

____

STEP 2. Enter your answer into the computer.

❻ RESULTS TABLE:

LNAME	DOB
SHIELD	59131
BURNS	11178

EXERCISE 15-8

❶ TOPIC: Connecting two subSELECTs with OR — additional search criteria

❷ EXERCISE 15-8. Find the LNAME, and DOB for the youngest and oldest managers in the company.

❸ INPUT TABLE:

EMPL

NBR	LNAME	FNAME	DOB	HIREDTE	PERF	JOB	DEPT	PROJ
01	LOWE	ROB	53012	85012	4	PROG	FIN	01
02	SHIELD	BROOKE	59131	87001	3	MAN	MKT	01
03	MOORE	ROGER	48111	86002	1	DIR	MKT	04
04	EASTWOOD	CLINT	41091	60120	3	PROG	FIN	03
05	MOSTEL	ZERO	21365	84211		PRES		
06	BURNS	GEORGE	11178	49001	2	SYS	FIN	01
07	O'NEIL	RYAN	42189	60121	3	DIR	ACC	05
08	MARVIN	LEE	32187	51876	2	VP	ACC	02
09	LANCASTER	BURT	41091	79092	1	MAN	R&D	02
10	BLAIR	LINDA	54013	85012	1	PROG	MKT	

❹ FORMAT

```
SELECT <column, column>
        FROM <table>
          WHERE <condition> AND <subquery condition>
      (SELECT FUNCTION FROM <table> WHERE <condition>)
          OR <subquery condition>
      (SELECT FUNCTION FROM <table> WHERE <condition>);
```

❺ STEP 1. ______________________________

Write your answer.

____

STEP 2. Enter your answer into the computer.

❻ RESULTS TABLE:

LNAME	DOB
SHIELD	59131
LANCASTER	41091

Hint: **You will need extra conditions in the outer-level SELECT and the subSELECT.**

15.14 THE STANDARD COMPARISON OPERATORS

As we learned previously, if your subSELECT returns multiple values, the standard comparison operators:

=

<

>

<=

>=

¬=

come up short because the value in the row we are comparing to the generated values may be equal to some and not equal to others, greater than some and less than others, etc.

We need to further qualify "how" equal to, greater than, or less than the row value must be for correct row SELECTion. There are several modifiers that can accomplish this. We'll start by learning how to use IN.

15.15 THE RELATIONAL MODIFIER — IN

IN is used to compare row values with a list of literals enclosed in parenthesis.

```
SELECT * FROM MOV WHERE RATING IN ('PG', 'R', 'PG13')
```

As translated by SQL, the above SELECT statement returns rows where the RATING value is equal to PG or R or PG13. In other words, where the row value is equal to *any one of the values contained in the list.*

When we use IN to do a subquery/subSELECT comparison, we tell SQL to SELECT the outer-level row, if the value in the comparison column is equal to any one of the values contained IN the list generated by the subSELECT.

Example:

Retrieve the LNAMEs of programmers who worked over 30 hours last week.

```
SELECT LNAME FROM EMPL WHERE JOB = 'PROG' AND NBR IN
       (SELECT NBR FROM PAY WHERE HOURS > 30);
```

LNAME
BLAIR

In executing the above query, SQL first executes the subSELECT and builds a list of employee NBRs from the PAY table, of all employees who worked over 30 hours. The list is then substituted as the outer-level SELECT's WHERE clause's search arguments, and the outer-level SELECT is executed. Rows are evaluated to see if the JOB column is equal to 'PROG' AND the NBR column is IN (equal to any of the NBRs in the generated list — (see the substitution query listed below):

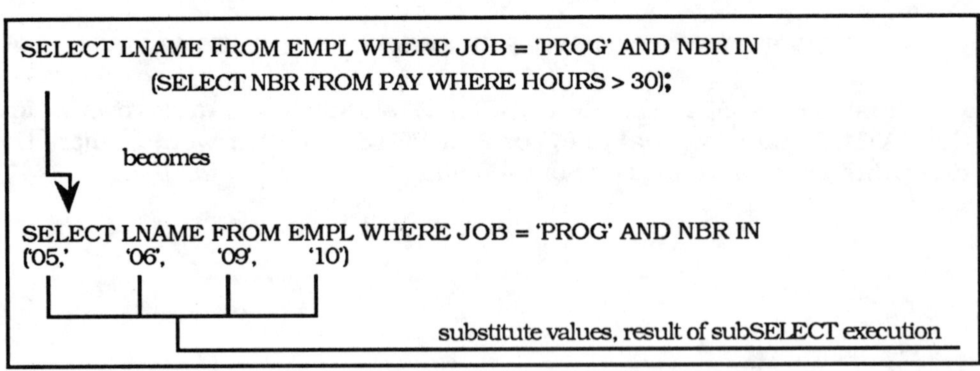

EXERCISE 15-9

❶ TOPIC: Subquery using IN.

❷ EXERCISE 15-9. Display all employee last names for employees who worked less than 30 hours last week.

❸ INPUT TABLE:

EMPL

NBR	LNAME	FNAME	DOB	HIREDTE	PERF	JOB	DEPT	PROJ

PAY

NBR	HOURS	RATE	DED	YTD

❹ FORMAT

```
SELECT   <column>
     FROM <table>
          WHERE  <subquery condition>
       (SELECT <column> FROM <table> WHERE <condition>);
```

❺ STEP 1. ______________________________
Write your answer.
____

STEP 2. Enter your answer into the computer.

❻ RESULTS TABLE:

LNAME
LOWE
SHIELD
MOORE
EASTWOOD
O'NEIL
MARVIN

Note: **How would you code this same statement as a table join.**

EXERCISE 15-10

❶ TOPIC: Subquery using NOT IN.

❷ EXERCISE 15-10. Find the LNAMES of all the employees NOT working on the SYSTEM R project.

❸ INPUT TABLE:

EMPL

NBR	LNAME	FNAME	DOB	HIREDTE	PERF	JOB	DEPT	PROJ

PROJ

NBR	NAME	DEPT	MAJPROJ

❹ FORMAT

```
SELECT   <column>
     FROM <table>
          WHERE <column> <operator>
       (SELECT <column> FROM <table> WHERE <condition>);
```

❺ STEP 1. ______________________________

Write your answer.

____

STEP 2. Enter your answer into the computer.

❻ RESULTS TABLE:

LNAME
LOWE
SHIELD
MOORE
BURNS
O'NEIL
MARVIN
LANCASTER

Note: **Blair and Mostel should be on this list. Why aren't they? Hint: Remember problems associated with searching for NULL fields? In section 15.23 we will learn how to satisfy this request (and get around these types of problems) using the EXISTS operator.**

EXERCISE 15-11

❶ TOPIC: Subquery using IN combined with table joins.

❷ EXERCISE 15-11. Display LNAMES and HOURS worked for programmers who worked as many hours as any of the directors.

❸ INPUT TABLE:

EMPL

NBR	LNAME	FNAME	DOB	HIREDTE	PERF	JOB	DEPT	PROJ

PROJ

NBR	NAME	DEPT	MAJPROJ

❹ FORMAT

```
SELECT    <column>
     FROM <table>
          WHERE <join condition> and <condition> and
               <subquery condition>
   (SELECT <column> FROM <table> WHERE < join condition>
          and <condition>);
```

❺ STEP 1. ______________________________

Write your answer.

____

STEP 2. Enter your answer into the computer.

❻ RESULTS TABLE:

LNAME	HOURS
EASTWOOD	26.75

15.16 UPDATING TABLES USING A SUBQUERY

One of the more important uses for subqueries is in update processing. Recall that you may not reference more than one table in the FROM clause of an UPDATE or DELETE statement. This restriction does not allow us to process requests with a single statement like the following:

> Delete all employees who worked less than 30 hours last week.
> (EMPL and PAY tables)
>
> Set the employee project number (PROJ) to '04' for all employees on the System R project.
> (EMPL and PROJ tables)
>
> Set the employee pay rate (RATE) up by 10% for all employees with a PERF greater than or equal to four.
> (EMPL and PAY tables)

Subqueries allow us to get around this restriction:

```
DELETE FROM PAY WHERE NBR IN
        (SELECT NBR FROM PAY WHERE HOURS < 30);
UPDATE EMPL SET PROJ = '04' WHERE PROJ IN
        (SELECT NBR FROM PROJ WHERE NAME = 'SYSTEM R');
UPDATE PAY SET RATE = RATE * 1.1 WHERE NBR IN
        (SELECT NBR FROM EMPL WHERE PERF > 4);
```

When SQL executes this statement, a list of Employee NBRs is generated and substituted back into the outer-level query as follows:

```
UPDATE PAY SET RATE = RATE * 1.1 WHERE NBR IN ('01', '06');
```

This becomes a perfectly legal update statement.

******* UPDATE EXERCISES NOTE *******

Please set your commit processing off and execute a ROLLBACK WORK statement after you've successfully finished Exercise 15.13. If you do not do this, and the row deletion becomes permanent, the rest of the results tables in this course will not match your results tables.

EXERCISE 15-12

❶ TOPIC: Deleting rows using a subquery.

❷ EXERCISE 15-12. Delete the employees from the employee table who worked under 40 hours last week.

❸ INPUT TABLE:

EMPL

NBR	LNAME	FNAME	DOB	HIREDTE	PERF	JOB	DEPT	PROJ

PAY

NBR	HOURS	RATE	DED	YTD

❹ FORMAT

```
DELETE FROM <table_name>
    WHERE <condition>
        (SELECT <col_n>
            FROM <table_name>
                WHERE <column> <comp op> value);
```

❺ STEP 1. ______________________________
Write your answer.
____

STEP 2. Enter your answer into the computer.

❻ RESULTS TABLE: Use SELECT * from EMPL to see what's left after DELETE operation.

NBR	LNAME	FNAME	DOB	HIREDTE	PERF	JOB	DEPT	PROJ
05	MOSTEL	ZERO	21365	84211		PRES		
09	LANCASTER	BURT	41091	79092	1	MAN	R&D	02

EXERCISE 15-13

❶ TOPIC: Updating rows using a subquery.

❷ EXERCISE 15-13. Update the PERF column on the employee table; set all the ratings to 5 for employees who worked over 40 hours last week.

❸ INPUT TABLES:

EMPL

NBR	LNAME	FNAME	DOB	HIREDTE	PERF	JOB	DEPT	PROJ

PAY

NBR	HOURS	RATE	DED	YTD

❹ FORMAT

```
UPDATE  <table_name>
    SET col_n = val_n  WHERE  <col_n>  IN
        (SELECT <col_n>
            FROM <table_name>
                WHERE <comp op> col_n);
```

❺ STEP 1. ____

Write your answer.

____

STEP 2. Enter your answer into the computer.

⑥ RESULTS TABLE: Use SELECT from EMPL to see what's left after DELETE operation.

NBR	LNAME	FNAME	DOB	HIREDTE	PERF	JOB	DEPT	PROJ
05	MOSTEL	ZERO	21365	84211	5	PRES		
09	LANCASTER	BURT	41091	79092	5	MAN	R&D	02

***Note:* Now issue the following statement to cancel your deletions:**

ROLLBACK;

15.17 MULTIPLE LEVEL SUBSELECTS

You may remember that up to 15 levels of subSELECTs may be coded in a single query. Given the complexity of subquery coding and testing, four levels of subSELECTs might be considered a more practical threshold for less than genius level IQs to comprehend. Let's step through a subquery with three levels of subSELECT and see how it would be executed by SQL. Figure 15.3 shows the subquery. What we are after is a list of customers who've rented a specific movie type. To do this, we must link up all the corresponding columns in the Video Store Database as shown below in Figure 15.4.

```
SELECT LNAME FROM CUST WHERE ID IN
   (SELECT CUSTID FROM INV WHERE NBR IN
         (SELECT I_NBR FROM I_M WHERE M_NBR IN
               (SELECT NBR FROM MOV WHERE RATING = 'R') ) );
```

Figure 15.3. Three level subSELECT — complete statement.

The order of the subSELECTs in a multiple level subquery is driven by the database design, table relationships and whatever literal search values are available. The outer-level SELECT will always retrieve the final report information. Our sole piece of search information is (WHERE RATING = 'R'). With this we begin selecting at the lowest level.

REQUEST: Retrieve the names of customers who rented 'R' rated movies.

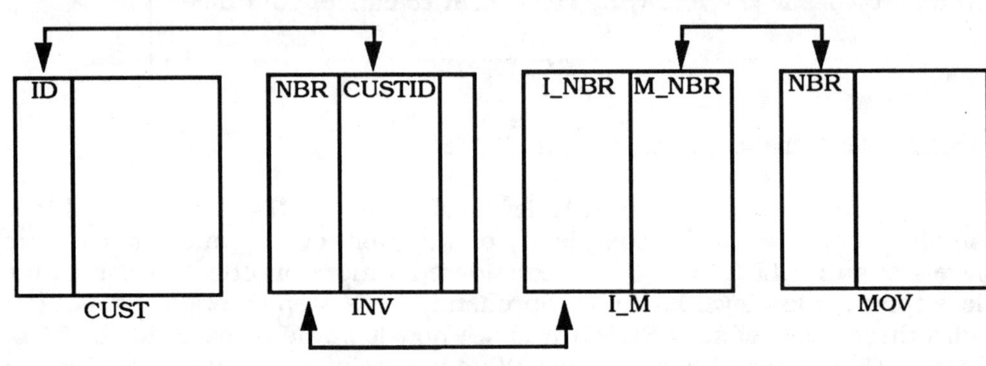

Figure 15.4. Video Store Database.

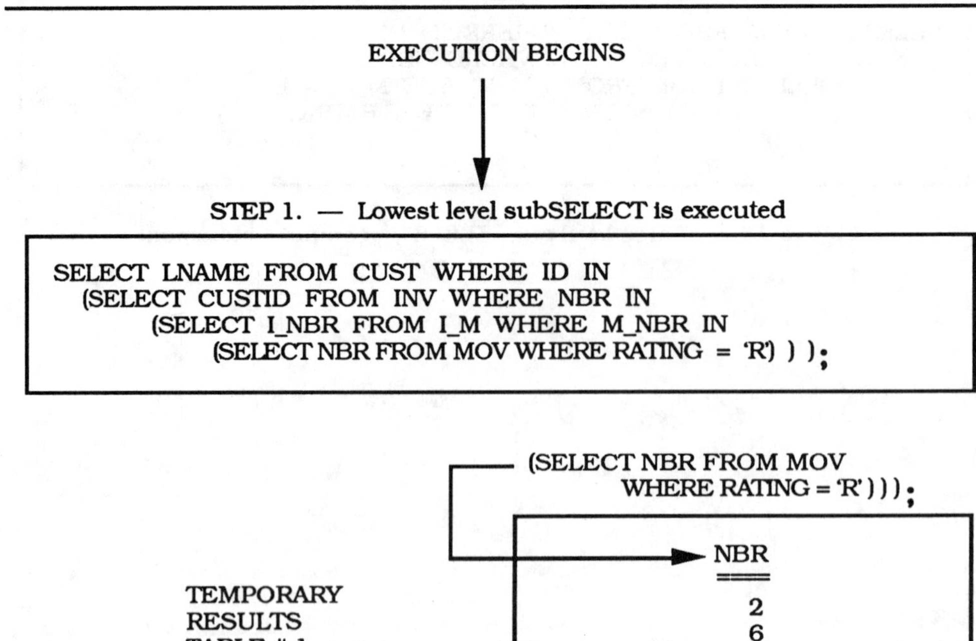

As you can see in Step 1, the lowest level subSELECT generates a list of values which is substituted in the higher level subSELECT's WHERE clause.

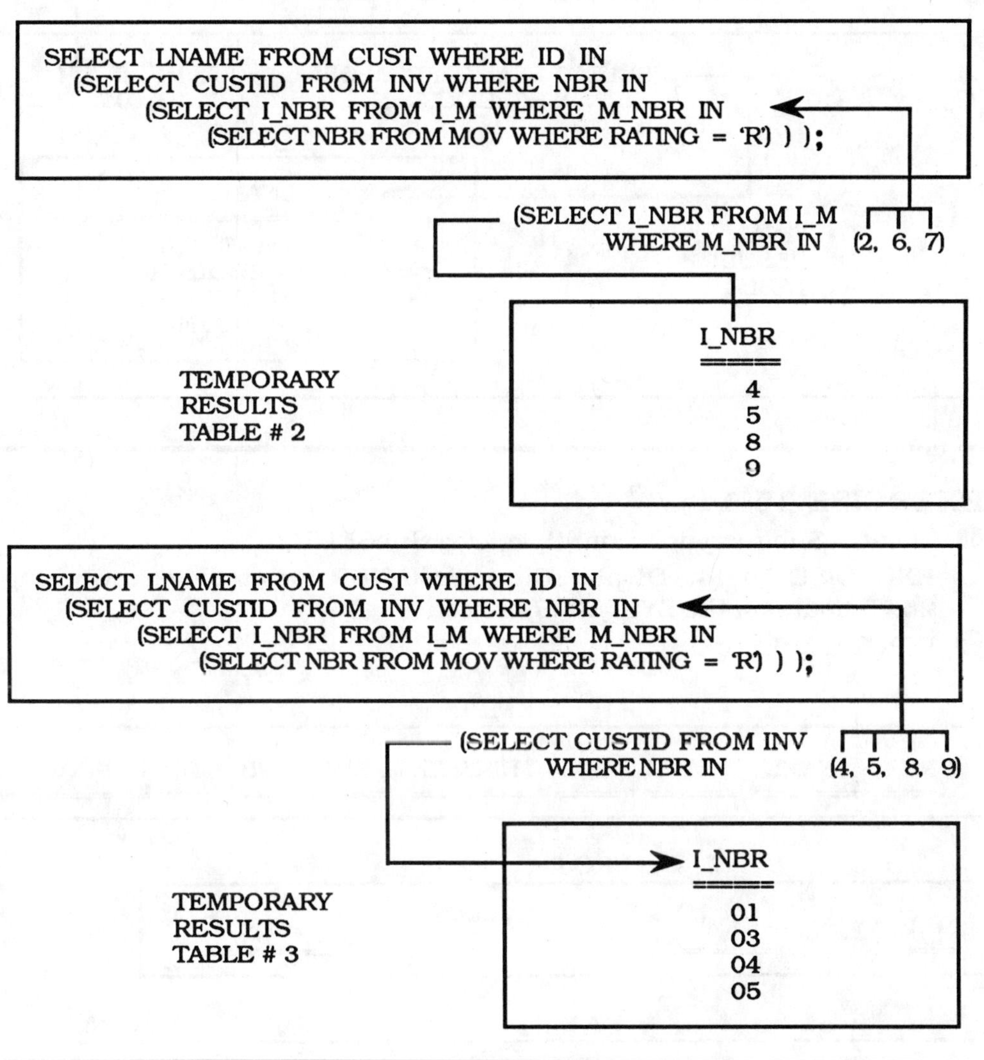

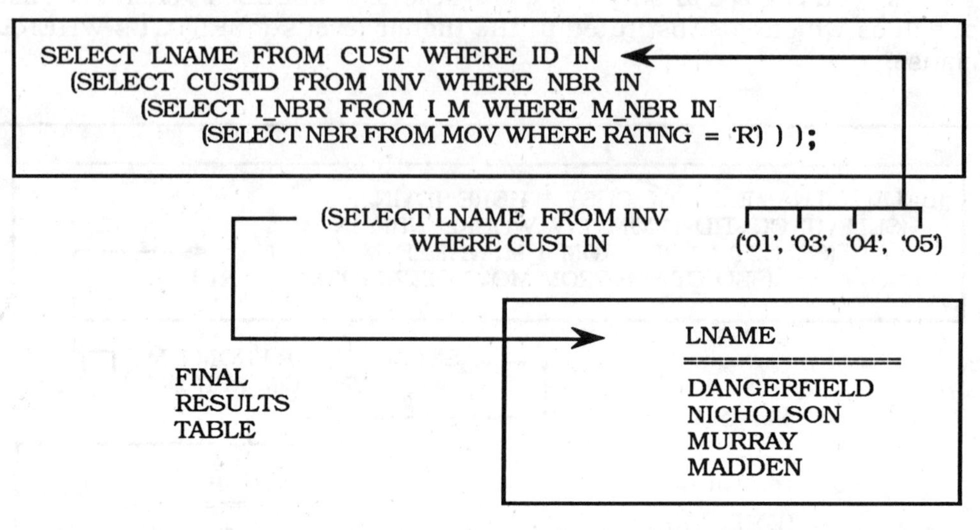

EXERCISE 15-14

❶ TOPIC: Subquery with multiple level SubSELECT.

❷ EXERCISE 15-14. Display the payroll NBR for employees working 26.75 hours on the SYSTEM R project.

❸ INPUT TABLES:

EMPL

NBR	LNAME	FNAME	DOB	HIREDTE	PERF	JOB	DEPT	PROJ

PROJ

NBR	NAME	DEPT	MAJPROJ

PAY

NBR	HOURS	RATE	DED	YTD

❹ STEP 1. ______________________________
Write your answer.
____

STEP 2. Enter your answer into the computer.

❺ RESULTS TABLE:

NBR
04

ADVANCED SUBQUERIES

As you can see from the rather exotic topic list at the beginning of this chapter, advanced subquery material is not for the faint-hearted. This is truly a technical section. It contains SQL coding techniques which, while they can be of real use in production SQL applications, are often left to the "SQL gurus" in your area. If you want to take the first step towards "guru-dom," then continue. Otherwise, skip ahead to Chapter 16.

15.18 SUBSELECTS WHICH PROCESS INDEPENDENTLY OF THE OUTER SELECT

Up to now, we've learned how to use subSELECTs to compute values which are substituted back in the outer-level SELECT. These subSELECTs can be thought of as independent SQL queries. They are stand-alone, SELECT statements, not related in any way to particular rows of the outer-level SELECT.

Example:

Find all employees whose performance evaluation is equal to the highest performance evaluation in the company.

```
SELECT * FROM EMPL WHERE PERF =
        (SELECT MAX(PERF) FROM EMPL);
```

The subSELECT in the illustration above is executed once and only once. It computes the MAX(PERF) for the entire EMPL table. Most business requests that are serviced by subqueries use standard subSELECTs.

15.19 SUBSELECTs WHICH ARE DEPENDENT ON AND RELATED TO A VALUE IN EACH ROW OF THE OUTER-LEVEL SELECT

On occasion, however, you may wish that you had a way to execute a subquery multiple times in a query to generate or compute values which depend on a field in each row of the outer-level SELECT table.

Example:

Find all employees whose performance evaluation is equal to the highest performance evaluation *in their respective departments.*

For this request, we need a WHERE condition in our subSELECT that computes a new MAX(PERF) for each DEPT value in the outer-level SELECT. In effect, the WHERE clause search argument value in the subSELECT must be variable. It must have a way to relate the DEPT value in each outer-level SELECT row with the WHERE DEPT = condition in the subSELECT used in computing MAX(PERF). In SQL we use a table ALIAS as a variable, and call it a correlation* variable. Subqueries which use correlation variables are called *correlated subqueries.*

```
SELECT * FROM EMPL CORRELATION_VAR WHERE PERF =
    (SELECT MAX(PERF) FROM EMPL
        WHERE EMPL.DEPT = CORRELATION_VAR.DEPT);
```

15.20 CORRELATED SUBQUERY EXECUTION

When SQL interprets an SQL statement like the one above and finds a correlation variable used to associate row values in the outer-level SELECT with a WHERE clause in the subSELECT, it reverses its usual execution order.

As shown in Figure 15.5, instead of executing the subSELECT first and only once per query, the subSELECT is executed one time for each row in the outer-level SELECT table. As each row becomes the current row, the two step process shown is repeated, until there are no more rows in the outer-level SELECT table.

Using correlated subqueries, the value generated by the subSELECT changes for each row and allows us to ask the types of questions shown in section 15.21.

---

* CORRELATION — a reciprocal or mutual relation — interrelated.

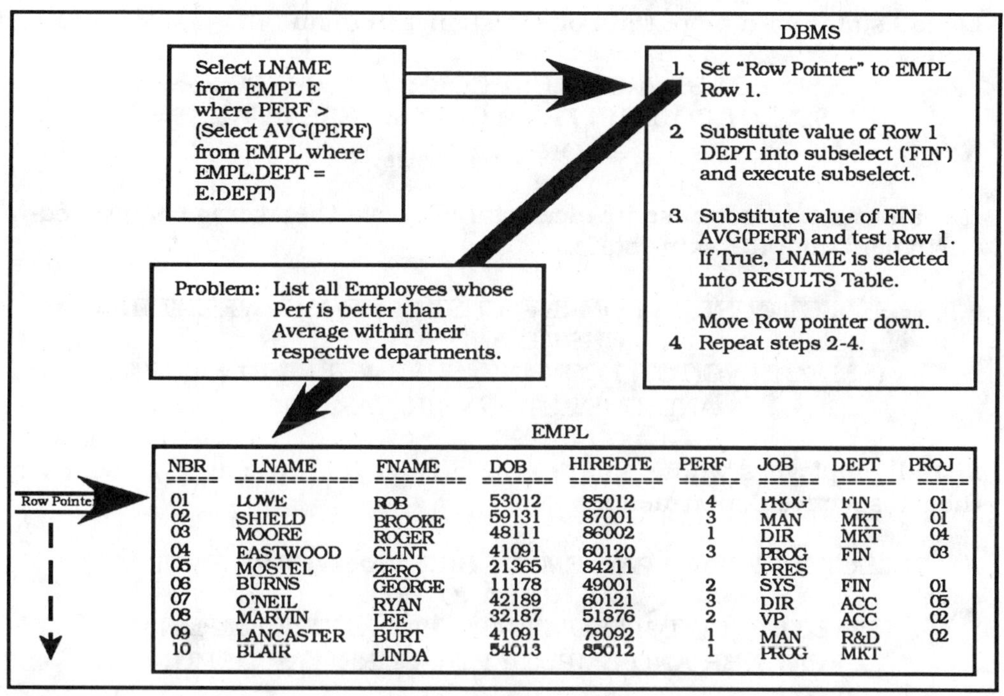

NBR	LNAME	FNAME	DOB	HIREDTE	PERF	JOB	DEPT	PROJ
01	LOWE	ROB	53012	85012	4	PROG	FIN	01
02	SHIELD	BROOKE	59131	87001	3	MAN	MKT	01
03	MOORE	ROGER	48111	86002	1	DIR	MKT	04
04	EASTWOOD	CLINT	41091	60120	3	PROG	FIN	03
05	MOSTEL	ZERO	21365	84211		PRES		
06	BURNS	GEORGE	11178	49001	2	SYS	FIN	01
07	O'NEIL	RYAN	42189	60121	3	DIR	ACC	05
08	MARVIN	LEE	32187	51876	2	VP	ACC	02
09	LANCASTER	BURT	41091	79092	1	MAN	R&D	02
10	BLAIR	LINDA	54013	85012	1	PROG	MKT	

Figure 15.5. Correlated Subquery Execution.

15.21 CORRELATED SUBQUERY EXAMPLES

Study these examples carefully. Type them in and try them on your system.

List all of the employees who have a performance evaluation higher than the average performance evaluation for their respective job positions.

```
SELECT * FROM EMPL CORR_VAR WHERE PERF >
(SELECT AVG(PERF) FROM EMPL WHERE EMPL.JOB =
CORR_VAR.JOB);
```

List all the more popular movie stars ("more popular" being defined as movie stars who starred in more than one movie).

```
SELECT STAR FROM MOV CORR_VAR WHERE 2 <=
(SELECT COUNT(*) FROM MOV WHERE MOV.STAR =
CORR_VAR.STAR);
```

List all states with more than one customer in them.

```
SELECT ST FROM CUST CORR_VAR WHERE 2 <=
(SELECT COUNT(*) FROM CUST WHERE CUST.ST =
CORR_VAR.ST);
```

List all customers whose invoice total is above the average compared to customers from the same state.

```
SELECT CUSTID FROM INV, CUST CORR_VAR WHERE ID =
CUSTID AND TOT >
(SELECT AVG(TOT) FROM INV, CUST WHERE ID = CUSTID
AND CUST.ST = CORR_VAR.ST);
```

List all employees whose salaries are greater than the average salary in their respective departments.

```
SELECT NBR FROM PAY, EMPL CORR_VAR WHERE EMPL.NBR =
PAY.NBR AND YTD>
(SELECT AVG(YTD) FROM EMPL, PAY WHERE EMPL.NBR =
PAY.NBR AND EMPL.DEPT = CORR_VAR.DEPT);
```

15.22 WRITING CORRELATED SUBQUERIES

The only difference between standard and correlated subqueries is the presence of the correlation variable. As shown below in Figure 15.5. the correlation variable is coded after a table name in the outer-level SELECT statement, and once again in the subSELECT's WHERE clause. In the WHERE clause it references the outer-level SELECT's column containing the value you wish to relate the outer-level SELECT and subSELECT together with (usually the column named in the specification requirements — "respective departments," etc.)

```
SELECT * FROM <outer-table CORR_VARIABLE>
  WHERE XCOLUMN <operator>
     (SELECT XCOLUMN FROM subselect_table WHERE
         subselect_table.YCOLUMN <operator> CORR_VARIABLE.YCOLUMN);
```

In Figure 15.5, XCOLUMN refers to the linking columns between the subSELECT and outer-level SELECT; YCOLUMN refers to the column used to relate the subSELECTs search to each row of the outer-level SELECT.

It is only necessary to qualify (subselect_table.YCOLUMN), as I have done in this example, when the table referenced in the outer-level SELECT and subSELECT are the same I show the table alias only to illustrate which column from which SELECT/TABLE is being compared.

Finally, correlation_variable can be any legal SQL TABLE name (up to 18 characters, beginning with a letter, etc.). They don't need to have the words correlation or variable anywhere in them.

Example: Retrieve the Research and Development employees whose performance evaluations are higher than average for their jobs.

```
SELECT * FROM EMPL E
  WHERE DEPT = 'R&D' AND PERF >
      (SELECT AVG(PERF) FROM EMPL WHERE
         EMPL.JOB = E.JOB) ;
```

EXERCISE 15-15

❶ TOPIC: Correlated subquery.

❷ EXERCISE 15-15. Display JOB, LNAME and DOB for employees who are older than the average employee within their JOB position.

❸ INPUT TABLES:

EMPL

NBR	LNAME	FNAME	DOB	HIREDTE	PERF	JOB	DEPT	PROJ

❹ FORMAT

```
SELECT <col_1, col_n>
  FROM <table_1, CORR VARR>
      WHERE <column> <relational operator>
         (SELECT <col_n> FROM <table_name>
           WHERE <col_n> <operator> <CORR VARR.col_n>) ;
```

❺ STEP 1. ______________________________

Write your answer.

____

STEP 2. Enter your answer into the computer.

❻ RESULTS TABLE:

JOB	LNAME	DOB
PROG	EASTWOOD	41091
DIR	O'NEIL	42189
MAN	LANCASTER	41091

EXERCISE 15-16

❶ TOPIC: Correlated subquery — With Table Join.

❷ EXERCISE 15-16. Find out which employees worked more than the average number of hours last week for their respective projects.

❸ INPUT TABLES:

EMPL

NBR	LNAME	FNAME	DOB	HIREDTE	PERF	JOB	DEPT	PROJ

PAY

NBR	HOURS	RATE	DED	YTD

❹ FORMAT

```
SELECT <col_1, col_n>
  FROM <table_1, CORR VARR>
      WHERE <column> <relational operator>
         (SELECT <col_n>  FROM  <table_name>
           WHERE <col_n> <operator> <CORR  VARR.col_n>);
```

❺ STEP 1. ______________________________
Write your answer.
____

STEP 2. Enter your answer into the computer.

❻ RESULTS TABLE:

NBR	LNAME	FNAME	DOB	HIREDTE	PERF	JOB	DEPT	PROJ
05	BURNS	GEORGE	11178	49001	2	SYS	FIN	01
09	LANCASTER	BURT	41091	79092	1	MAN	R&D	02

15.23 EXISTS — TESTING FOR THE EXISTENCE OF A VALUE

The SQL existence function (EXISTS), is used to find out whether or not an outer-level SELECT row value (the value brought down using the correlation variable) tests true in the subSELECT. In other words, where there is at least one row examined in the subSELECT that passes the conditional test established by:

```
WHERE   subselect_table.column   = COR_VAR.column
        (subselect_table.column  = correlation value)
```

If the test is true, it "proves the existence" of the value in at least one row of the subSELECT table. For some query processing specifications, existence may be all we need to know.

Example: Retrieve the names of all customers who have rented movies.

```
SELECT LNAME FROM CUST CORR_VAR WHERE EXISTS
   (SELECT * FROM INV
       WHERE INV.CUSTID = CORR_VAR.ID);
```

In this example what we're asking SQL to do is to (using CORR_VAR.ID) pass each customer's identification number — CUST table, ID column) down into the subSELECT, and try to find a match on the corresponding column from the INV table (CUSTID). The EXISTS function (along with its associated SELECT * from the subSELECT), acts like an on/off switch, returning a HIT/NO HIT condition to the outer-level SELECT. If the value contained in the correlation variable EXISTS (HIT), the row containing that value is SELECTed in the outer-level SELECT. If the value contained in the correlation variable tests false (NOT HIT), the row is not SELECTed.

15.24 NOT EXISTS — TESTING FOR THE NON-EXISTENCE OF A VALUE

You may be asking yourself, "Why did we use a correlated subquery with EXISTS to find out which customers rented movies?" After all, you can accomplish the same result with a simple table join. Fair enough. But what if you wanted to find which customers have not rented movies. Table joins come up short here. You use NOT EXISTS (non-existence tests) to accomplish these.

```
SELECT LNAME FROM CUST CORR_VAR WHERE NOT EXISTS
    (SELECT * FROM INV WHERE INV.CUSTID = CORR_VAR.ID);
```

In this example the test for NOT EXISTS checks to see if the outer-level SELECT's row values, passed into the subSELECT using CORR_VAR.ID, tested false. The outer-level row is SELECTed when this occurs. Put another way, if the value compared in the subSELECT using the correlation variable (CUST table ID value) is not found anywhere on the INV table CUSTID column, that value does not exist, and the row is SELECTed.

NOT EXISTS allows us to track down and test for NULL fields. * Recall from Exercise 15.9 that there were two rows omitted form our final result table due to the presence of NULLs. If we'd used the NOT EXISTS test instead of NOT IN, we would have retrieved those rows.

---

* Note: Because NULLS represent a logical "absence of value" (the "maybe" condition), one could argue that NULLs testing true or false for EXISTS is not a feature of IBM SQL, but a bug! This matter can only be resolved by an "SQL Standards" committee.

NOT EXISTS also allows us to do exception type reporting and what are called outer joins.

15.25 OUTER JOINS

Outer joins are queries that join two or more tables and retrieve those rows which do not co-exist between the two tables — rows where there are values in one table that do not match any value in the other table.

Example:

Find all movies that have not been rented.

```
SELECT TITLE FROM MOV CORR_VAR WHERE NOT EXISTS
  (SELECT * FROM I_M WHERE M_NBR = CORR_VAR.NBR);
```

TITLE
COCOON

Recall that the existence or presence of the movie number (M_NBR) in the I_M table means that the movie has been rented. This is a typical relational database design strategy. Instead of using artificial columns in tables that describe events such as MOVIE_HAS_BEEN_RENTED, or CUSTOMER_HAS_ORDERS, relational database designers let the data speak for itself. If a customer has rented a movie, the customer's identification number is in the CUSTID column of INV. If a particular movie has been rented, the movie number value is in the I_M table, M_NBR column.

This design strategy makes perfect sense from a business perspective. Unfortunately, it makes our processing a bit more complex, as we must learn how to use outer joins to return exception conditions that exist in our database.

15.26 RULES FOR CODING EXISTS/NOT EXISTS

The rules for using these operators are fairly easy:

- EXISTS is always the linking condition between the outer-level SELECT and the subSELECT.
- EXISTS will always be part of a correlated subquery.
- To test for non-existence, preface EXISTS with the logical NOT operator — NOT EXISTS.
- No column names are allowed in the existence tests:
  ** wrong ** WHERE <colm> NOT EXISTS . . . etc.
- Use SELECT * in the subSELECT following EXISTS and NOT EXISTS.

EXERCISE 15-17

❶ TOPIC: Correlated subquery using EXISTS.

❷ EXERCISE 15-17. Display the name of all projects that have employees working on them.

❸ INPUT TABLES:

EMPL

NBR	LNAME	FNAME	DOB	HIREDTE	PERF	JOB	DEPT	PROJ

PAY

NBR	HOURS	RATE	DED	YTD

❹ FORMAT

```
SELECT <col_1, col_n>
   FROM <table_1, CORR VARR>
      WHERE EXISTS
         (SELECT * FROM <table2>
            WHERE <col_n> <operator> <CORR VARR.col_n>);
```

❺ STEP 1. ______________________________

Write your answer.

____

STEP 2. Enter your answer into the computer.

❻ RESULTS TABLE:

NAME
PHASERS
SYSTEM X
SYSTEM R
LASERS
R*

EXERCISE 15-18

❶ TOPIC: Correlated subquery using NOT EXISTS.

❷ EXERCISE 15-18. Display the name of any project that has no employees working on it.

❸ INPUT TABLES:

EMPL

NBR	LNAME	FNAME	DOB	HIREDTE	PERF	JOB	DEPT	PROJ

PAY

NBR	HOURS	RATE	DED	YTD

❹ FORMAT

```
SELECT <col_1, col_n>
   FROM <table_1, CORR VARR>
        WHERE NOT EXISTS
            (SELECT * FROM <table2>
               WHERE <col_n> <operator> <CORR VARR.col_n>);
```

❺ STEP 1. ______________________________
Write your answer.
____

STEP 2. Enter your answer into the computer.

❻ RESULTS TABLE:

```
NAME
==========
NEW PROJ
```

15.27 USING A SUBSELECT TO FILTER OUT GROUPS OF ROWS

Previously we learned how to use the HAVING clause to filter out summary rows that had been generated using GROUP BY.

Example:

Summarize the Departmental salary information by department. Only produce a report for departments with an average salary greater than $30,000.

```
SELECT DEPT, MIN(YTD), MAX(YTD), AVG(YTD) FROM EMPL, PAY
   WHERE EMPL.NBR = PAY.NBR
         GROUP BY DEPT
         HAVING AVG(YTD) > 30000;
```

Using the HAVING clause with a literal search argument like this will probably suffice for most reports requiring GROUP BY/HAVING processing. But what if we wanted to produce the same summary shown above for only those departments whose salary average is greater than the company average as a whole? Again, we could run separate queries to accomplish this, or simply code a subSELECT after the HAVING clause to compute the company average.

```
SELECT DEPT, MIN(YTD), MAX(YTD), AVG(YTD) FROM EMPL, PAY
   WHERE EMPL.NBR = PAY.NBR
        GROUP BY DEPT
        HAVING AVG(YTD) >
           (SELECT AVG(YTD) FROM PAY);
```

The above query first computes the company AVG(YTD) from the PAY table, then uses that value in the conditional selection of the GROUPed rows.

EXERCISE 15-19

❶ TOPIC: HAVING using a subSELECT.

❷ EXERCISE 15-19. Display the DEPTs and average PERF rating for all departments having an average or higher than average rating for their employees.

❸ INPUT TABLES:

EMPL

NBR	LNAME	FNAME	DOB	HIREDTE	PERF	JOB	DEPT	PROJ

PAY

NBR	HOURS	RATE	DED	YTD

❹ FORMAT

```
SELECT <col_1, col_n>
    FROM <table_1>
        GROUP BY <col_n>
            HAVING FUNC operator
              (SELECT <FUNC(col_n)>
                  FROM <table_name>);
```

❺ STEP 1. ______________________________

Write your answer.

____

STEP 2. Enter your answer into the computer.

❻ RESULTS TABLE:

DEPT	AVG(PERF
ACC	2.5
FIN	3

15.28 USING SUBSELECTS AND HAVING TO SIMULATE A FUNCTION OF A FUNCTION

As of this writing, SQL does not allow us to nest COLUMN functions — to apply one function on top of another. This restriction means that we cannot retrieve the highest average hours MAX(AVG(HOURS)), or the lowest total pay MIN(SUM(YTD))) from our database.

If your processing calls for these types of statistical summaries, you can simulate functions of functions by combining GROUP BY/HAVING and subSELECTs.

Example:

Return the DEPT and average salary for the department with the *highest average* salary.

```
SELECT DEPT, AVG(YTD) FROM EMPL, PAY
   WHERE EMPL.NBR = PAY.NBR
         GROUP BY DEPT HAVING AVG(YTD) > = ALL
              (SELECT AVG(YTD) FROM EMPL,PAY
                   WHERE EMPL.NBR = PAY.NBR
                        GROUP BY DEPT);
```

In this query:

The subSELECT first generates a list of average YTDs, one for each DEPT in the EMPL table. (Note that this is not a correlated subquery.)

The outer-level SELECT then generates the summary rows with DEPT number and AVG(YTD) one for each DEPT in the EMPL table.

Finally, the HAVING clause in the outer-level SELECT examines each summary row produced by the outer-level SELECT to see if the AVG(YTD) in the row is greater than or equal to the highest number from the list returned by the subSELECT.

The relational modifier ALL is shown here. It basically dictates:

Don't SELECT this row unless the value being compared is greater than/less than/equal to *ALL* values in the list.

ALL specifies "SELECT this row only if the condition tested for is true for all values returned by the subSELECT." In our example we tested the AVG(YTD) (generated for each DEPT) against the list of ALL AVG(YTD)s for DEPTs. We SELECTed only the ones equal to the highest value in the list.

EXERCISE 15-20

❶ TOPIC: Using HAVING to obtain the Maximum Average.

❷ EXERCISE 15-20. Display the DEPTs and average DOB for the department having the youngest average employee age.

❸ INPUT TABLES:

EMPL

NBR	LNAME	FNAME	DOB	HIREDTE	PERF	JOB	DEPT	PROJ

❹ FORMAT

```
SELECT <col_1, col_n>
   FROM <table_1>
       GROUP BY <col_n>
           HAVING  <condition>
               (SELECT  <FUNC(col_n)>
                  FROM  <table_name>
                      GROUP BY  <col_n>);
```

❺ STEP 1. ______________________________
Write your answer.
____

STEP 2. Enter your answer into the computer.

❻ RESULTS TABLE:

DEPT	AVG(DOB)
MKT	53751

15.29 SUMMARY SUBQUERY COMPARISON

One of the less sanquine aspects of the SQL language's flexibility is its redundancy. There are usually several different ways to formulate the same request specification, each way performing better or worse than its sibling. In the chart below I have attempted to give you some "rules-of-thumb" regarding advanced SQL coding techniques. These suggested constructs show which techniques to use under which circumstances. They are "rules-of-thumb" only, and other factors such as table size and database design, index placement, etc., could override any of them on occasion.

	SPECIFICATION REQUEST	SQL CODING TECHNIQUE
1.	COLUMN STATISTICS (MIN, MAX, AVG,SUM OR COUNT) THE NUMBER OF ROWS	USE THE SQL COLUMN FUNCTIONS
2.	SUMMARIZE TABLE INFORMATION AND OBTAIN STATISTICS BROKEN DOWN BY DIFFERENT COLUMN VALUES	USE GROUP BY
3.	REMOVE ROWS FROM SUMMARY REPORTS NOT MEETING STATISTICAL CONDITIONS	USE GROUP BY/HAVING
4.	REMOVE ROWS FROM SUMMARY REPORTS NOT MEETING STATISTICAL CONDITIONS WHICH ARE COMPUTED BY A SUBSELECT	USE GROUP BY/HAVING WITH A SUBSELECT AFTER HAVING

5.	COMBINE OR MERGE COLUMNS FROM TWO OR MORE TABLES IN A SINGLE QUERY	JOIN TABLES
6.	EXECUTE A QUERY THAT DEPENDS ON A SINGLE COMPUTED VALUE — A VALUE WHICH YOU DON'T WANT TO RUN A SEPARATE QUERY FOR — TO OBTAIN	STANDARD SUBQUERY USING (>, <, =, <=, >=, =) AS COMPARISON OPERATORS — SUBSELECT SELECTS A SINGLE VALUE COLUMN FUNCTION OR A TABLE JOIN
7.	EXECUTE A QUERY THAT DEPENDS ON SEVERAL COMPUTED VALUES — VALUES WHICH YOU DON'T WANT TO RUN A SEPARATE QUERY FOR — TO OBTAIN	STANDARD SUBQUERY MODIFYING THE STANDARD COMPARISON OPERATORS WITH ALL/ANY/IN OR A TABLE JOIN
8.	EXECUTE A QUERY IN WHICH YOU RELATE A ROW VALUE IN EACH ROW OF THE OUTER-LEVEL SELECT WITH COMPARISON VALUES IN THE SUBSELECT	CORRELATED SUBQUERY
9.	EXECUTE A QUERY WHICH TESTS FOR THE EXISTENCE OR NON-EXISTENCE OF VALUES IN A TABLE	CORRELATED SUBQUERY USING EXISTS OR NOT/EXISTS

15.30 ANSWERS TO SELECTED QUESTIONS

There may be more than one answer to an exercise. However, the one given emphasizes the topic discussed.

EXERCISE 15-1 SELECT LNAME, DOB FROM EMPL WHERE DOB = (SELECT DOB FROM EMPL WHERE LNAME = 'LANCASTER' AND FNAME = 'BURT');

EXERCISE 15-2 SELECT NBR FROM EMPL WHERE PROJ = (SELECT PROJ FROM EMPL WHERE LNAME = 'BURNS' AND FNAME = 'GEORGE');

EXERCISE 15-3 SELECT LNAME FROM EMPL WHERE PROJ = (SELECT NBR FROM PROJ WHERE NAME = 'SYSTEM R');

EXERCISE 15-4 SELECT NBR FROM PAY WHERE YTD = (SELECT MAX(YTD) FROM PAY);

EXERCISE 15-5 SELECT LNAME FROM EMPL WHERE JOB = 'PROG' AND PERF > (SELECT MAX(PERF) FROM EMPL WHERE JOB = 'DIR');

EXERCISE 15-6 SELECT LNAME, YTD FROM EMPL,PAY WHERE EMPL.NBR = PAY.NBR AND YTD = (SELECT MAX(YTD) FROM PAY);

EXERCISE 15-7 SELECT LNAME, DOB FROM EMPL WHERE DOB = (SELECT MAX(DOB) FROM EMPL) OR DOB = (SELECT MIN(DOB) FROM EMPL);

EXERCISE 15-8 SELECT LNAME, DOB FROM EMPL WHERE JOB = 'MAN' AND (DOB = (SELECT MAX(DOB) FROM EMPL WHERE JOB = 'MAN') OR DOB = (SELECT MIN(DOB) FROM EMPL WHERE JOB = 'MAN'));

EXERCISE 15-9 SELECT LNAME FROM EMPL WHERE NBR IN (SELECT NBR FROM PAY WHERE HOURS < 30);

EXERCISE 15-10 SELECT LNAME FROM EMPL WHERE PROJ IN (SELECT NBR FROM PROJ WHERE NAME ! = 'SYSTEM R');
OR
SELECT LNAME FROM EMPL WHERE PROJ NOT IN (SELECT NBR FROM PROJ WHERE NAME = <> 'SYSTEM R');

EXERCISE 15-11 SELECT LNAME, HOURS FROM EMPL E, PAY P WHERE E.NBR = P.NBR AND JOB = 'PROG' AND HOURS IN (SELECT HOURS FROM EMPL E, PAY P WHERE E.NBR = P.NBR AND JOB = 'DIR');

EXERCISE 15-12 DELETE FROM EMPL WHERE NBR IN (SELECT NBR FROM PAY WHERE HOURS < 40);

EXERCISE 15-13 UPDATE EMPL SET PERF = 5 WHERE NBR IN (SELECT NBR FORM PAY WHERE HOURS > 40);

EXERCISE 15-14 SELECT NBR FROM PAY WHERE HOURS = 26.75 AND NBR IN (SELECT NBR FROM EMPL WHERE PROJ = (SELECT NBR FROM PROJ WHERE NAME = 'SYSTEM R'));

EXERCISE 15-15 SELECT X.JOB, X.LNAME, X.DOB FROM EMPL X WHERE X.DOB < (SELECT AVG(DOB) FROM EMPL WHERE JOB = X.JOB);

EXERCISE 15-16 SELECT * FROM EMPL L_V, PAY WHERE L_V.NBR = PAY.NBR AND HOURS > (SELECT AVG(HOURS) FROM PAY, EMPL WHERE EMPL.NBR = PAY.NBR AND EMPL.PROJ = L_V.PROJ);

EXERCISE 15-17 SELECT NAME FROM PROJ X WHERE EXISTS (SELECT * FROM EMPL WHERE X.NBR = PROJ);

EXERCISE 15-18 SELECT NAME FROM PROJ X WHERE NOT EXISTS (SELECT * FROM EMPL WHERE X.NBR = PROJ);

EXERCISE 15-19 SELECT DEPT, AVG(PERF * 1.0) FROM EMPL GROUP BY DEPT HAVING AVG(PERF * 1.0) >= (SELECT AVG(PERF * 1.0) FROM EMPL);

EXERCISE 15-20 SELECT DEPT, AVG(DOB) FROM EMPL GROUP BY DEPT HAVING AVG(DOB) >= ALL (SELECT AVG(DOB) FROM EMPL GROUP BY DEPT);

Chapter 16
SQL VIEWS

Here you will learn how to code data definition language statements to create and drop VIEWs. You will also learn more about what VIEWs are, why and when to use them, and how SQL processes a request against a VIEW.

The following topics are covered:

16.1 What is a VIEW?
16.2 Why use VIEWs?
16.3 Creating VIEWs
16.4 Creating DBA VIEWs
16.5 Creating VIEWs with the CHECK OPTION
16.6 Using VIEWs and the USER keyword to control data access
16.7 DROP VIEW
16.8 How SQL processes SQL requests against VIEWs
16.9 Answers to Questions

This material discusses VIEWs from an application and Database Administrator perspective. Many companies adhere to strict guidelines on the creation and use of VIEWs. You should obtain documentation on relational DBMS standards and procedures at your installation to see if:

1. VIEWs are used at your installation at all.
2. You are qualified to create or use them.

If after doing some reading in the standards manual on VIEWs you find that the material here is not relevant to you, feel free to skip ahead.

16.1 WHAT IS A VIEW?

There are three types of tables in relational databases:

- Base Tables
- Results Tables
- Views

A base table is a defined table that physically stores data

A results table is a temporary, undefined table created by the execution of a SELECT, that reflects data values stored in a base table or view — at the moment of execution

A VIEW is a stored SELECT statement that *materializes* into a results table — see Figure 16.1.

When you create a table with the CREATE TABLE statement, you are defining an SQL object that can store data. Another term for the type of table created by the CREATE TABLE statement is a BASE TABLE.

Statements to create a BASE TABLE

```
CREATE TABLE MOV (NBR SMALLINT, TITLE CHAR (30),
TYPE CHAR(6), RATING CHAR(4), STAR CHAR (10),
QTY SMALLINT, PRICE DECIMAL(5,2));
```

If I load this MOV base table up with our data and execute a SELECT * against it, the "results table" reflects the data values physically stored in the table.

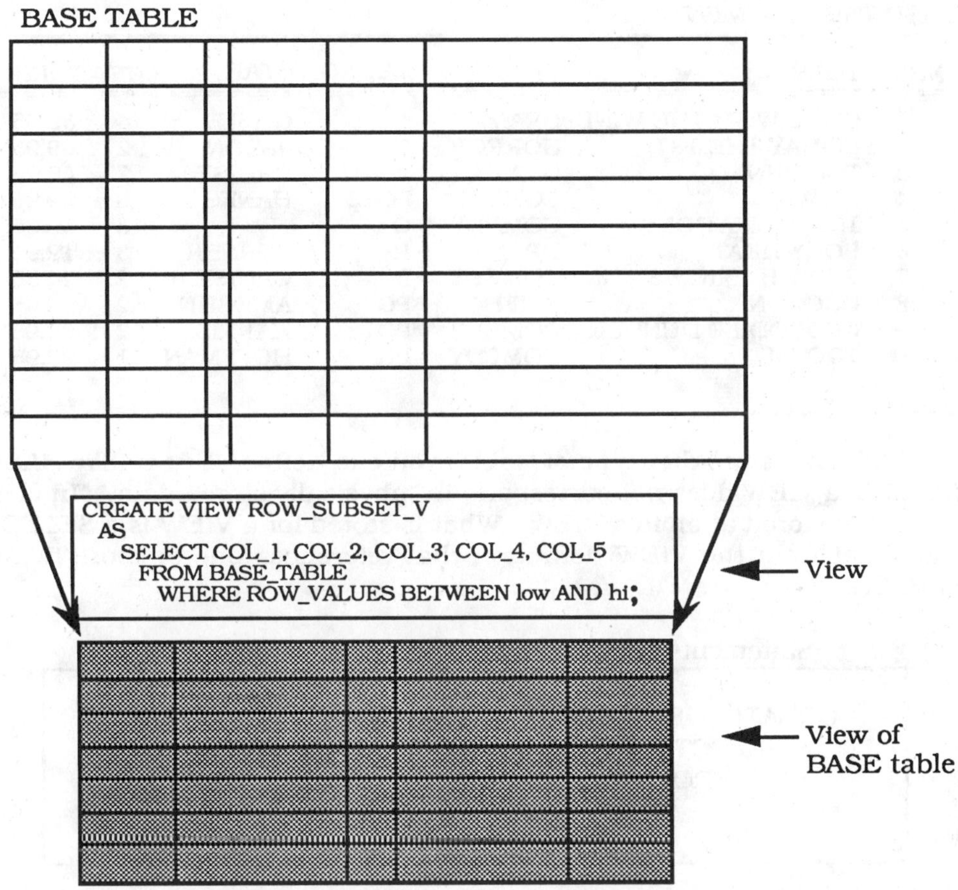

Figure 16.1. Stored SELECT against base table.

BASE TABLE — MOV:

NBR	TITLE	TYPE	RATING	STAR	QTY	PRICE
1	GONE WITH THE WIND	DRAMA		GABLE	4	39.95
2	FRIDAY THE 13TH	HORROR	R	JASON	2	69.95
3	TOP GUN	DRAMA	PG	CRUISE	7	49.95
4	SPLASH	COMEDY	PG13	HANKS	3	29.95
5	101 DALMATIONS	COMEDY	G		3	59.95
6	BODY HEAT	DRAMA	R	TURNER	3	19.95
7	RISKY BUSINESS	COMEDY	R	CRUISE	2	44.55
8	COCOON	SCIFI	PG	AMECHE	2	
9	CROCODILE DUNDEE	COMEDY	PG13	HARRIS	2	69.95
10	TOOTSIE	COMEDY	PG	HOFFMAN	1	29.95

A VIEW is another type of "table" you can define. But unlike BASE TABLES, a VIEW does not correspond to "physically stored data." In fact, no data is stored at all in a VIEW. What is stored for a VIEW is a SELECT statement from the VIEW definition that references one or more BASE TABLES.

Statements to create a VIEW on the MOV base table

```
CREATE VIEW CHEAP_MOVIE AS
    SELECT TITLE, TYPE, RATING, QTY, PRICE
        FROM MOV
            WHERE PRICE < 40.00;
```

If I process a SELECT * against the CHEAP_MOVIE VIEW, the results table reflects the values stored in the MOV table, and the results table looks like this:

VIEW CHEAP_MOVIE:

TITLE	TYPE	RATING	STAR	QTY	PRICE
GONE WITH THE WIND	DRAMA		GABLE	4	39.95
SPLASH	COMEDY	PG13	HANKS	3	29.95
BODY HEAT	DRAMA	R	TURNER	3	19.95
TOOTSIE	COMEDY	PG	HOFFMAN	1	29.95

The CHEAP_MOVIE VIEW shown above is a row/column subset of the MOV table. It looks like a base table, it works like a base table. In fact, a user can process it as though it were a base table. But it's defined to SQL as a VIEW of a base table. It is an *image* or reflection of the data in the MOV table — some of the columns and some of the rows.

What would happen to the CHEAP_MOVIE VIEW if the MOV table changed? The CHEAP_MOVIE VIEW would change with it to reflect the values in the MOV table. If the price of "Gone with the Wind" was updated to $50 in the MOV table, we would no longer see its row if we SELECTed against the CHEAP_MOVIE VIEW.

What kinds of VIEWS can you create?

Just about any legal SELECT statement can become a VIEW. You can create a VIEW that is a column-subset view:

BASE TABLE — MOV:

NBR	TITLE	TYPE	RATING	STAR	QTY	PRICE
1	GONE WITH THE WIND	DRAMA		GABLE	4	39.95
2	FRIDAY THE 13TH	HORROR	R	JASON	2	69.95
3	TOP GUN	DRAMA	PG	CRUISE	7	49.95
4	SPLASH	COMEDY	PG13	HANKS	3	29.95
5	101 DALMATIONS	COMEDY	G		3	59.95
6	BODY HEAT	DRAMA	R	TURNER	3	19.95
7	RISKY BUSINESS	COMEDY	R	CRUISE	2	44.55
8	COCOON	SCIFI	PG	AMECHE	2	
9	CROCODILE DUNDEE	COMEDY	PG13	HARRIS	2	69.95
10	TOOTSIE	COMEDY	PG	HOFFMAN	1	29.95

COLUMN SUBSET VIEW DEFINITION

```
CREATE VIEW COLUMN_SUBSET AS
    SELECT TITLE, RATING, TYPE, STAR
           FROM MOV;
```

SELECT * FROM COLUMN_SUBSET:

TITLE	RATING	TYPE	STAR
GONE WITH THE WIND		DRAMA	GABLE
FRIDAY THE 13TH	R	HORROR	JASON
TOP GUN	PG	DRAMA	CRUISE
SPLASH	PG13	COMEDY	HANKS
101 DALMATIONS	G	COMEDY	
BODY HEAT	R	DRAMA	TURNER
RISKY BUSINESS	R	COMEDY	CRUISE
COCOON	PG	SCIFI	AMECHE
CROCODILE DUNDEE	PG13	COMEDY	HARRIS
TOOTSIE	PG	COMEDY	HOFFMAN

Note, in this example, that you can rearrange the default table/column ordering with views.

ROW SUBSET VIEW:

You can create row subset views.

BASE TABLE — MOV:

NBR	TITLE	TYPE	RATING	STAR	QTY	PRICE
1	GONE WITH THE WIND	DRAMA		GABLE	4	39.95
2	FRIDAY THE 13TH	HORROR	R	JASON	2	69.95
3	TOP GUN	DRAMA	PG	CRUISE	7	49.95
4	SPLASH	COMEDY	PG13	HANKS	3	29.95
5	101 DALMATIONS	COMEDY	G		3	59.95
6	BODY HEAT	DRAMA	R	TURNER	3	19.95
7	RISKY BUSINESS	COMEDY	R	CRUISE	2	44.55
8	COCOON	SCIFI	PG	AMECHE	2	
9	CROCODILE DUNDEE	COMEDY	PG13	HARRIS	2	69.95
10	TOOTSIE	COMEDY	PG	HOFFMAN	1	29.95

ROW SUBSET VIEW

```
CREATE VIEW ROW_SUB AS
    SELECT *
        FROM MOV
            WHERE RATING IN ('R', 'PG');
```

SELECT * FROM ROW_SUB:

NBR	TITLE	TYPE	RATING	STAR	QTY	PRICE
2	FRIDAY THE 13TH	HORROR	R	JASON	2	69.95
3	TOP GUN	DRAMA	PG	CRUISE	7	49.95
6	BODY HEAT	DRAMA	R	TURNER	3	19.95
7	RISKY BUSINESS	COMEDY	R	CRUISE	2	44.55
8	COCOON	SCIFI	PG	AMECHE	2	
10	TOOTSIE	COMEDY	PG	HOFFMAN	1	29.95

ROW/COLUMN SUBSET VIEW:

You can create views that materialize into row and column subsets of a BASE TABLE.

BASE TABLE — MOV:

NBR	TITLE	TYPE	RATING	STAR	QTY	PRICE
1	GONE WITH THE WIND	DRAMA		GABLE	4	39.95
2	FRIDAY THE 13TH	HORROR	R	JASON	2	69.95
3	TOP GUN	DRAMA	PG	CRUISE	7	49.95
4	SPLASH	COMEDY	PG13	HANKS	3	29.95
5	101 DALMATIONS	COMEDY	G		3	59.95
6	BODY HEAT	DRAMA	R	TURNER	3	19.95
7	RISKY BUSINESS	COMEDY	R	CRUISE	2	44.55
8	COCOON	SCIFI	PG	AMECHE	2	
9	CROCODILE DUNDEE	COMEDY	PG13	HARRIS	2	69.95
10	TOOTSIE	COMEDY	PG	HOFFMAN	1	29.95

ROW/COLUMN SUBSET VIEW

```
CREATE VIEW ROW_COL_SUB AS
     SELECT  TITLE, RATING, TYPE, STAR
       FROM MOV
             WHERE RATING IN ('PG', 'R');
```

SELECT * FROM ROW_COLUMN_SUB:

TITLE	RATING	TYPE	STAR
FRIDAY THE 13TH	R	HORROR	JASON
TOP GUN	PG	DRAMA	CRUISE
BODY HEAT	R	DRAMA	TURNER
RISKY BUSINESS	R	COMEDY	CRUISE
COCOON	PG	SCIFI	AMECHE
TOOTSIE	PG	COMEDY	HOFFMAN

You can also create views that join tables.

BASE TABLE — MOV:

NBR	TITLE	TYPE	RATING	STAR	QTY	PRICE
1	GONE WITH THE WIND	DRAMA		GABLE	4	39.95
2	FRIDAY THE 13TH	HORROR		JASON	2	69.95
3	TOP GUN	DRAMA	R	CRUISE	7	49.95
4	SPLASH	COMEDY	PG	HANKS	3	29.95
5	101 DALMATIONS	COMEDY	PG13		3	59.95
6	BODY HEAT	DRAMA	G	TURNER	3	19.95
7	RISKY BUSINESS	COMEDY	R	CRUISE	2	44.55
8	COCOON	SCIFI	R	AMECHE	2	
9	CROCODILE DUNDEE	COMEDY	PG	HARRIS	2	69.95
10	TOOTSIE	COMEDY	PG13	HOFFMAN	1	29.95
			PG			

JOIN TABLE VIEW

```
CREATE VIEW JOIN TABLE      AS
    SELECT NBR, TITLE, I_NBR
        FROM MOV, I_M
            WHERE RATING IN ('PG', 'R')
                AND M_NBR = NBR;
```

BASE TABLE — I_M:

I_NBR	M_NBR
1	4
1	5
2	3
3	1
3	10
4	6
4	3
4	5
4	1
5	1
5	7
7	1
7	3
7	5
8	4
8	7
9	9
9	2
10	4
10	4
10	1
10	5
10	4

You can create views of tables that invoke the column statistics and GROUP BY clause.

BASE TABLE — MOV:

NBR	TITLE	TYPE	RATING	STAR	QTY	PRICE
1	GONE WITH THE WIND	DRAMA		GABLE	4	39.95
2	FRIDAY THE 13TH	HORROR	R	JASON	2	69.95
3	TOP GUN	DRAMA	PG	CRUISE	7	49.95
4	SPLASH	COMEDY	PG13	HANKS	3	29.95
5	101 DALMATIONS	COMEDY	G		3	59.95
6	BODY HEAT	DRAMA	R	TURNER	3	19.95
7	RISKY BUSINESS	COMEDY	R	CRUISE	2	44.55
8	COCOON	SCIFI	PG	AMECHE	2	
9	CROCODILE DUNDEE	COMEDY	PG13	HARRIS	2	69.95
10	TOOTSIE	COMEDY	PG	HOFFMAN	1	29.95

STATISTICAL SUMMARY VIEW*

```
CREATE VIEW STAT_SUMMARY
    (TYPE, AVG_PRICE, SUM_PRICE)
        AS SELECT TYPE, AVG(PRICE),
            SUM(PRICE) FROM MOV
              GROUP BY TYPE;
```

SELECT * FROM STAT_SUMMARY:

TYPE	AVG_PRICE	SUM_PRICE
COMEDY	46.87	234.35
DRAMA	36.6166667	109.85
HORROR	69.95	69.95
SCIFI		

Notice in our last example how I created new column names for the columns in the statistical summary VIEW.

You can rename columns in any VIEW definition if you want. You do this by naming the columns after the CREATE VIEW statement — before the AS SELECT clause. Most of the time you will be content to accept the existing column names from the SELECTed base table to be

* Note that once this view is built, you can then manipulate the data through the view in ways not possible with standard SQL. For example, I can

```
SELECT MAX(AVG_PRICE)
        FROM STAT_SUMMARY;
```

your column names. However, if you define a VIEW that SELECTs an arithmetic expression or function (any type of derived field), you will need to define names for the VIEW columns.

VIEW definitions, like all other SQL object definitions are stored in the System Catalog. Let's summarize what we've learned about VIEWs before we go on:

VIEWs:

- Are Stored SELECTs;
- Do not physically store data;
- Are processed like base tables (most of the time);
- Can *extend* the users perception of their data (particularly VIEWs
- that join tables or create summary extracts);
- Can *filter* the users perception of their data (particularly VIEWs
- that are row/column subsets of a single table).

16.2 WHY USE VIEWS?

VIEWs are used for four main purposes in relational database systems:

Security — By customizing particular users' VIEW of data to only those data elements they *need* to see, we can implement tight control on data access of our tables.

Performance: — By customizing particular users' VIEW of data to only those data elements they *need* to see, we can ensure that users (including application programs) do not access more data than is necessary for their processing purposes.

Simplicity: — By customizing particular users' VIEW of data to only those data elements they *want* to see, we can offload some of the SQL statement complexity by storing it in the VIEWs SELECT FROM WHERE definition. This has the added advantage of centralizing our application's processing specifications. The rules for an application can be stored in the System Catalog as VIEWs as opposed being to stored in separate programs as logic constructs.

Flexibility: By customizing particular users' VIEW of data to only those data elements they *need* to see, we allow for greater program/data independence. Database Administrators may define and manage data in base tables one way, and define different user VIEWs as needed for application processing another way.

16.3 CREATING VIEWS

To define views, we execute a CREATE VIEW statement. However, instead of specifying column definitions as we would if we were defining a base table, we specify a subselect

SELECT columns FROM table WHERE condition

statement that — in effect — becomes the VIEW definition.

```
CREATE VIEW view_name
     (optional column names)    AS
          SELECT columns
               FROM table(s)
                    WHERE condition;
```

A few other coding rules apply:

VIEWs use the same naming conventions as base tables.

The VIEW definition cannot contain:
- The UNION operator,
- The ORDER BY operator.

You can define VIEWs on top of existing VIEWs.

If no column names are specified, the VIEW columns will be named the same as the underlying base table's.

The VIEW columns will assume the underlying base table column's attributes.

- VIEWs may hold only 140 columns (as opposed to 300 for a base table
- Some VIEWs are read-only (non-updatable). Such VIEWs are defined with:
  - DISTINCT
  - GROUP BY/HAVING
  - BUILT IN FUNCTIONS
  - MORE THAN ONE TABLE (JOINED TABLES)
  - SELECTED ARITHMETIC EXPRESSIONS
  - SELECTED CONSTANTS
  - SUBQUERIES

To simply this list, VIEWs are updateable ONLY if the VIEW definition is a row/column subset of a single table.*

EXERCISE 16-1

❶ TOPIC: Creating a Column subset of a table through a VIEW.

❷ EXERCISE 16-1. Create a view called PROJECT_VIEW that contains the NBR, LNAME, FNAME, DEPT, JOB, PERF, PROJ columns from the EMPL table.
Rename the LNAME and FNAME columns to LAST_NAME and FIRST_NAME respectively. Then issue a select * against it to verify your work.

❸ INPUT TABLE:

EMPL

NBR	LNAME	FNAME	DOB	HIREDTE	PERF	JOB	DEPT	PROJ

As a point of interest, recall that there is no way to specifically rename table columns using SQL. It is, however, possible to rename columns through VIEW definitions.

Also, it is not possible to delete columns from an existing base table without dropping and recreating the entire table. This rather complex and arduous task may be avoided by defining VIEWs of tables (although some update restrictions may apply).**

---

* Interestingly, if you try to update a non-updateable view, DB2 disallows the oepration with a -551 Authorization Privilege Failure.

** If you need to update or insert rows through a view, you cannot leave a column out of the view definition that is defined as NOT null.

❹ FORMAT

```
CREATE VIEW <viewname>
    (<colname, colname, . . .) as
            SELECT <columns> from <tablename>
```

❺ STEP 1. ______________________________
Write your answer.
____

STEP 2. Enter your answer into the computer.

EXERCISE 16-2

❶ TOPIC: Creating a Row subset of a table through a VIEW.

❷ EXERCISE 16-2. Create a VIEW called MKTFIN that contains all columns in the EMPL table but on those from the marketing and finance departments. Issue a SELECT to verify your answer.

❸ INPUT TABLE:

EMPL

NBR	LNAME	FNAME	DOB	HIREDTE	PERF	JOB	DEPT	PROJ

As a point of interest, if you define a VIEW as SELECT * FROM a base table, then later ALTER the base table to add new columns, those new columns WILL NOT appear in the VIEW. You will have to delete and re-define the VIEW to pick up the new table definition.

❹ FORMAT

```
CREATE VIEW <viewname>
    (<colname, colname, . . .) as
            SELECT <columns> from <tablename>
```

❺ STEP 1. ______________________________
Write your answer.
____

STEP 2. Enter your answer into the computer.

EXERCISE 16-3

❶ TOPIC: Creating a a VIEW of joined tables.

❷ EXERCISE 16-3. Create a VIEW called PAYCHECK that contains the employee net pay from the PAY table (HOURS * RATE – DED), and employee LNAME and FNAME from the EMPL table. Issue a SELECT * to verify your answer.

❸ INPUT TABLES;

EMPL

NBR	LNAME	FNAME	DOB	HIREDTE	PERF	JOB	DEPT	PROJ

PAY

NBR	HOURS	RATE	DED	YTD

Hint: **Under what circumstances do you have to name your VIEW's columns?**

❹ FORMAT

```
CREATE VIEW <viewname>
    (<colname, colname, . . .) as
        SELECT <columns> from <tablename>
```

❺ STEP 1. ______________________________
Write your answer.
____

STEP 2. Enter your answer into the computer.

EXERCISE 16-4

❶ TOPIC: Creating a Statistical Summary VIEW.

❷ EXERCISE 16-4. Create a VIEW called DEPTSTATS containing the AVG(PERF), MAX(PERF) and MIN(PERF) from the EMPL table, broken down by JOB within DEPT. Check your answer with a SELECT *.

❸ INPUT TABLE:

EMPL

NBR	LNAME	FNAME	DOB	HIREDTE	PERF	JOB	DEPT	PROJ

❹ FORMAT

```
CREATE VIEW <viewname>
    (<colname, colname, . . .) as
            SELECT <columns> from <tablename>
               GROUP BY <condition>
```

❺ STEP 1. ______________________________
Write your answer.
____

STEP 2. Enter your answer into the computer.

16.4 CREATING DBA VIEWS

For reasons of performance or administration, DBAs may need flexibility in managing the physical data in relational tables. Without affecting existing application programs or queries, DBAs might want to:

- Split a table horizontally and separate the rows onto separate devices;
- Split a table vertically and separate the columns onto separate devices;
- Rename columns;
- Delete columns.

This type of program/data independence is only possible currently through the use of VIEWs. And not just any VIEWs, but VIEWs of VIEWs. These are sometimes called DBA VIEWs. Let's see how DBA VIEWs can improve program/data independence.

Example:

You have a VIEW that is a join of two tables:

```
CREATE VIEW TAB_JOIN AS
        SELECT * FROM TABLE1, TABLE2
                WHERE CONDITION;
```

This creates the following User VIEW of the data:

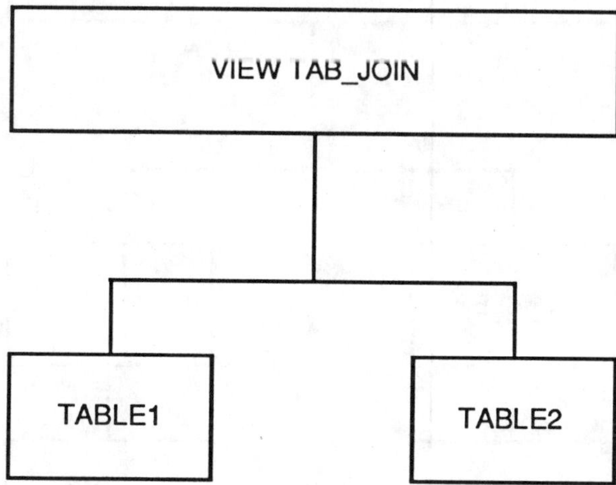

What would happen if TABLE2 grew very large, and it was determined that it should be split vertically into two new tables and joined when necessary?

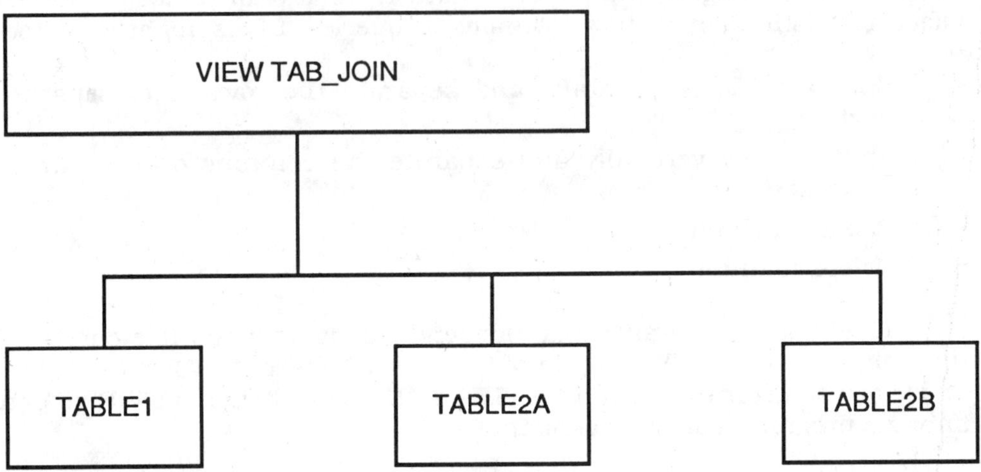

This is fine for the base tables involved, but now all queries, QMF PROCS, application programs and VIEWs of the old TABLE2 must be modified to reflect the change.

A way to avoid this task would be to define VIEWs on all the base tables involved.

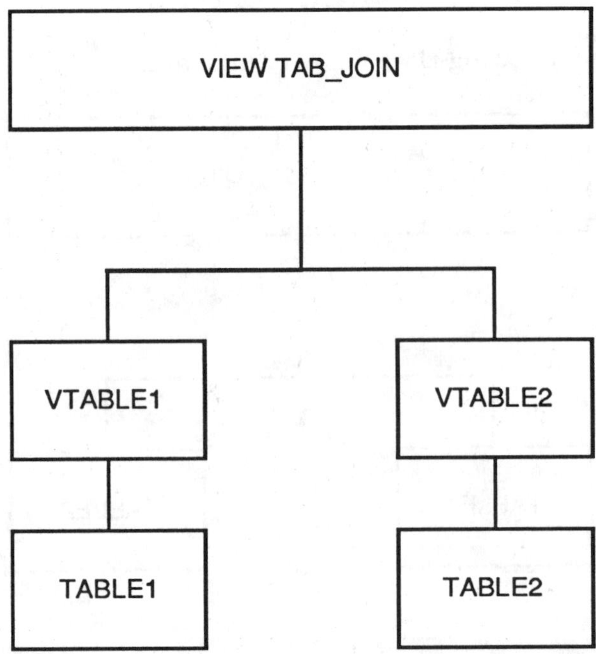

Now if TABLE2 is modified in any way, we need only change the VIEW definition for VTABLE2. We do not have to change the User VIEWs on the table!

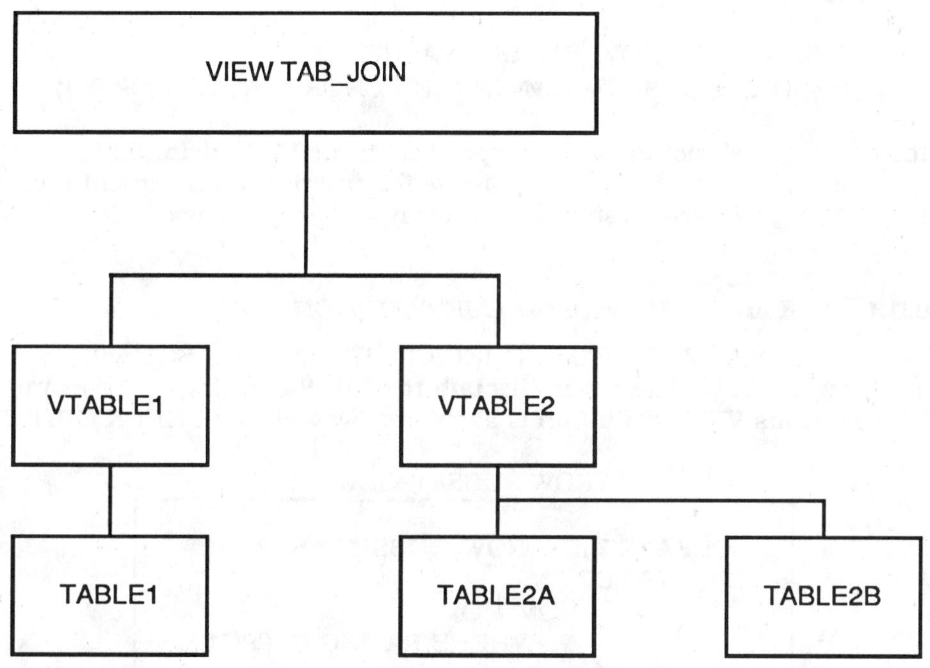

16.5 CREATING VIEWS WITH THE CHECK OPTION

The VIEW definitions we have looked at and written so far all suffer from one rather dramatic shortcoming. Users could defeat the purpose of the VIEW on row update processing.

Example:

Here's an example of a VIEW definition from earlier in this chapter:

ROW SUBSET VIEW

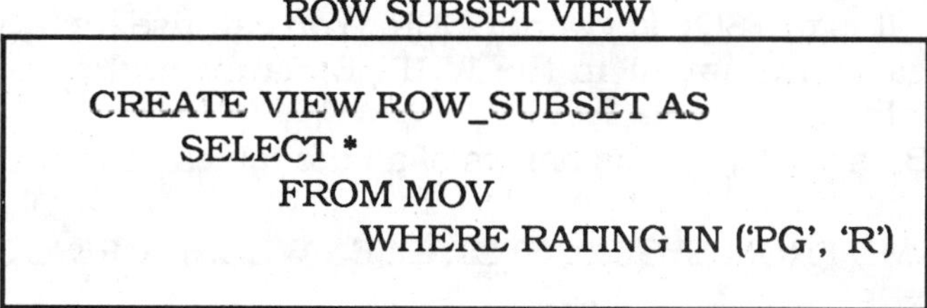

```
CREATE VIEW ROW_SUBSET AS
    SELECT *
        FROM MOV
            WHERE RATING IN ('PG', 'R')
```

Since this VIEW is a row/column subset of a single table we are allowed to update the MOV table through it. As a matter of fact, we could update through this VIEW as defined here with rows that defeat the VIEW definition. We could:

```
INSERT INTO ROW_SUBSET VALUES
      (12, 'BAD SITUATION INDEED', 'HORROR', 'X', 1,99.99);
```

with a rating of X, not PG or R as specified in the VIEW definition.

This is a particularly unattractive feature of the implementation of VIEWs in SQL based systems. As always, there is a way around this problem.

16.5.1 Defining VIEWs with the CHECK OPTION

If you define your VIEW with a CHECK OPTION clause, SQL will enforce data integrity in update mode through the VIEW definition. An example of the previous VIEW definition is shown below with the CHECK OPTION.

ROW SUBSET VIEW

```
CREATE VIEW ROW_SUBSET AS
      SELECT *
            FROM MOV
                  WHERE RATING IN ('PG', 'R')
WITH CHECK OPTION;
```

The CHECK OPTION will enforce all VIEW definition rules in update processing. It probably should be considered a corporate standard for critical production tables.

16.6 USING VIEWS AND THE USER KEYWORD TO CONTROL DATA ACCESS

There is a system reserved word that equates to each individuals User-ID or Logon ID. It is the SQL keyword USER. You can use the USER keyword both as a search argument in the WHERE clause and as SELECTed or INSERTed data.

In a DB2 system, get the names of all the tables you have created:

```
SELECT NAME FROM SYSIBM.SYSTABLES WHERE CREATOR = USER;
```

Notice that because USER is an SQL reserved word, we do not enclose it with single quotes the way we would normally enclose a character search argument.

You can implement the USER keyword effectively in security and data access mechanisms. For User-sensitive data, create a column in each table to hold the User-ID of the person allowed to access the row. Create a VIEW of the table like the following:

```
CREATE VIEW view_name AS SELECT only_designated_columns
   FROM table_name WHERE User-ID_column_name = USER;
```

This will restrict data access of the table to only those rows that the users inserted.

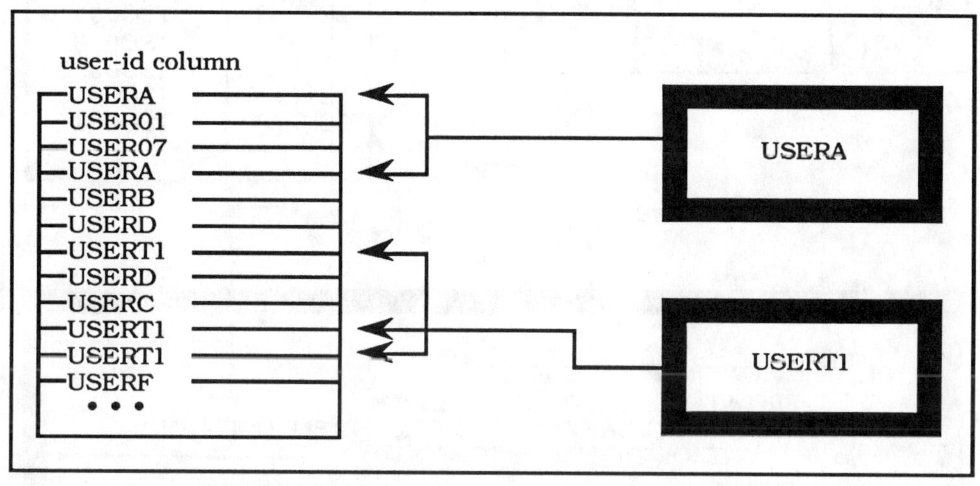

Figure 16.3 VIEWS used to restrict access.

You can also create a more flexible scheme by:

- Creating codes that designate groups of users;
- Populating the base table rows with the appropriate codes;
- Create a cross reference table made up of two column, code and User-ID;
- Populating this table with rows assigning each User-ID one or more codes as necessary;
- Creating a VIEW of the table as follows;

```
CREATE VIEW view_name AS SELECT stuff_you_want FROM TABLE A,
                    CONTROL_TABLE
      WHERE TABLEA.CODE = CONTROL_TABLE.CODE AND
            CONTROL_TABLE.User_ID = USER;
```

This scheme is illustrated below:

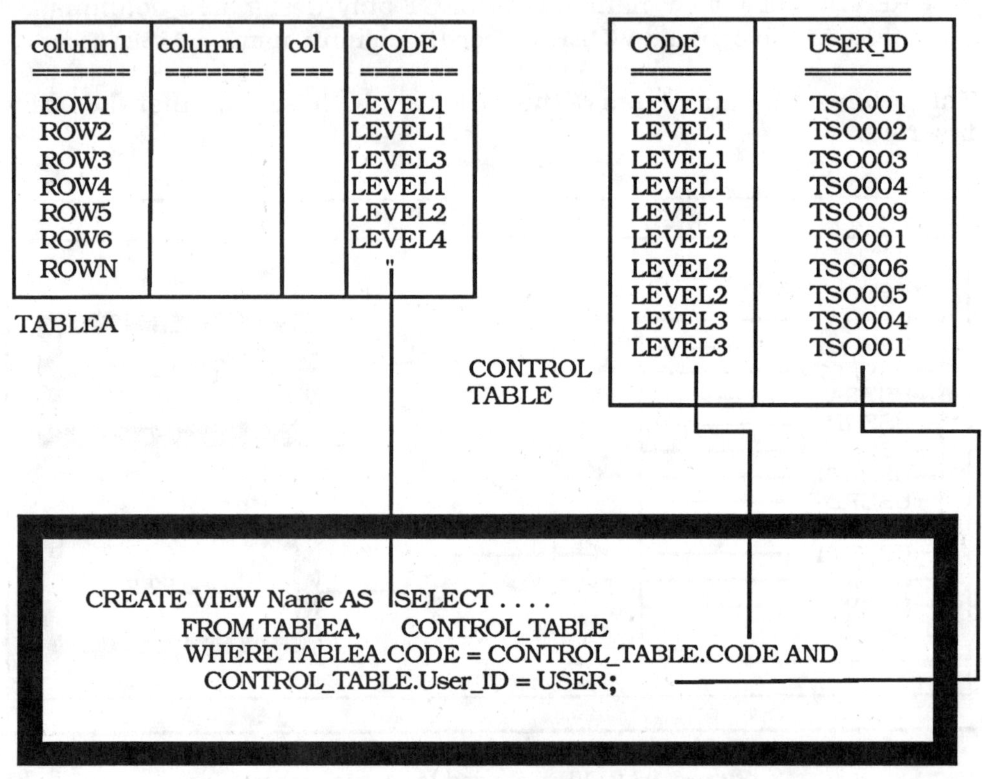

Figure 16.4. Using a VIEW to enforce Row-level security.

16.7 DROP VIEW

To remove a User VIEW on a base table simply execute the statement:

```
DROP VIEW <view_name>;
```

Example:

Drop the ROW_SUBSET VIEW defined earlier in the chapter:

```
DROP VIEW ROW_SUBSET;
```

After successful execution of this statement, the VIEW and all its entries in the System Catalog are deleted. This includes all permissions to use the VIEW (next chapter) and all VIEWs that are dependent on the VIEW.

You can only DROP VIEWs that you have created unless you have a very high level of system authorization.*

16.8 HOW SQL PROCESSES SQL REQUESTS AGAINST VIEWS

A final look at VIEWs will demonstrate the way that SQL executes SELECTs against VIEWs.

Given the following VIEW of our EMPL table:

ROW/COLUMN SUBSET VIEW

```
CREATE VIEW ROW_COLUMN_SUB AS
    SELECT TITLE, RATING, TYPE, STAR
        FROM MOV
            WHERE RATING IN ('PG', 'R')
```

And given the following SQL SELECT against that VIEW:

```
SELECT TITLE, RATING, TYPE
    FROM ROW_COLUMN_SUB
        WHERE TYPE = 'DRAMA'
ORDER BY TITLE;
```

here's what happens when you execute the SELECT against the ROW_COLUMN_SUB VIEW.

SQL builds a composite SELECT statement from both the VIEW definition (as obtained from the system catalog), and the "run-time" SELECT statement from the terminal. This is shown in Figure 16.5.

---

* In DB2, you must have SYSADM authority to drop views.

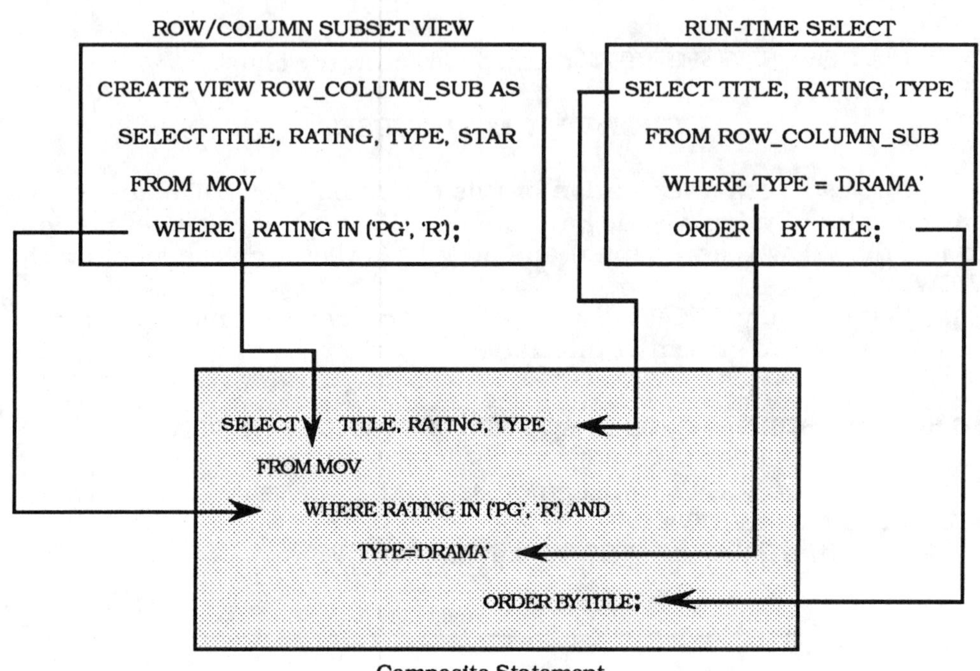

Figure 16.5. Composite Executable Statement (text merge operation).

Notice that the:

"Run-time" SELECT overrode the VIEW's column definition;

VIEW Definition's RATING specifications were merged in with the "run-time" SELECT's TYPE = 'DRAMA' requirement;

ORDER BY was obtained from the run-time SELECT.

***Note:* Now cancel all your work by entering:**

```
ROLLBACK WORK;
```

16.9 ANSWERS TO QUESTIONS

EXERCISE 16-1 CREATE VIEW PROJECT_VIEW (NBR, LAST_NAME, FIRST_NAME, DEPT, JOB, PERF, PROJ) AS SELECT NBR, LNAME, FNAME DEPT, JOB, PERF, PROJ, FROM EMPL;

EXERCISE 16-2 CREATE VIEW MKTFIN AS SELECT*FROM EMPL WHERE DEPT IN ('MKT', 'FIN');

EXERCISE 16-3 CREATE VIEW PAYCHECK (LNAME, FNAME, HOURS, RATE, NET_PAY) AS SELECT LNAME, FNAME, HOURS, RATE, (HOURS*RATE) - DED FROM EMPL, PAY WHERE EMPL.NBR=PAY.NBR;

EXERCISE 16-4 CREATE VIEW DEPTSTATS (DEPT, JOB, AVG_PERF, MAX_PERF, MIN_PERF) AS SELECT DEPT, JOB, AVG(PERF), MAX(PERF), MIN(PERF) FROM EMPL GROUP BY DEPT, JOB;

Chapter 17
DATA CONTROL STATEMENTS

Here you will learn how to code data control language statements to define and modify a security and authorization subsystem to control access to your data.

Topics include:

This material discusses Data Control Language (DCL) statements and security from an application and Database Administrator perspective. Most companies adhere to strict policy guidelines on the creation and administration of security and data access control. The student is encouraged to read the appropriate standards and procedures manuals and internal documentation on the subject, either before or after completing the exercises.

INTRODUCTION

Decisions on how to implement security are not usually black and white choices. There are many options:

1. You can use VIEWs as part of a Security mechanism.
2. You can define the Security mechanism with Data Control Statements
3. You can mix the above options

DB2 and SQL/DS have various access controls, levels and functions that can be defined within its internal security system using the Data Control Language statements we will be learning. This internal security system uses works in cooperation with your mainframe system sign-on and security package* to permit or restrict what you can do or not do in the DB2 and SQL/DS world. As a student, you may not be able to execute all the statements shown. This doesn't mean that you can't try them out though.

If after executing a statement, you receive an error message to the effect that you do not have the necessary authorization to do so, this may be all right. You probably entered the statement in a syntactically correct format because DB2 and SQL/DS do syntax checking before they do authorization checking. Consider this successful execution and continue on to the next exercise.

Note that if you are taking this course on a PC using SQLBase you should have no trouble with the types of internal DB2 restrictions discussed above. You will have to be concerned about them, however, when you start working on DB2 or SQL/DS at a mainframe terminal.

You will be given an opportunity to define a security mechanism for the EMPL database.

17.1 INTERNAL VS. EXTERNAL SECURITY IN DB2 AND SQL/DS

Access to DB2 and SQL/DS is usually controlled externally by the use of the operating system security package you have installed in your installation — or on using the standard CP password protection or unsecure. These packages allow you to sign-on to the mainframe only if you type in the correct password associated with your User-ID. SQL/DS depends on these external security systems to identify you, and to verify that "you are who you say you are." In addition, in SQL/DS, your DBA may have to grant your ID "connect" authorization.** This allows you to access ISQL, DBSU or other SQL/DS interfaces such as QMF or REXX.***

Finally, when you execute an SQL statement to access SQL/DS data, DB2 obtains your User-ID (called an Authorization ID in the IBM manuals) from the above mentioned external security system and searches the appropriate System Catalog tables to see if you have permission to execute the SQL statement. This is illustrated in Figure 17.1.

---

* Like RACF (Resource Access Control Facility), Top Secret, or ACF2.

** If the "ALL USERS" connect from the SQL/DS install is revoked.

*** Some SQL/DS shops also revoke the default grant to public on ISQL, DBSU and QMF. In this case you must be granted explicit run privileges on these modules.

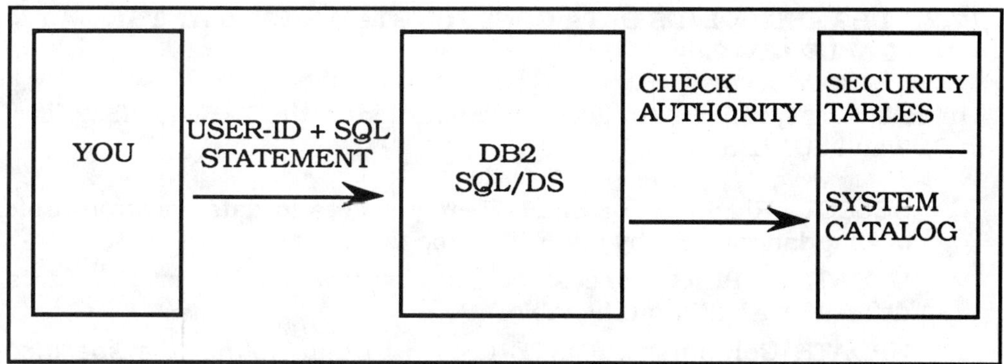

Figure 17.1. DB2 and SQL/DS Use System Catalog Security tables to check access authority.

The authorizations contained in the System Catalog security tables come in two varieties:

Specific Table Level Privileges — An authorization to perform a specific type of activity (like SELECT, INSERT, UPDATE, etc.) against a specific TABLE or VIEW.

Generic Privilege "Bundles" — A group of authorizations at the system-wide level. These privilege bundles allow users to perform, administer and manage activities on all objects contained in SQL/DS.

The system-wide level of authorizations are beyond the scope of this course. Refer to the *IBM System Planning and Administration for VM/SP — SH09-8018* for information on these topics.

The same basic sign-on and system access controls are set up in DB2. Access to DB2I/SPUFI and QMF are granted to either the public or individual system users.

We continue now with a look at the table level privileges.

17.2 DB2 AND SQL/DS DATA CONTROL STATEMENTS AT THE TABLE LEVEL

The following types of data access are granted at the TABLE/VIEW level in DB2 and SQL/DS:

SELECT: Read-only or SELECT-only access to data within a table or that data defined by a VIEW on the table.

UPDATE: UPDATE access to data within a table or that data defined by a VIEW on the table.

UPDATE Columns: UPDATE access to data *within the specified columns* of a Table or that data defined by a VEW on the table.

DELETE: DELETE access allows users to remove existing rows from a table or from a table through a VIEW defined on the table.

INSERT: INSERT access allows users to INSERT new rows into a TABLE or into a table through a VIEW defined on the table.

ALTER: ALTER access allows users to ALTER the table by adding additional columns.

ALL: ALL allows users to use the table with all of the above access levels.

These access levels are given to either factors:

- Individual authorization IDs (User-IDs)
- PUBLIC — All User-IDs able to accessDB2 data

with the GRANT statement, and taken away with the REVOKE statement.

17.3 WRITING DATA CONTROL STATEMENTS — GRANT

GRANT follows the format illustrated below:

```
GRANT <access_level(s)>
    ON <object_type>  <object_name>
        TO  <User-id_list/PUBLIC>
            WITH GRANT OPTION;
```

Example 1:

Grant SELECT, UPDATE and DELETE authorizations on the EMPL table to User-IDs TS0001, TS0009 and TS013.

```
GRANT SELECT, UPDATE, DELETE
    ON TABLE EMPL
        TO TS0001, TS0009, TS0013;
```

Example 2:

Grant all authorizations on the ROW_SUBSET VIEW to all DB2 users.*

```
GRANT ALL
    ON TABLE ROW_SUBSET
        TO PUBLIC;
```

As you can see, there is nothing particularly difficult about writing Data Control Statements. The work is all in the careful planning and definition of a flexible, manageable and administrable system-wide security and access mechanism. Such a mechanism will adequately protect system resources, allow system users to accomplish their work and be manageable by the security and administration group.

Let's look at how to REVOKE previously GRANTed privileges.

17.4 WRITING DATA CONTROL STATEMENTS — REVOKE

REVOKE confiscates privileges from User-IDs, and follows the format illustrated below:

```
REVOKE <access_level(s)>
    ON <object_type>  <object_name>
        FROM  <User-id_list/PUBLIC>
```

* Note: DB2 considers views to be tables.

Example 1:

Revoke the SELECT, UPDATE and DELETE authorizations previously GRANTed on the EMPL table from User-IDs TS0001, TS0009 and TS0013.

```
REVOKE SELECT, UPDATE, DELETE
    ON TABLE EMPL
        FROM TS0001, TS0009, TS0013;
```

Example 2:

Revoke all authorizations on the ROW_SUBSET VIEW from all DB2 users.

```
REVOKE ALL
    ON TABLE ROW_SUBSET
        FROM PUBLIC;
```

Again, as with GRANT, there is nothing particularly difficult about writing REVOKE statements. Let's look at another GRANT option.

17.5 HIERARCHICAL PRIVILEGES — WITH GRANT OPTION

The GRANT statement has an additional clause that allows users who are GRANTed authorization levels to pass those authorizations on to other DB2 users. The clause is the WITH GRANT OPTION, and it is coded at the end of a GRANT statement as illustrated below:

```
GRANT SELECT ON TABLE PAY TO TS0003
    WITH GRANT OPTION;
```

After successful execution of the above statement, User-ID TS0003 can retrieve data from all rows and columns in the PAY table. They are also free to pass the SELECT privilege to any other DB2 system user. This has the effect of establishing a hierarchy of authorities as illustrated below in Figure 17.2.

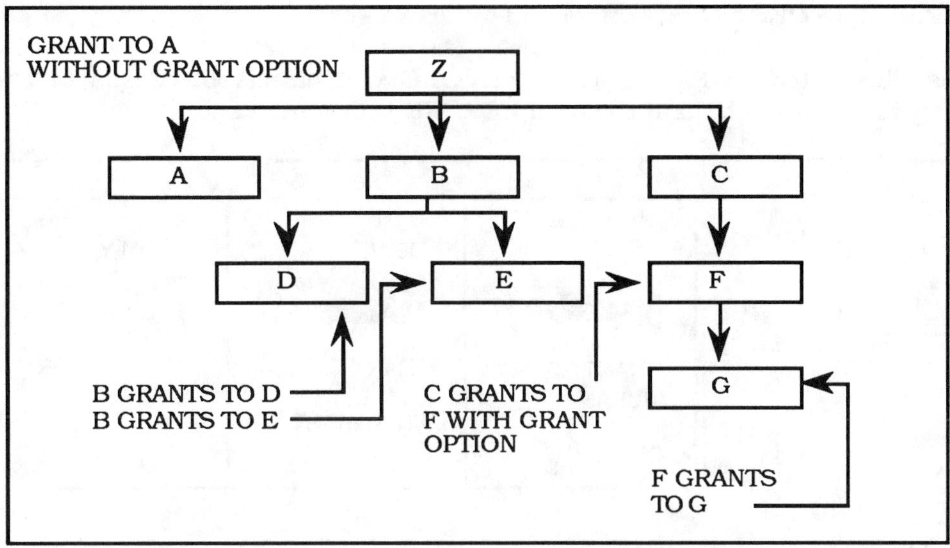

Figure 17.2. Hierarchy of Grant Authority.

Such a mechanism can tend to be difficult to administer, and many installations choose not to use the GRANT OPTION at all.

Let's see how REVOKE works with privileges passed on with the WITH GRANT OPTION.

17.6 REVOKE's Effect on Passed-on Privileges

As illustrated in Figure 17.3, REVOKE will cancel privileges in the specific GRANT chain at the level REVOKE is executed on.

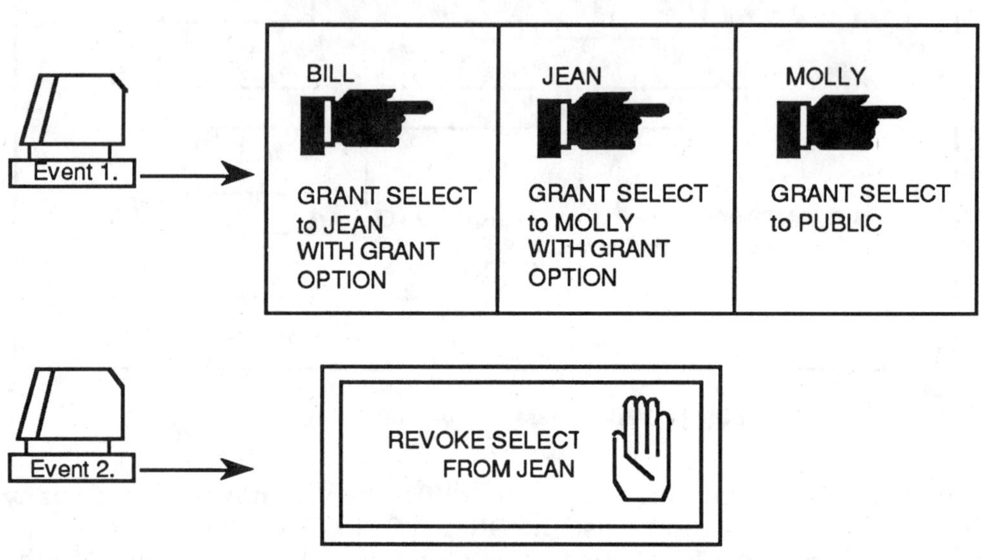

Figure 17.3. Revoking Privileges.

The example shows User-ID BILL granting SELECT to JEAN with the GRANT OPTION. JEAN then grants SELECT to MOLLY with the GRANT OPTION. MOLLY grants SELECT to the PUBLIC.

If at a later time BILL REVOKEs SELECT from JEAN, that REVOKE "cascades" through the GRANT chain terminating SELECT privileges from MOLLY and the PUBLIC.

17.7 WORKSHOP EXERCISE

Prepare a Security and Authorization scheme for an application area, and its users on the EMPL and PAY tables in the Employee database. Decide *who* should have Table-level Authorizations for the following access and *why* they should have them:

SELECT

UPDATE/DELETE/INSERT

ALTER

Do not GRANT ALL on the tables, and do not grant any of the access levels TO PUBLIC.

Decide when and where to use VIEWs as a part of your Authorization scheme. Recall that the Data Control Language statements control access to your data at the *table* and *column* level. VIEWs can control access to your data at the *row* level. When would you choose one, the other or both, and why?

After you have planned your authorization scheme, write and execute the necessary Data Control Language statements to implement it.

USER-IDS	JOB	DESCRIPTION
TS0001	SR. PROGRAMMER/ANALYST	MANAGE ENTIRE DEVELOPMENT EFFORT
TS0002	APPLICATION PROGRAMMER	RESPONSIBLE FOR MONITORING USER DATA ON EMPL TABLE
TS0003	APPLICATION PROGRAMMER	RESPONSIBLE FOR MONITORING USER DATA ON PAY TABLE
TS0005	END-USER	USES EMP, PAY — READ ONLY
TS0006	END-USER	USES PAY — READ ONLY, UPDATES ALL COLUMNS OF EMPL
TS0007	END-USER	UPDATES EMPLOYEE NUMBER COLUMN ON BOTH TABLES

17.7.1 Workshop Solutions

```
GRANT ALL ON TABLE EMPL, PAY TO TS0001;
GRANT SELECT ON EMPL TO TS0002, TS0005, TS0006, TS0007;
GRANT SELECT ON PAY TO TS0003, TS0005, TS0006, TS0007;
GRANT UPDATE ON EMPL TO TS0006;
GRANT UPDATE ON EMPL.NBR, PAY.NBR TO TS0007;
```

Chapter 18
THE DB2 CATALOG TABLES

Here you will learn about the system catalog tables: what they do, how DB2 uses them, and how to write queries that access the information in them.

Topics include:

WHO SHOULD BENEFIT FROM THIS MATERIAL?

The DB2 Catalog tables are system owned and maintained tables that contain data regarding all system and user defined data objects, and the privileges on and processing against those objects. This type of data is called "Metadata" — data about data — and is really only of use to three types of system users:

Database Administrators (DBA's) — will need an in-depth knowledge of all of the above types of information. In fact, DBAs will need to get information out of the catalog tables and run queries that go beyond the scope presented here.

Applications Personnel — will need some knowledge of the Systems Catalog tables. More specifically, this material has been designed around Applications Personnel and their particular needs.

Systems Administrators — will need to know about DB2 privilege dispersion, and need to access the Security and Access Catalog Tables. Some of those types of queries are contained here.

If you don't fall into one of the above categories, feel free to simply read this material without doing the exercises, or skip.

18.1 WHAT IS THE SYSTEM CATALOG?

The DB2 System Catalog is a collection of thirty or so tables, each containing information on some part of the DB2 system. These catalog tables act as a functioning data dictionary, and are used by DB2 to direct and control the operation of the system. The catalog tables break down into four separate groups as shown in Figure 18.1. Those relate to:

- System data objects and data object dependencies — such as SYSTABLES, which contains one row for each table or VIEW defined to DB2; SYSVIEWDEP, which shows the relationship of a VIEW to each separate table or VIEW it is dependent on (i.e. is built on).
- Application Plans and plan/data object dependency — tables which contain information on compiled and linked SQL statements (SQL queries that have been embedded in application programs).
- User privilege tables — which contain information on all privileges granted to users and other information on specific user-IDs and their privilege classes.
- Internal DB2 tables — which DB2 uses for its own internal processes to effectively process using the System Catalog tables.

18.2 HOW DOES DB2 USE THE SYSTEM CATALOG?

DB2 uses the System Catalog in many activities. It writes to the appropriate catalog table when new data objects are added or existing ones are altered, it reads from the System Catalog when it executes dynamic SQL queries (from QMF, SPUFI etc.), and certain DB2 utility programs use the System Catalog Tables. The updating of the catalog tables is illustrated for you in Figure 18.3. Here you can see that when a new table is added to the DB2 system, a row is inserted into the

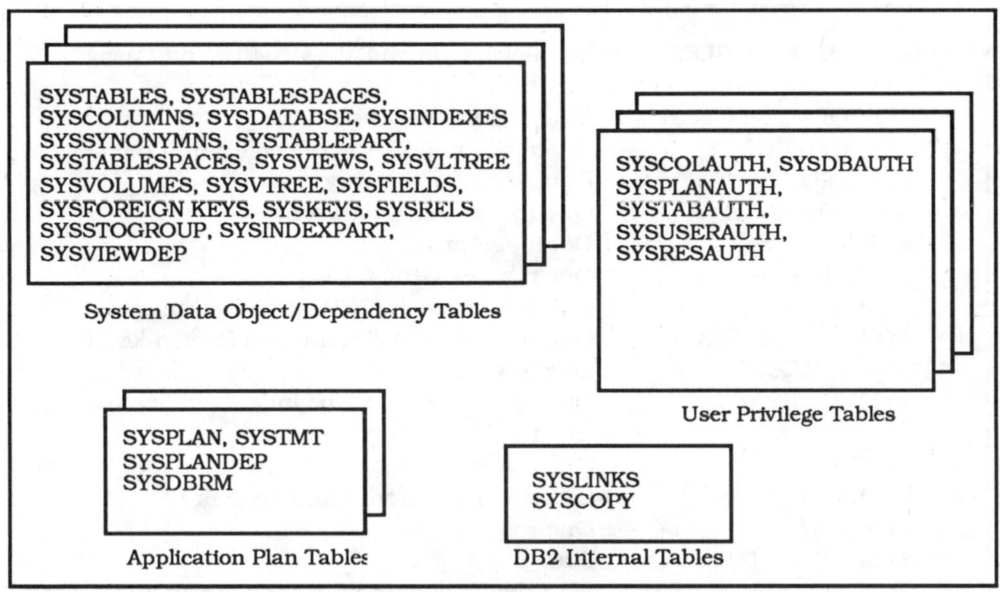

Figure 18.1. The DB2 Catalog tables.

SYSTABLES catalog table. SYSTABLES describes the general features of the new table — how many columns, the table name, an internal DB2 identifier, the number of active pages of data, the Database in which the table resides, whether the table is a base table or a VIEW, etc. Also, one row is inserted into SYSCOLUMNS, for each column in the new table, describing each column's characteristics — datatype attributes, NULLs allowed, Primary/Foreign Key, etc. Other catalog tables are also updated.

18.3 WHAT IS IN THE DB2 CATALOG TABLES?

Figure 18.2 below shows the general contents of the DB2 Catalog tables. I have marked with an asterisk the ones I feel applications personnel should know about. Note that this is very much my personal opinion, and your installation will almost definitely have different standards in this regard. You should realize that like all other DB2 tables, the privilege of using the System Catalog tables is not automatic. Your Systems Administrators must either:

DB2 CATALOG TABLES — Store information on DB2 structure and usage

*	SYSCOLAUTH	—	UPDATE privileges on table or view columns
*	SYSCOLUMNS	—	One row for each existing column in the system
	SYSCOPY	—	Data used for recovery
	SYSDATABASE	—	One row for each database
	SYSDBAUTH	—	Database privileges
*	SYSDBRM	—	One row per DBRM
	SYSFIELDS	—	One row for each FIELDPROC column
*	SYSFOREIGN KEYS	—	One row for each column of a foreign key
*	SYSINDEXES	—	One row per index
	SYSINDEXPART	—	One row per each simpe index and one row for each partition
*	SYSKEYS	—	One row per indexed column
*	SYSLINKS	—	One row per parent-children link
*	SYSPLAN	—	One row per plan
*	SYSPLANAUTH	—	Plan privileges
*	SYSPLANDEP	—	Plan dependencies on tables, views, table space and indexes
*	SYSRELS	—	One row per link
	SYSRESAUTH	—	System privileges
	SYSSTMT	—	One row per each SQL
	SYSSTOGROUP	—	One row per storage group
*	SYSSYNONYMS	—	One row per synonyum
*	SYSTABAUTH	—	Tables or views privileges
	SYSTABLEPART	—	One row per each simple table space and one row per each partition
*	SYSTABLES	—	One row per table or view
*	SYSTABLESPACE	—	One row per table space
*	SYSUSERAUTH	—	System privileges
*	SYSVIEWDEP	—	Views dependencies on tables, views
*	SYSVIEWDEP	—	Views dependencies on tables, views
*	SYSVIEWS	—	One (more) row per view
	SYSVLTREE	—	Pars tree representation of views
	SYSVOLUMES	—	A row per storage group volume
	SYSVTREE	—	Pars tree of a view

Figure 18.2. General contents of System Catalog.

*Important tables for application personnel.

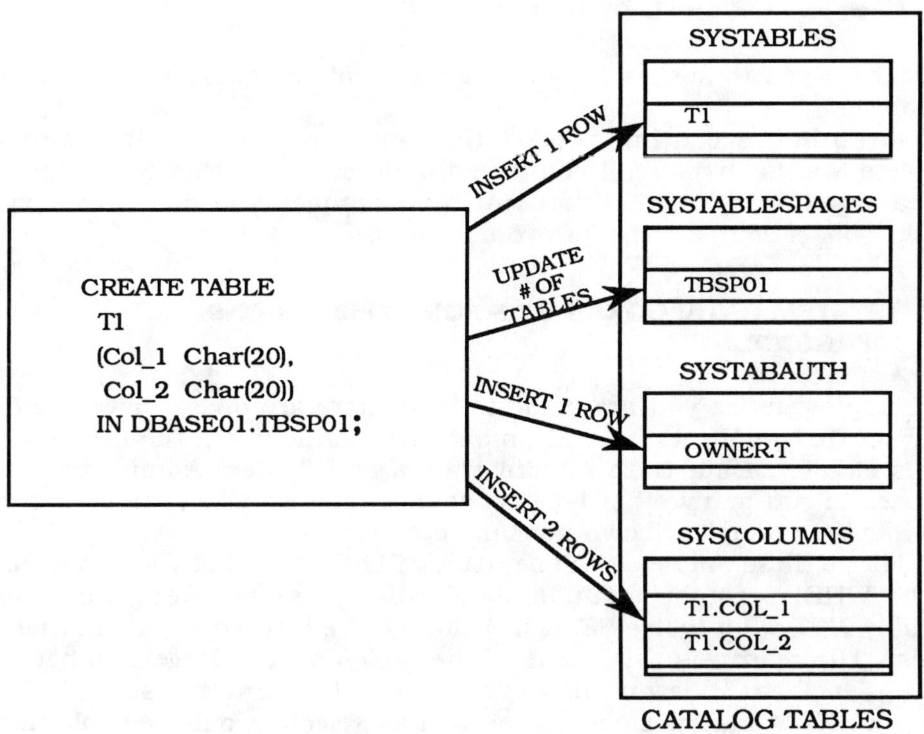

Figure 18.3. How DB2 updates the System Catalog Tables.

Grant you specific privileges to see specific tables, or

Have granted privileges to the PUBLIC

in order for you to successfully execute the catalog queries in the next section.

Regarding executing queries there are no exercises for you to complete in this chapter. If you have the necessary authority to access the catalog tables, type in and run the examples shown in the text. Results will, of course, vary with your system.

18.4 SYSTEM CATALOG QUERIES FOR APPLICATIONS PERSONNEL

The sample queries you are going to learn here are ones I have found useful in DB2 application development. As mentioned at the beginning of this chapter, Data Base Administrators and System Administrators will need to run many other types of catalog queries. These include the ones like the examples shown here and others.

For all these queries, I will be using SELECT * so that you can focus on the WHERE clause to understand what it is I'm after. You will probably *NOT want to use SELECT ** after you've learned what's in these tables. The reason is that some of the tables have a large number of columns in them. Viewing this information at a terminal is unwieldy. Also, SELECT * may not perform as well as selecting only the columns you need to see.

We'll be taking a look at the contents of various catalog tables, then seeing how they can be used to verify, or administer, DB2 data objects and authorizations.

18.4.1 SYSIBM.SYSTABLES Table

This table contains one row for each table or view defined to DB2. It carries general information describing the table — who created it, when, what database it lives in, how big the table is, whether referential integrity is to be enforced on it, and so on.

Column Name	Data Type	Description
NAME	VARCHAR(18)	Name of the table or view.
CREATOR	CHAR(8)	Authorization ID of the owner of the table or view.
TYPE	CHAR(1)	Whether the row describes a table or view: T Table; V View.
DBNAME	CHAR(8)	For a table, or a view of tables, the name of the database that contains the table space name in TSNAME. For a view of a view, the value is DNSDB06.
TSNAME	CHAR(8)	For a table, or a view of one table, the name of the table space that contains the table. For a view of more than one table, the name of a tablespace that contains one of the tables. For a view of a view, the value is SYSVIEWS.
DBID	SMALLINT	Internal identifier of the database; 0 if the row describes a view.
OBID	SMALLINT	Internal identifier of the table; 0 if the row describes a view.
COLCOUNT	SMALLINT	Number of columns in the table or view.
EDPROC	CHAR)(8)	Name of the edit procedure; blank if the row describes a view or a table without an edit procedure.
VALPROC	CHAR(8)	Name of the validation procedure; blank if the row describes a view or a table without validation procedure.
CLUSTERTYPE	CHAR(1)	Not used; all values are blank.
CLUSTERRID	INTEGER	Not used; all values are 0.
CARD	INTEGER	Total number of rows in the table ; –1 if statistics not gathered or the row describes a view. This is an updateable column.
NPAGES	INTEGER	Total number of pages on which rows of the table appear; –1 if statistics not gathered or the row describes a view; This is an updateable column

Column Name	Data Type	Description
PCTPAGES	SMALLINT	Percentage of total pages of the table space that contain rows of the table. If the tablespace is segmented, the percentage of total pages in the set of segments assigned to the table. –1 if statistics not gathered or the row describes a view. This is an updateable column.
IBMREQD	CHAR(1)	Whether the row came from the basic machine-readable material (MRM) tape: N, No; Y, Yes; B, Release 3 dependency indicator, not from MRM tape; C, Version 2 dependency indicator, not from table is a dependent.
CHILDREN	SMALLINT	The number of relationships in which the table is a parent.
KEYCOLUMNS	SMALLINT	The number of columns in the table's primary key.
RECLENTH	SMALLINT	The maximum length of any record in the table, including all control fields.
STATUS	CHAR(1)	I, the table's definitions incomplete because it lacks a primary index; X, the table has a primary index blank Blank if the table has no primary key, or is a catalog table.
KEYOBID	SMALLINT	Internal DB2 identifier of the index or link that enforces uniqueness of the table's primary key; 0 if not applicable.
LABEL	VARCHAR(30)	The table's label as given by a LABEL ON statement; otherwise an empty string.
CHECKFLAG	CHAR(1)	C, the table space containing the table is in CHECK PENDING mode and there are rows in the table that may violate referential constraints. Blank, the Table does not contain rows that violate referential constraints.

Column Name	Data Type	Description
CHECKRID	CHAR(4)	CHECKRID is blank if the table or partition is not in a check pending state (the CHECKFLAG is blank) or if the tablespace is partitioned. Otherwise, it is the RID of the first row of the table space partition that may violate referential constraints or the value is X'00000000' indicating that any row may violate referential constraints.
AUDITING	CHAR(1)	Value of the audit option: A, AUDIT ALL; C, AUDIT CHANGE; BLANK, AUDIT NONE.
CREATEDBY	CHAR(8)	Primary authorization ID of the user who created the table.
LOCATION	CHAR(16)	Not used; all values are blank.
TBCREATOR	CHAR(16)	Not used; all values are blank.
TBNAME	VARCHAR(18)	Not used; all values are blank.
IBMREQD	CHAR(1)	Whether the row came from the basic machine-readable material (MRM) tape: N, no; Y, yes; C, version 2 dependency indicator; not from MRM tape.
ROOTNAME	VARCHAR(18)	Unqualified name of the root table of the space. Blank if it is not a structured table space.
ROOTCREATOR	CHAR(8)	Authorization ID of the owner of the root table.
SEGSIZE	SMALLINT	The number of pages in each segment of a segmented table space.
CREATEDBY	CHAR(8)	Primary authorization ID of the user who created the table space.

18.4.2 SYSIBM.SYSTABLES

Here are some sample queries you can execute against the SYSTABLES catalog table.

List all base tables you own.

```
SELECT * FROM SYSIBM.SYSTABLES WHERE OWNER = USER
AND TYPE = 'T';
```

List all views you own.

```
SELECT * FROM SYSIBM.SYSTABLES WHERE OWNER = USER
AND TYPE = 'V';
```

List all tables in your application's database.

```
SELECT * FROM SYSIBM.SYSTABLES WHERE DBNAME =
<dbname>;
```

List all tables with over 10,000 rows.

```
SELECT * FROM SYSIBM.SYSTABLES WHERE CARD > 10000;
```

List all dependent tables in your application's database (see referential integrity description — Chapter 10).

```
SELECT * FROM SYSIBM.SYSTABLES WHERE PARENTS > 0;
```

List all parent tables in your application's database (see referential integrity description — Chapter 10).

```
SELECT * FROM SYSIBM.SYSTABLES WHERE CHILDREN > 0;
```

List all tables in your application (your application has a high-level qualifier of XYZ in the third, fourth and fifth positions).

```
SELECT * FROM SYSIBM.SYSTABLES WHERE NAME LIKE
'__XYZ%';
```

18.4.3 SYSIBM.SYSCOLUMNS Table

The SYSIBM.SYSCOLUMNS table contains one row for every column of each table and view. It carries detail information on what type of column, how big the column can be, it allows whether nulls, and even statistics on what is in the column (see COLCARD, HIGH2KEY, LOW2KEY).

Column Name	Data Type	Description
NAME	VARCHAR(18)	Name of the Column.
TBNAME	VARKCHAR(18)	Name of the table or view which contains the column.
TBCRATOR	CHAR(8)	Authorization ID of the owner of the table or view that contains the column.
COLNO	SMALLINT	Numerical plan of the column in the table or view; for example 4 (out of 10).
COLTYPE	CHAR(8)	Type of Column: INTEGER, large integer; SMALLINT, small integer; FLOAT, floating point; CHAR, fixed-length character string; VARCHAR, variable-length character string; LONGVAR. varying-length character string; DATE,date; TIME, time; TIMESTMP, timestamp.
LENGTH	SMALLINT	The length attribute of the column; or, in the case of a decimal column, its precision. The number does not include the internal prefixes used to record actual length and null state where applicable.

Column Name	Data Type	Description
SCALE	SMALLINT	Scale of decimal data; 0 if not a decimal column.
NULLS	CHAR(1)	Whether the column can contain null values.
COLCARD	INTEGER	Number of distinct values in the column; −1 if statistics have not been gathered. This is an updateable column
HIGH2KEY	CHAR(8)	Second highest value of the column; blank if statistics have not been gathered. If the key has a non-character data type, the data may not be printable. This is an updateable column
LOW2KEY	CHAR(8)	Second lowest value of the column; blank if statistics have not been gathered. If the key has a non-character data type, the data may not be printable. This is an updateable column
UPDATES	CHAR(1)	Whether the column can be updated: N, no; Y, yes. (The value is N only if the column is part of the key of a partitioned index or is derived from a function or arithmetic expression. Thus, the value can be Y for columns of read-only view. N also applies to catalog columns that are not updateable.
IMBREQD	CHAR(1)	Whether the row came from the basic machine-readable material (MRM) tape: N, no; Y, yes.

Column Name	Data Type	Description
REMARKS	VARCHAR(254)	A character string provided by the user with the COMMENT ON statement.
DEFAULT	CHAR(1)	Whether the column has a default value (null or non-null): N, no; Y, yes.
KEYSEQ	SMALLINT	The column's numerical position within the table's primary key; 0 if it is not part of a primary key.
FOREIGNKEY	CHAR(1)	Contains B if the column contains bit data. All other values indicate that the column does not contain bit data. This is an updateable column.
FLDPROC	CHAR(1)	Whether the column has a field procedure: N, no; Y, yes. This field is blank for views.
LABEL	VARCHAR(30)	The column label as given by a LABEL ON statement; otherwise an empty string.

18.4.4 SYSIBM.SYSCOLUMNS

Some sample queries to find out information about your table columns:

List all columns in table XYZ_123 (you created this table) — in their default create order.

```
SELECT * FROM SYSIBM.SYSCOLUMNS WHERE TBCREATOR =
    USER AND
      TBNAME = 'XYZ_123'
            ORDER BY COLNO;
```

List the number of *different* values in the PART columns of the PARTS_INV table which is owned by "PROD_DBA".

```
SELECT COLCARD FROM SYSIBM.SYSCOLUMNS WHERE
TBCREATOR = 'PROD_DBA' AND TBNAME = 'PARTS_INV' AND
NAME = 'PART';
```

List all tables that have VARCHAR COLUMNS.

```
SELECT T.* FROM SYSIBM.SYSCOLUMNS C,
SYSIBM.SYSTABLES T
      WHERE T.NAME = C.TBNAME AND COLTYPE = 'VARCHAR';
```

List all tables that have a columnname similar to NBR.

```
SELECT * FROM SYSIBM.SYSCOLUMNS
      WHERE NAME LIKE '%NBR%';
```

18.4.5 SYSIBM.SYSSYNONYMS Table

The SYSIBM.SYSSYNONYMS table contains one row for each synonym of a table or view.

Column Name	Data Type	Description
NAME	VARCHAR(18)	Synonym for the table or view.
CREATOR	CHAR(8)	Authorization ID of the owner of the synonym.
TBNAME	VARCHAR(18)	Name of the table or view.
TBCREATOR	CHAR(8)	Authorization ID of the owner of the table or view.
IBMREQD	CHAR(1)	Whether the row came from the basic machine-readable (MRM) tape: N, no; Y, yes.
CREATEDBY	CHAR(8)	Primary authorization ID of the user who created the synonym.

18.4.6 SYSIBM.SYSSYNONYMS

List all SYSNONYMS created for table XYZ_123 (you created this table).

```
SELECT * FROM SYSIBM.SYSSYNONYMS WHERE TBCREATER =
USER AND
   TBNAME = 'XYZ_123';
```

List all table names for SYNONYM "S123456".

```
SELECT * FROM SYSIBM.SYSSYNONYMS WHERE NAME = '123456';
```

18.4.7 SYSIBM.SYSVIEWS Table

The SYSIBM.SYSVIEWS table contains one or more rows for each view. Recall that a VIEW is nothing more than a stored SELECT statement. Here (among other places) is where the original SELECT statement is stored. (see TEXT column)

Column Name	Data Type	Description
NAME	VARCHAR(18)	Name of the view.
CREATOR	CHAR(8)	Authorization ID of the owner of the view.
SEQNO	SMALLINT	Sequence number of this row; the first portion of the view is on row 1 and successive rows have increasing values of SEQNO.
CHECK	CHAR(1)	Whether the CHECK option was specified in the CREATE VIEW statement: N, no; Y, yes. Note: The value will be N if the view has no WHERE clause.
IBMREQD	CHAR(1)	Whether the row came from the basic machine-readable material (MRM) tape: N, no; Y, yes. B, Release 3 dependency indicator; not from MRM tape. C, Version 2 dependency indicator; not from MRM tape.
TEXT	VARCHAR(254)	The text portion of the text of the CREATE VIEW statement.

18.4.8 SYSIBM.SYSVIEWS

List the SELECT statement that defines the VIEW "V123456".

```
SELECT TEXT FROM SYSIBM.SYSVIEWS WHERE NAME =
  'V123456';
```

18.4.9 SYSIBM.SYSVIEWDEP Table

The SYSIBM.SYSVIEWDEP table records the dependencies of views on tables and other views.

Since views are built on other DB2 objects (base tables and other VIEWs) DB2 keeps track of all the object dependencies in this table.

Column Name	Data Type	Description
BNAME	VARCHAR(18)	Name of a table or view the view is dependent on.
BCREATOR	CHAR(8)	Authorization ID of the owner of BNAME.
BTYPE	CHAR(1)	Type of object BNAME: T, table; V, view.
DNAME	VARCHAR(18)	Name of the view.
DCREATOR	CHAR(8)	Authorization ID of the owner of the view.
IBMREQD	CHAR(1)	Whether the row came from the basic machine-readable material (MRM tape): N, no; Y, yes.

18.4.10 SYSIBM.SYSVIEWDEP

List all views dependent on Table XYZ_123 (you created this table).

```
SELECT * FROM SYSIBM.SYSVIEWDEP WHERE BCREATOR =
USER AND
      BNAME = 'XYZ_123';
```

18.4.11 SYSIBM.SYSTABAUTH Table

The SYSIBM.SYSTABAUTH table records the privileges held by users on tables and views.

Recall from the material on SQL Data Control Language statements that one way to implement a security mechanism is through explicit GRANT and REVOKE statements. All such activity is recorded in this table.

Column Name	Data Type	Description
GRANTOR	CHAR(8)	Authorization ID of the user who granted the privileges.
GRANTEE	CHAR(8)	Authorization ID of the user who holds the privileges or the name of a plan that uses the privileges.
GRANTEETYPE	CHAR(1)	Meaning: blank, GRANTEE is an authorization ID; P, GRANTEE is an application plan.
DBNAME	CHAR(8)	If the privileges were received from a user with DBADM, DBCTRL, or DBMAINT authority, DBNAME is the name of the dbase on which the GRANTOR has that authority. Otherwise, DBNAME is blank.
SCREATOR	CHAR(8)	If the row of SYSIBM.SYSTABAUTH was created as a result of a CREATE VIEW statement, SCREATOR is the authorization ID of the owner of a table or view referenced in the CREATE VIEW statement. Otherwise, SCREATOR is the same as TCREATOR.
STNAME	VARCHAR(18)	If the row of SYSIBM.SYSTABAUTH was created as a result of a CREATE VIEW statement, STNAME is the name of a table or view referenced in the CREATE VIEW statement. Otherwise, SCREATOR is the same as TCREATOR.
TCREATOR	CHAR(8)	Authorization ID of the owner of the table or view.
TTNAME	VARCHAR(18)	Name of the table or view.
AUTHHOWGOT	CHAR(1)	Authorization level of the user from whom the privileges were received. Note: This is the authorization level used by the authorization sub-component; It is not necessarily the highest authorization level of the grantor. Blank, not applicable; S, SYSADM; D, DBADM; C, DBCTL; M,DBMAINT.

Column Name	Data Type	Description
TIMESTAMP	CHAR(12)	Time at which the privileges were granted (internal timestamp format).
DATEGRANTED	CHAR(6)	Date the privileges were granted, in the form yymmdd.
TIMEGRANTED	CHAR(8)	Time the privileges were granted, in the form hhmmssth.
UPDATECOLS	CHAR(1)	The value of this column is blank if the value of UPDATEAUTH applies uniformly to all columns of the table or view. The value is an asterisk(*) if the value of UPDATEAUTH applies to some columns but not to others. In this case, rows will exist in SYSIBM.SYSCOLAUTH with matching timestamps which list the columns on which update option. Y, privilege is held without the GRANT option.
DELEAUTH	CHAR(1)	Whether the GRANTEE can delete rows from the table or view: blank – privilege is not held G – privilege is held with the GRANT option Y – privilege is held without the GRANT option
INSERTAUTH	CHAR(1)	Whether the GRANTEE can insert rows into the table or view: blank, privilege is not held; G, privilege is held with the GRANT option; Y, privilege is held without the GRANT option.
SELECTAUTH	CHAR(1)	Whether the GRANTEE can select rows from the table or view: blank, privilege is not held; G, privilege is held with the ; GRANT option; Y, privilege is held without the GRANT option.

Column Name	Data Type	Description
UPDATEAUTH	CHAR(1)	Whether the GRANTEE can update rows from the table or view: blank, privilege is not held; G, privilege is held with the GRANT option; Y, privilege is held without the GRANT option.
IBMREQD	CHAR(1)	Whether the row came from the basic machine-readable material (MRM) tape: N, no; Y, yes.
GRANTELOCATION	CHAR(16)	Not used.

18.4.12 SYSIBM.SYSTABAUTH

Extract the user privileges of table 'T123456' created by DBA001.

```
SELECT * FROM SYSIBM.SYSTABAUTH WHERE TTNAME =
'T123456' AND
        TCREATOR = 'DBA001';
```

User TS0001 has left the company, find out which tables or views they had privileges against.

```
SELECT * FROM SYSIBM.SYSTABAUTH
WHERE GRANTEE = 'TS0001';
```

List all privileges granted by TS0001, on tables or views that TS001 owned after July 4, 1989.

```
SELECT * FROM SYSIBM.SYSTABAUTH WHERE GRANTOR =
'TS001'
        AND DATEGRANTED > '890704' AND TBCREATOR =
        'TS0001';
```

18.5 SYSTEM CATALOG TABLE REVIEW

You have examined the DB2 System Catalog tables. The data you saw (if you were able to execute the statements on a mainframe) described the DB2 data objects, their dependencies and authorizations:

Base Tables and VIEWS
Databases
Table Columns
Table/View authorities

Figure 18.4 illustrates the dependent nature of DB2 data objects. Figures 18.5, 18.6, and 18.7 illustrate the way these dependencies are materialized in the various system catalog tables. Figure 18.8 illustrates how DB2 updates the system catalog Tables during SQL DDL execution.

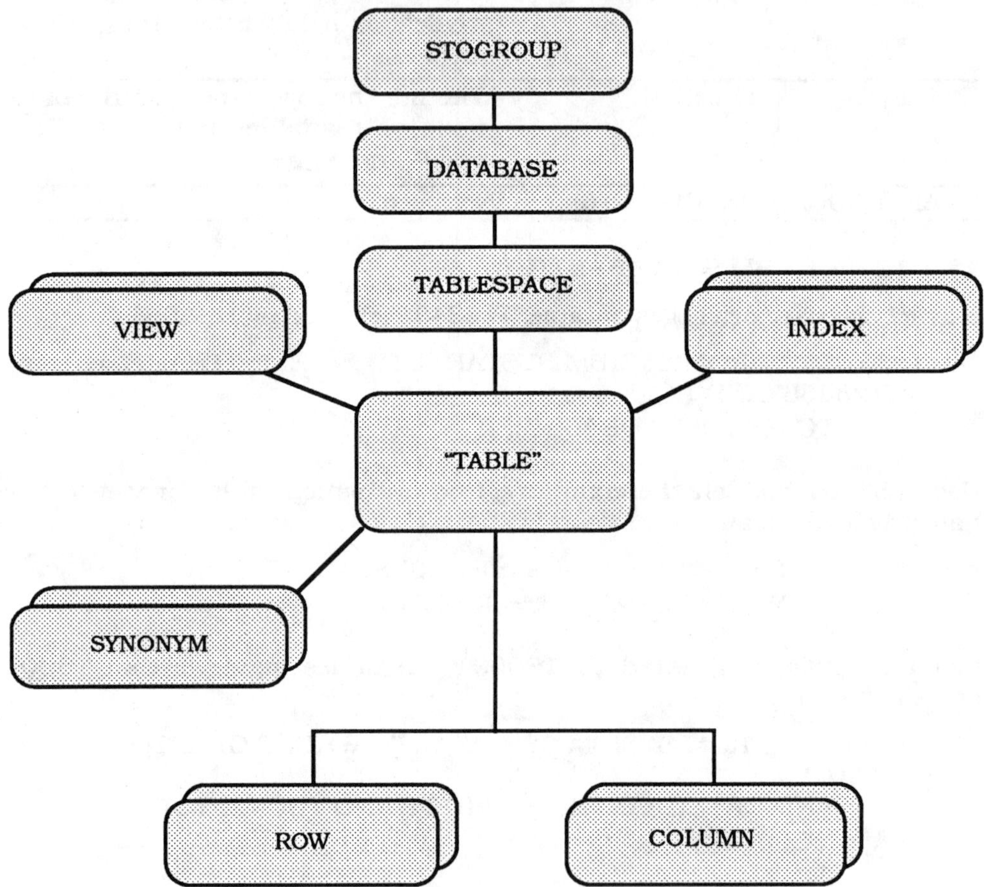

Figure 18.4. DB2 data object dependencies.

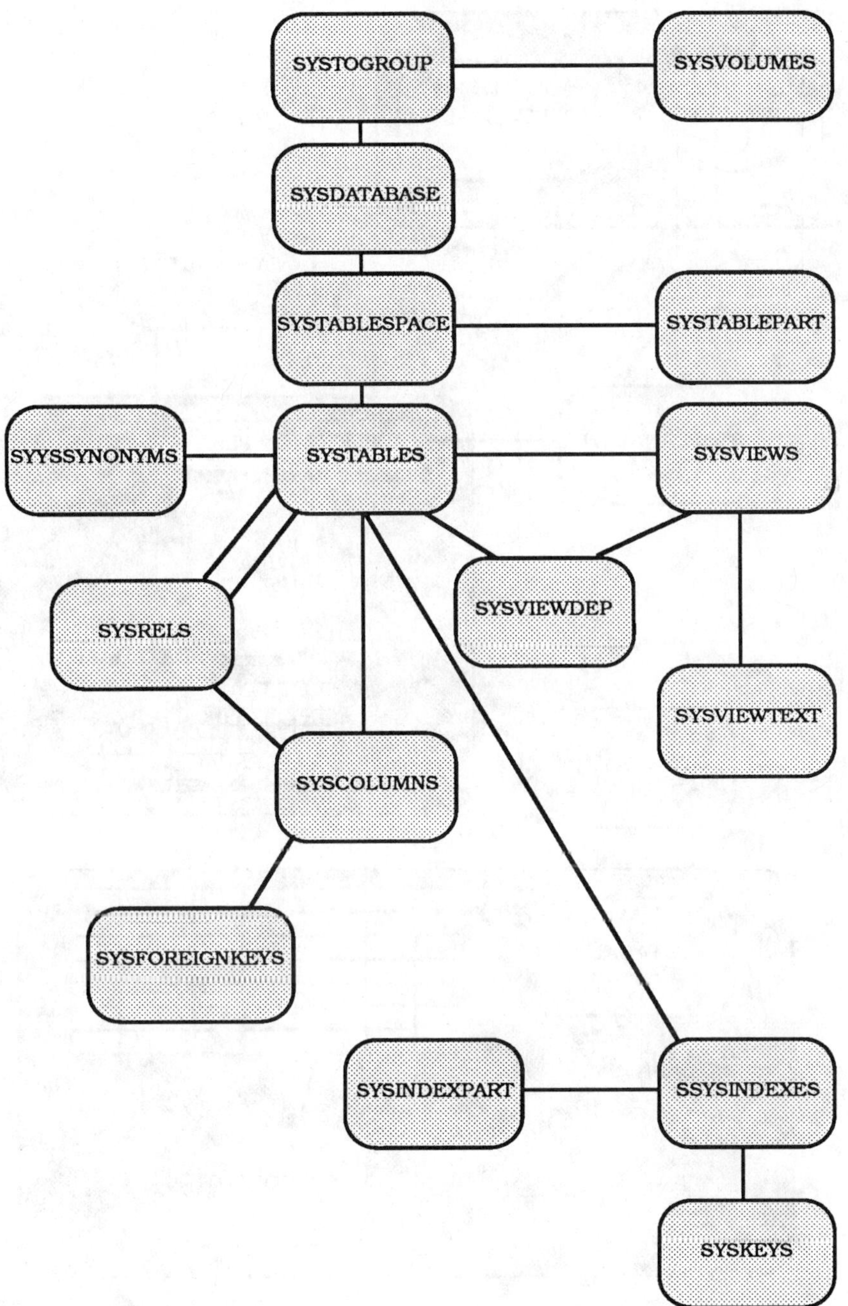

Figure 18.5. DB2 Data Object Catalog Tables.

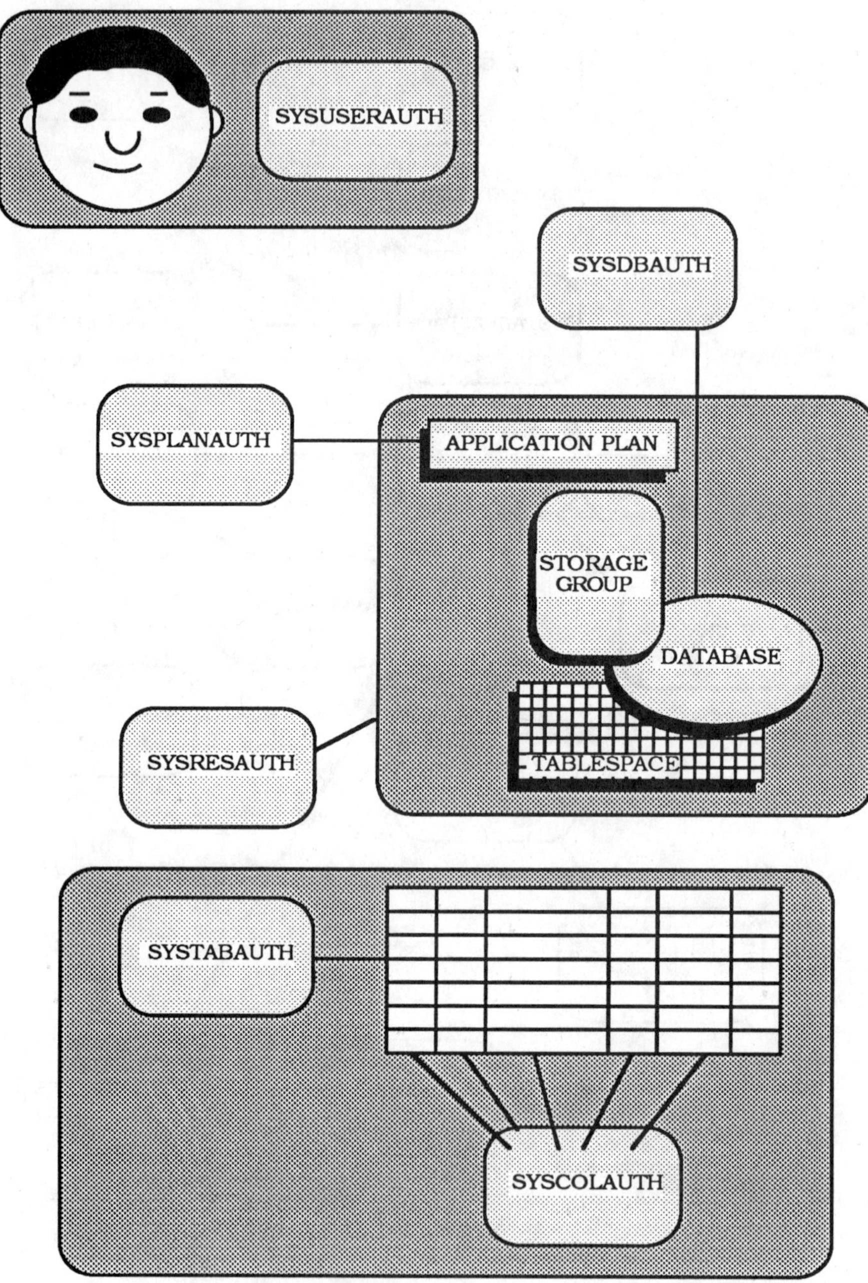

Figure 18.6. DB2 Authorization Catalog Tables.

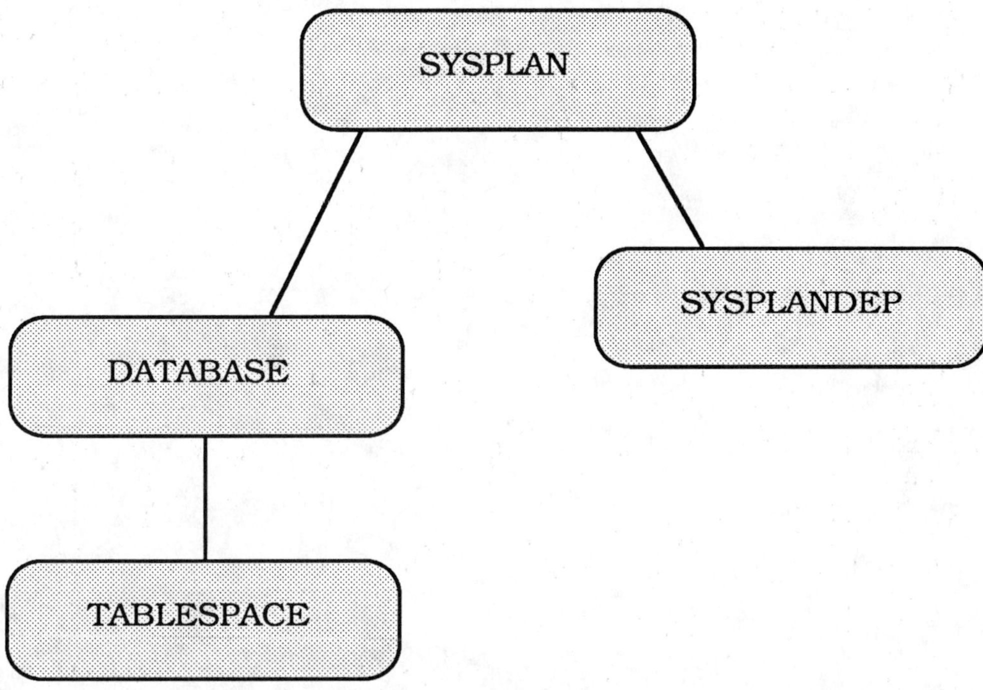

Figure 18.7. DB2 Application Plan Catalog Tables.
(Courtesy of Johnston Education Services.)

Chapter 19
OPTIMIZING YOUR QUERIES IN DB2 OR SQL/DS

Here you will learn how to write SQL queries for optimum performance. You will learn how DB2 chooses to access your data and ways to take advantage of existing indexes through proper SQL coding techniques.

Up to now, you haven't been looking at what DB2 does "under the covers" to access data. You've been concentrating on learning SQL from a syntactical and a logical standpoint. You have not been concerned with *HOW* DB2 got your data. This is fine for querying the small Employee and Video-store Databases consisting of ten to twenty rows per table. However, in the real world you will be writing queries that access tables of thousands, hundreds of thousands, perhaps even millions of rows. Will you be able to remain unconcerned over coding efficient SQL queries that access such large amounts of data? Experience has proved that you will have to be concerned with SQL query performance once you start working with large "real-world" volumes of data.

This is important information. Relational databases such as DB2 are notorious for poor response time and heavy system overhead. If you learn to use the techniques and coding strategies presented in this chapter, you will be saving your company a lot of money in the near and far term. And you will find that your queries run faster, thus saving you a lot of time in the near and far term.

One final point of interest before we begin. DB2, or more specifically, the DB2 Optimizer, chooses how best to access your data. You can only "indirectly influence"* the optimizer's decisions by using the strategies shown in this chapter. You could almost say that the goal of this chapter is to teach you the mandatory SQL "maneuvers" to outwit the Optimizer. Sound like fun? Let's start.

---

* DB2 Version 2.1 contains features that allow Database and Systems Administrators a more "direct influence" on the DB2 Optimizer's choice of data access. They can directly update certain DB2 System Catalog tables which hold statistics used by the Optimizer to calculate search algorithms and weigh alternatives.

19.1 WHEN DO YOU HAVE TO CONSIDER OPTIMIZING YOUR QUERIES?

A good rule of thumb regarding when you must consider optimizing your queries is if the SQL statement:

- Involves a table join;
- Involves a subquery;
- Accesses over 10,000 rows;
- Involves sorting of over 1,000 rows.

When does DB2 sort your data? DB2 invokes a sort if your statement contains any of the following:

- ORDER BY
- GROUP BY
- UNION
- DISTINCT
- Certain table joins

Sorts of queries using UNION or DISTINCT are particularly expensive, as DB2 must evaluate the entire row within the sort.

Now let's go "under-the-covers" and learn how DB2 accesses physical stored data. If you need a refresher on tablespaces and pages, please return to chapter 2 at this point and reread the information on these topics.

19.2 DB2 ACCESS PATH SELECTION

In accessing data in tables, DB2 chooses between two methods:

- Using an index to access the data
- Reading all pages of a TABLESPACE/DBSPACE*

Notice that I said reading all pages of a TABLESPACE/DBSPACE — not table. This is not a typo and is illustrated for you in Figure 19.1:

---

* With DB2 Version 2.1 you are able to define segmented tablespaces. This allows DB2 to access only the pages corresponding with the tables referenced in your query.

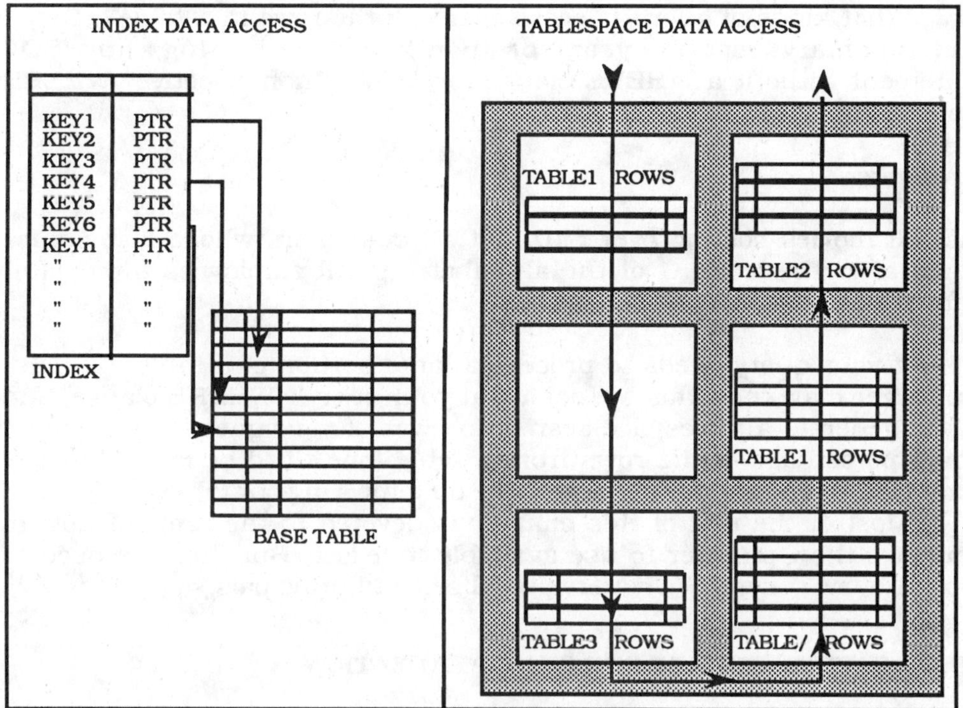

Figure 19.1. DB2 data access.

As illustrated above, when DB2 uses an index to access your data, it is able to read a sorted file of keys (indexed columns), and use an associated pointer with the key to go directly to the row requested. When DB2 is forced to perform a full tablespace scan, it must retrieve each and every page in the tablespace in order to examine the rows in the page.

19.3 TABLESPACE/DBSPACE SCAN VERSUS INDEX SCAN

If by the nature of the business request, your query will be accessing over one-third of the rows in a table, then a tablespace scan will be more efficient than an index scan. The reason for this is that in accessing data through an index, DB2 and SQL/DS may need to perform two physical I/O operations for each row:

- The first I/O operation to read the Index and get the row pointer.
- The second I/O operation to access the row on the base table.

What kinds of queries access over half the rows in a table? Detail and summary reports, year- or month-end processing, any SQL statement without a WHERE clause, or with a "non-selective" WHERE clause.

Example:

List all movies costing over $10, list all customers whose last names begin after the letter G of the alphabet, list all employees hired after January 1, 1940, etc.

If your query needs to process a large portion of a table, the DB2 Optimizer may know this by looking at your query's WHERE clause, and it will generate a tablespace scan. However, if your query needs only a small percentage of the rows from a table (one third or less), you will want DB2 to use an index to access the data rows in your table.*

Most of the rest of this chapter is devoted to the topic of how to convince the Optimizer to use available indexes. But first, we need to look at a few general SQL query guidelines and principals.

19.4 SAYLES' RULE #1 FOR SQL OPTIMIZATION

The most important single thing you can do to optimize your queries is to be as *selective* as possible regarding the amount of data accessed by DB2. This selectivity is both vertical and horizontal.

By vertical selectivity I mean "only retrieve the columns you need." Avoid SELECT * — especially in combination with DISTINCT or UNION.

By horizontal selectivity I mean try to pinpoint with your WHERE clause only the rows you need to satisfy your business request. Understand the specification completely, and be sure that you have the necessary and sufficient additional search arguments AND'd or OR'd together to access only the rows you need.

19.5 DON'TS FOR QUERIES THAT SHOULD USE INDEXES

As mentioned in 19.1, unless your specification requires you to access a majority of the rows in a table, you should give DB2 the "opportunity" to use an index. Such an opportunity arises from two conditions:

- An index exists on a column you referenced as part of your WHERE clause ,

---

* As an aside, note that as of this writing, SQL/DS is limited to using at most one index per table per SELECT. DB2 version 2.2 is not bound by this limitation.

- You code your WHERE clause using the predicates shown in Figure 19.3, and not using the types of predicates outlined in Figure 19.2 below.

DB2 and SQL/DS *will not* consider using an index in the following predicates:

```
COLUMN NOT BETWEEN LOVALUE AND HIVALUE
COLUMN NOT IN (LIST_OF_VALUES)
COLUMN NOT LIKE 'SEARCH_STRING'
NOT COLUMN most predicates
COLUMN = arithmetic expression
COLUMN = datatype mismatch.*
COLUMN ¬ = value
COLUMN IN (SUBQUERY)
TABLE1.COLUMNA = TABLE1.COLUMNB
COLUMN1 = x or COLUMN2 = y**
COLUMN > x OR COLUMN > y**
COLUMN < x OR COLUMN < y**
```

```
UPDATE SET COLUMNA = NEW VALUE WHERE COLUMNA search condition
```

* By datatype mismatch, I refer to search argument values that do not match precisely their table/column definition in the WHERE clause
Example:

PERF is defined as SMALLINT. This query would not use an index on PERF even if one existed:

```
SELECT * FROM EMPL WHERE PERF = 1.0;
```

You can also get a datatype mismatch in string padding.
Example:

DEPT is defined as CHAR(3). This query would not use an index on DEPT even if one existed:

```
SELECT * FROM EMPL WHERE DEPT = 'ACC'; (note extra blanks)
```

** DB2 version 2.2 can use an index on multiple OR conditions.

Figure 19.2. SQL Non-indexable predicates.

Realize that DB2 and SQL/DS may also choose not to use an index based on other factors such as:

- The number of rows in the table. The more rows, the better chance DB2 and SQL/DS will utilize existing indexes.

- The type of indexes available. A non-unique index would *not* be a good bet as a performer, while a unique clustered index would be an excellent bet as a performer.[*]

19.6 DOs FOR QUERIES THAT SHOULD USE INDEXES

Below I have listed specific WHERE conditions that give DB2 the opportunity to use an existing index on a column.

```
DB2 and SQL/DS will consider using an index in the following predicates:

        COLUMN BETWEEN LOVALUE AND HIVALUE
        COLUMN LIKE 'SEARCH_STRING'
        COLUMN > , <, =, <=, >= value
        COLUMN =, <, >, (SUBQUERY)
        COLUMN IN (LIST_OF_VALUES)
        TABLE1.COLUMN = TABLE2.COLUMN
        EXISTS (SUBQUERY)
        COLUMN1 = x OR COLUMN1 = y
        COLUMN  = x AND COLUMN = y
```

Figure 19.3. SQL Indexable predicates.

Again you should realize that DB2 may choose not to use an existing index based on the types of circumstances discussed in 19.4.

In spite of this, you should attempt to solve your business requests using SQL statements that conform to the above list of WHERE predicates, and stay away from the predicates in Figure 19.2.

Let's take a look at some more advanced SQL optimization concepts.

19.7 SUBSTITUTE PREDICATES — UNION FOR OR[**]

If you read Figures 19.2 and 19.3 carefully, you probably noticed that DB2 will only use an index with OR if the OR:

- refers to the same column, and
- compares the column with an equal sign.

---

[*] A cluster index is a special type of index that gives very efficient sequential row access.

[**] DB2 version 2.2 will use available indexes or multiple dissimilar conditions connected with OR. We do not recommend this union substitute technique with DB2 v2.2.

Example 1: OR using index:

```
SELECT * FROM EMPL WHERE PERF = 2 OR PERF = 4;
```

Example 2: OR not using index:

```
SELECT * FROM EMPL WHERE PERF = 2 OR DEPT = 'FIN';
```

Example 3: OR not using index:

```
SELECT * FROM EMPL WHERE PERF < 2 OR PERF > 5;
```

This is an alarmingly limiting aspect of the current state of DB2 affairs! However, the solution to the above problems is to recode the query using the UNION operator.[*]

Example 4 : OR replaced with UNION ALL — does not use index:

```
SELECT * FROM EMPL WHERE PERF < 2
UNION ALL
SELECT * FROM EMPL WHERE PERF > 5;
```

Example 5: OR replaced with UNION — uses index:

```
SELECT * FROM EMPL WHERE PERF = 2
UNION
SELECT * FROM EMPL WHERE DEPT = 'FIN';
```

The substitution shown in Example 5, would not only allow DB2 to use an index on PERF if it existed; DB2 could also choose to use an index and an additional index if one existed on the DEPT column. Recall from Section 19.2 that DB2 is limited to using one index per table *per SELECT*. If indexes existed on the PERF and DEPT columns for the EMPL table, the way to allow DB2 to utilize both is to cast the query in UNION format.

19.8 SUBSTITUTE PREDICATES — JOINS FOR SUBQUERIES

Again, if you read figures 19.2 and 19.3 carefully, you probably noticed that DB2 will not use an index for a subquery connected by IN.

Example 6: subquery not using index that exists on the NBR column:

```
SELECT * FROM EMPL WHERE NBR IN (SELECT NBR FROM PAY
WHERE HOURS > 30;)
```

---

[*] Note: The difference between UNION and UNION ALL is that UNION invokes a DB2 sort to remove duplicates; UNION ALL does not. Use UNION ALL if possible.

This is another restriction we'd like to avoid. Simply recode the query as a table join and DB2 will be able to choose the index on the NBR column.

Example 7: join using index on NBR column:

```
SELECT E.* FROM EMPL E, PAY P
        WHERE E.NBR = P.NBR AND HOURS > 30;
```

As a rule of thumb substitute a join for a subquery whenever possible.

19.9 AVOIDING TRUNCATION AND PADDING PROBLEMS

If the values you specify in your search arguments do not match the datatype of the corresponding columns, DB2 will not use an index if one exists for that column in evaluating your SELECT.

Example:

QTY from the MOV table has an index on it, and the column has a SMALLINT (small integer) datatype. This query will not allow DB2 to use the index:

```
SELECT * FROM MOV WHERE QTY BETWEEN 1.0 AND 5.0;
```

These problems should be avoided altogether by coding the WHERE clause as

```
WHERE QTY BETWEEN 1 AND 5
```

If you don't know the datatype for a particular column (say the system documentation got shredded last week, or has pages missing), extract the column datatype information from the System Catalog* then code the query correctly.

The same principal holds true for CHAR or character datatypes. Don't specify your search arguments with the incorrect lengths.

19.10 NON-SQL CODING STRATEGIES

Here is a list of non-SQL strategies (database design, implementation, etc.) for you to think about. They are only rules of thumb and may be broken at any time for specific reasons, but you might find it helpful to learn if you are first-time DB2 users.

---

* OR in QMF issue the ===> DRAW Tablename (TYPE = INSERT command).

1. Match data to end-user needs by creating VIEWs of:
    - Permanent Table Joins
    - Summary data
    - Row/Column subset tables
    - Complicated queries
2. Indexes should be created on columns commonly used in:
    - WHERE Conditions
    - Table Joins/Subqueries
    - GROUP BY/ORDER BY
    - Built in Functions
3. Because of the limitation of one index per SELECT/TABLE* multi-column indexes may be more efficient than single column indexes
4. In order to give the DB2 and SQL/DS Optimizer the most information on which to base its access strategy, you must periodically execute a DB2 and SQL/DS statistics gathering utility.
5. Prototype your database and tables using realistic production scale models. And stress test your queries using realistic production scale models (within 10% of the estimated final production volume).
6. Use the DB2 or SQL/DS EXPLAIN facility for all production queries.

Figure 19.4. Non SQL-coding performance strategies.

19.11 REVIEW EXERCISES

The following queries contain SQL DON'Ts. Recode the queries using the SQL optimization techniques discussed in this chapter.

Indexes exist on the NBR, PERF, LNAME, DEPT and PROJ columns.

* DB2 version 2.2 does not have this limitation.

1. Retrieve the employee last names for employees with performance evaluations greater than 1, and also those employees from the Finance department:

```
SELECT * FROM EMPL WHERE PERF > 1 OR DEPT = 'FIN';
```

2. Retrieve the employee last names for employees who work on a project without a name:

```
SELECT * FROM EMPL WHERE PROJ IN
    (SELECT NBR FROM PROJ WHERE NAME IS NULL);
```

3. Retrieve all employee numbers and hire dates for recently hired employees:

```
SELECT * FROM EMPL WHERE HIREDTE > 87001.0;
```

4. Retrieve the employee last names for employees whose performance evaluations are equal to one or six and seven:

```
SELECT * FROM EMPL WHERE PERF NOT BETWEEN 2 AND 5;
```

5. Retrieve the employee last names for employees whose performance evaluations are equal to four or higher:

```
SELECT * FROM EMPL WHERE NOT PERF < 4;
```

6. Retrieve all employee information for employees working in the R&D department:

```
SELECT * FROM EMPL WHERE DEPT = 'R&D';
```

Appendix 1
DB2I AND SPUFI FOR DB2

When you purchase the DB2 DBMS package, IBM throws in a group of ISPF/PDF panels to help you work with DB2 interactively and be more productive. These panels offer the typical "IBM/ISPF-ish," menu-driven, fill-in-the-blanks approach to getting work accomplished, and include such activities as:

- Running DB2 Utilities for table and file management
- Issuing DB2 Commands for system and transaction management
- Preparing programs for execution
- Setting your session defaults
- BINDing (compiling) SQL embedded in application programs
- Executing SQL statements interactively

We are interested mainly in the last activity, how to execute SQL statements. The DB2I option to execute SQL statements is called SPUFI (actually pronounced — no giggles please — spoofy).

OVERVIEW OF SPUFI

1. Edit a TSO file, and fill it with SQL statements
2. Submit that file to SPFUI — which processes (prepares and executes) each statement in your file invoking DB2
3. Browse the results of your statement's execution

This is shown in Figure A.1:

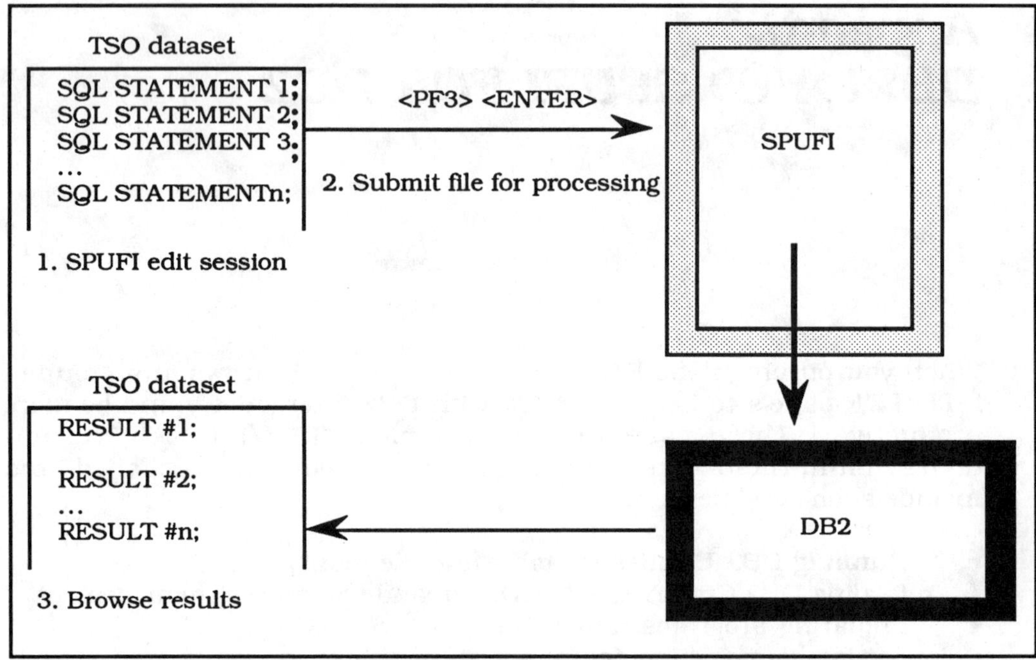

Figure A.1. SPUFI process and execution.

In step 1, (Edit a TSO dataset), you can enter one or many SQL statements at a time. If a statement fails execution, SPUFI processing halts, and you will be brought into Browse mode on the results TSO dataset that holds your DB2 results table and some run-time comments and error messages.

USING SPUFI

SPUFI is no harder to use than any other ISPF/PDF panel. In fact, it is much easier than some. There are just a few simple steps:

Step #1 — Allocate a SPUFI file for your statements

The file that will hold your SQL statements must exist before you enter a SPUFI session. Create this file as a partitioned dataset (library) using ISPF option 3.2. Use the file characteristics listed below in Figure A.2.

SPACE	==>	10	TRKS PRIMARY
		5	TRKS SECONDARY
		10	DIRECTOR BLOCKS
MODE	==>	FB	(Fixed Block)
LRECL	==>	80	
BLKSIZE	==>	3200	

Figure A.2. File characteristics for SPUFI dataset.

Step #2 — Getting to SPUFI

This step will be different at every installation, so I can't give you exact instructions on how to get to SPUFI. However, the problem is obvious. Before you can use the SPUFI panels to process the statements in this book, you have to find them.*

SPUFI is option 1 from the DB2I Main Menu panel, and DB2I is usually found as an option on the ISPF Main Menu screen (although many installations require you to execute a CLIST which allocates the DB2 and DB2I datasets to your TSO session).

If DB2I is not an observable option from your ISPF Main Menu screen, you will have to contact your Database Administration or Help desk personnel to find out how to connect up with it.

Once you arrive at the DB2I Primary Option Menu (shown in Figure A.3), type 1 in the option line for SPUFI and press <ENTER>.

---

* You may need to logon to TSO with a special DB2/TSO logon procedure that allocates necessary DB2 files to your TSO session.

```
                    DB2I PRIMARY OPTION MENU
===>

Select one of the following DB2 functions and press ENTER.

  1.  SPUFI                   (Process SQL statements)
  2.  DCLGEN                  (Generate SQL and source language
                              declarations)
  3.  PROGRAM PREPARATION     (Prepare a DB2 and application program to
                              run)
  4.  PRECOMPILE              (Invoke DB2 precompiler)
  5.  BIND/REBIND/FREE        (BIND, REBIND, or FREE application
                              plans)
  6.  RUN                     (RUN an SQL program)
  7.  DB2 COMMANDS            (Issue DB2 commands)
  8.  UTILITIES               (Invoke DB2 utilities)
  D   DB2I DEFAULTS           (Set global parameters)
  X   EXIT                    (Leave DB2I)

PRESS:  END to exit           HELP  for more information
```

Figure A.3. SPUFI primary option menu.

Step #3 — Enter your TSO dataset names

In this step you will type in the names of the dataset you want to use for your input file of SQL statements (1. on the SPUFI Panel), and your output browse dataset (4. on the SPUFI panel). Since this is the first time you've used SPUFI, we will also take a brief look at the CURRENT SPUFI DEFAULTS screen.

Figure A.4 shows the SPUFI main panel, set up with input and output datasets, and with the appropriate processing options specified:

- Enter the name of the file you allocated for your SPUFI statements on line one — INPUT DATA SET. This is still ISPF so you don't need your high-level TSO-ID qualifier in front of the dataset name. Don't forget to specify a member name — I've chosen to specify a single PDS member for each exercise, but, you might want to simply specify a single PDS member for each chapter. You can comment out exercises you have completed within the dataset.

- Enter any legal TSO name for the output data set on line two. SPUFI will allocate and manage it for you. If you want to use one that you've already allocated, specify it correctly.
- Enter YES on all of the processing options
- Press <ENTER>

```
                                  SPUFI
===>
  Enter the input data set name:                (Can be sequential or partitioned)
     1  DATA SET NAME          ===>  SPUFI.CNTL (CHP3EX01)
     2  VOLUME SERIAL          ===>             (Enter if not cataloged)
     3  DATA SET PASSWORD      ===>             (Enter if password protected)

  Enter the output data set name:               (Must be a sequential dataset)
     4  DATA SET NAME . . .    ===>  ANY.OLD.NAME.DATA

  Specify processing options:
     5  CHANGE DEFAULTS        ===>     YES  (Y/N — Display SPUFI defaults panel?)
     6  EDIT INPUT             ===>     YES  (Y/N — Enter SQL statements?)
     7  EXECUTE                ===>     YES  (Y/N — Execute SQL statements?)
     8  AUTOCOMMIT             ===>     YES  (Y/N — Commit after successful run?)
     9  BROWSE OUTPUT          ===>     YES  (Y/N — Browse output dataset?)

  PRESS:  ENTER to process     END to exit     HELP for more information
```

Figure A.4. SPUFI main panel.

Step #4 — Changing the SPUFI Defaults

This panel allows you to tailor your SPUFI output processing. I suggest you use my defaults as shown in Figure A.5. These will be more than acceptable for this course, and should not conflict with any installation standards or protocol.

If in the future you need to change your defaults, for example to view more than 250 lines of output — say for system testing purposes — then the CURRENT SPUFI DEFAULTS screen allows you to adjust SPUFI output processing (line 2, MAX SELECT LINES).

By the way, all of the SPUFI panels have a decent HELP facility, available at any time by pressing <PF11>.

Upon <PF3> from the SPUFI Defaults screen, you should be brought into an ISPF Edit session on the input dataset name you entered in on the previous SPUFI panel.

CURRENT SPUFI DEFAULTS

```
===>
===>
Enter the following to control your SPUFI session:
  1  ISOLATION LEVEL      ===>  CS      (RR=Repeatable Read, CS=Cursor
                                         Stability)
  2  MAX SELECT LINES     ===>  250     (Maximum number of lines to be
                                         returned from a SELECT)

Output data set characteristics:
  3  RECORD LENGTH        ===>  4092    (LRECL=Logical record length)
  4  BLOCK SIZE           ===>  4096    (Size of one block)
  5  RECORD FORMAT        ===>  VB      (REDFM=F, FB, FBA, V, VB, or VBA)
  6  DEVICE TYPE          ===>  SYSDA   (Must be DASD unit name)

Output format characteristics:
  7  MAX NUMERIC FIELD    ===>  20      (Maximum width for numeric
                                         fields)
  8  MAX CHAR FIELD       ===>  80      (Maximum width for character
                                         fields
  9  COLUMN HEADING       ===>  NAMES   (NAMES, LABELS, ANY or BOTH)
```

Figure A.5. Current SPUFI defaults panel.

Step #5 — Editing SQL Statements

SPUFI edit is *no different* from standard ISPF edit. The same full-screen edit, scrolling, search/replace (RFIND/RCHANGE) and external copy features are available so if you know ISPF, you should feel right at home.

Type your SQL statements anywhere in the file. Try not to word-wrap around, and if you are going to process more than one SQL statement in a file, you must end each statement with a semi-colon (;).

An empty dataset is shown in Figure A.6, and a dataset with SQL statements is shown in Figure A.7. Note the use of comments (prefixed by two dashes) for documenting your queries.

When you are finished editing, press <PF3>. This will bring you back to the SPUFI panel, and a message instructing you to "PRESS ENTER TO CONTINUE" is displayed (as shown in Figure A.8). If you press enter, SPUFI will execute your file of statements, however, there are two other alternatives for you at this point:

1.	Press <PF3> from the SPUFI panel	This will bring you back to the DB2I Primary Option Menu — without executing your statements
2.	Change your processing options on the SPUFI panel	You can re-edit your input file by changing your Asterisk to a "Y" — without executing your statements

Both of these alternatives may be useful sometime in the future, so salt them away carefully.

```
EDIT - - - -  TSOID.SPUFI.CNTL(CHP3EX01) — 01.00 - - - - - - - - - - - - - -COLUMNS 001 072
COMMAND ===>                                                  SCROLL ===> PAGE
****** ********************************* TOP OF DATA **********************************
''''''
''''''
''''''
''''''
''''''
''''''
''''''
''''''
''''''
''''''
''''''
''''''
''''''
''''''
''''''
''''''
''''''
''''''
''''''
''''''
****** ******************************** BOTTOM OF DATA ********************************
```

Figure A.6. Beginning SPUFI edit session.

```
EDIT - - - -  TSOID.SPUFI.CNTL(CHP3EX01) — 01.00 - - - -  - - - - - - - - -COLUMNS 001 072
COMMAND ===>                                               SCROLL ===> PAGE
****** ********************************** TOP OF DATA **********************************
 000001   SELECT * FROM MOV;  - -  THIS WILL SELECT ALL COLUMNS AND
 000002                       - -  ROWS FROM THE MOVIE TABLE
 000003   SELECT * FROM INV;  - -  BY THE WAY, THE DOUBLE DASHES MAKE ALL THIS
 000004                       - -  "NON-SQL" TEXT, A COMMENT
 ''''''
 ''''''
 ''''''
 ''''''
 ''''''
 ''''''
 ''''''
 ''''''
 ''''''
 ''''''
 ''''''
 ''''''
 ''''''
 ''''''
 ''''''
 ''''''
****** ********************************** BOTTOM OF DATA **********************************
```

Figure A.7. Input dataset with SQL statements.

```
                                   SPUFI
===>
DSNE808A  EDIT SESSION HAS COMPLETED.  PRESS ENTER TO CONTINUE
Enter the input data set name:                   (Can be sequential or partitioned)
      1   DATA SET NAME          ===> SPUFI.CNTL(CHP3EX01
      2   VOLUME SERIAL          ===>             (Enter if not catalogued)
      3   DATA SET PASSWORD      ===>             (Enter if password protected)

Enter the output data set name:                   (Must be a sequential data set)
      4   DATA SET NAME          ===> ANY.OLD.NAME.DATA

Specify processing options:
      5   CHANGE DEFAULTS        ===>    *        (Y/N — Display SPUFI defaults
                                                  panel?)
      6   EDIT INPUT             ===>    *        (Y/N — Enter SQL statements?)
      7   EXECUTE                ===>    YES      (Y/N — Execute SQL statements?)
      8   AUTOCOMMIT             ===>    YES      (Y/N) — Commit after successful
                                                  run?)
      9   BROWSE OUTPUT          ===>    YES      ((Y/N — Browse output data set?)

PRESS:  ENTER to process         END to exit      HELP for more information
```

Figure A.8. SPUFI main panel (with prompt).

Step #6 — Executing SQL Statements

Before you can actually execute an SQL statement against a DB2 table, you will have to check with the course administrator to be sure that all the necessary authorizations for your TSO-ID are in place. If you haven't already done so, you may not be able to successfully execute your statements.

While SPUFI is processing your statements and sending them to DB2 for execution, you will see the SPUFI panel. When execution is complete you are brought automatically into browse mode on the output dataset (in my example (ANY.OLD.NAME.DATA).

Step #7 — Viewing Your Results

When SPUFI and DB2 are finished with your SQL file, you will automatically enter a SPUFI Browse session. Again, the same full-screen browse, scrolling and searching are available so you should feel right at home. This screen is shown in Figure A.10. After you finish browsing press PF3 to back out to the SPUFI Main Menu (Figure A.9).

Note in Figure A.9, that I have changed the SPUFI processing options to:

```
===> NO
===> YES
===> YES
===> YES
```

This is because I no longer wish to change my SPUFI defaults — nor do I wish to revisit the CURRENT SPUFI DEFAULTS screen anymore, en route to my edit session.

```
BROWSE - -    TSOID.ANY.OLD.NAME.DATA - - - - - - - - - - - - LINE 00000000 COL 001 080
COMMAND ===>                                            SCROLL ===> PAGE
************************************ TOP OF DATA ****************************************
--------- + ---------- + --------- + ---------- + --------- + ---------- + --------- + ---------- + -------
SELECT * FROM MOV; -- THIS WILL SELECT ALL COLUMNS AND ROWS
--------- + ---------- + --------- + ---------- + --------- + ---------- + --------- + ---------- + -------
  NBR   TITLE                TYPE      RATING    STAR       QTY   PRICE
--------- + ---------- + --------- + ---------- + --------- + ---------- + --------- + ---------- + -------
    10  TOOTSIE              COMEDY    PG        HOFFMAN    1     29.95
     9  CROCODILE DUNDEE     COMEDY    PG13      HARRIS     2     69.95
     8  COCOON               SCIFI     PG        AMECHE     2     ------
     7  RISKY BUSINESS       COMEDY    R         CRUISE     2     44.55
     6  BODY HEAT            DRAMA     R         TURNER     3     19.95
     5  101 DALMATIONS       COMEDY    G         ------     3     59.95
     4  SPLASH               COMEDY    PG13      HANKS      3     29.95
     3  TOP GUN              DRAMA     PG        CRUISE     7     49.95
     2  FRIDAY THE 13TH      HORROR    R         JASON      2     69.95
     1  GONE WITH THE WIND   DRAMA     ------    GABLE      4     39.95
DSNE610I  NUMBER OF ROWS DISPLAYED IS 10
DSNE616I  STATEMENT EXECUTION WAS SUCCESSFUL, SQLCODE IS 100
--------- + ---------- + --------- + ---------- + --------- + ---------- + --------- + ---------- + -------
                    -- ROWS FROM THE MOVIE TABLE
SELECT * FROM INV;   -- BY THE WAY, THE DOUBLE DASHES MAKE ALL THIS
--------- + ---------- + --------- + ---------- + --------- + ---------- + --------- + ---------- + -------
  NBR   CUSTID     TOT       RENT      RETURN
--------- + ---------- + --------- + ---------- + --------- + ---------- + --------- + ---------- + -------
    10   02         7.55     87030               87060
     9   05        10.35     87052             -- -----
     8   04        12.55     87052               87053
     7   04        10.35     86355             --------
     6   02         9.75     87047               87051
     5   03        12.55     87045               87051
     4   01        11.25     87042               87072
     3   02        21.01     87041               87042
     2   02        13.55     87030             --------
     1   01         9.55     87055             --------
DSNE610I  NUMBER OF ROWS DISPLAYED IS 10
DSNE616I  STATEMENT EXECUTION WAS SUCCESSFUL, SQLCODE IS 100
--------- + ---------- + --------- + ---------- + --------- + ---------- + --------- + ---------- + -------
                    -- "NON-SQL" TEXT, A COMMENT
--------- + ---------- + --------- + ---------- + --------- + ---------- + --------- + ---------- + -------
DSNE617I  COMMIT PERFORMED,  SQLCODE IS 0
DSNE616I  STATEMENT EXECUTION WAS SUCCESSFUL,  SQL CODE IS 0
--------- + ---------- + --------- + ---------- + --------- + ---------- + --------- + ---------- + -------
DSNE601I  SQL STATEMENTS ASSUMED TO BE BETWEEN COLUMNS 1 AND 72
DSNE620I  NUMBER OF SQL STATEMENTS PROCESSED IS 2
  BROWSE -- TSOID.ANY.OLD.NAME.DATA - - - - - - - - - LINE 00000044 COL 001 080
  COMMAND ===>                                      SCROLL ===> PAGE
DSNE621I  NUMBER OF INPUT RECORDS READ IS 4
DSNE622I  NUMBER OF OUTPUT RECORDS WRITTEN IS 45
  ************************************ BOTTOM OF DATA ****************************************
```

Figure A.9. Browse session on output dataset.

```
                                 SPUFI
===>
Enter the input data set name:                    (Can be sequential or partitioned)
     1  DATA SET NAME          ===>  SPUFI.CNTL(CHP4EX01
     2  VOLUME SERIAL          ===>                (Enter if not catalogued)
     3  DATA SET PASSWORD      ===>                (Enter if password protected)

Enter the output data set name:                    (Must be a sequential data set)
     4  DATA SET NAME          ===>  ANY.OLD.NAME.DATA

Specify processing options:
     5  CHANGE DEFAULTS        ===>    NO          (Y/N — Display SPUFI defaults
                                                   panel?)
     6  EDIT INPUT             ===>    YES         (Y/N — Enter SQL statements?)
     7  EXECUTE                ===>    YES         (Y/N — Execute SQL statements?)
     8  AUTOCOMMIT             ===>    YES         (Y/N) — Commit after successful
                                                   run?)
     9  BROWSE OUTPUT          ===>    YES         (Y/N — Browse output data set?)

PRESS:  ENTER to process       END to exit         HELP for more information
```

Figure A.10. SPUFI main panel (note processing options).

Step #8 — Ending a SPUFI Session

To end SPUFI press <PF3> from the SPUFI Main Menu. This will bring you back out to the DB2I Menu Panel. To end DB2I press <PF3)> (or in some installations, X to exit) and this will bring you out of DB2I and back to ISPF.

Appendix 2
DB2 PERFORMANCE MANAGEMENT FOR DB2 *

The area of performance management as it applies to DB2 spans many areas and software products because of the multitude of ways in which DB2 can be used and accessed. The factors that need to be considered include DB2 table access paths, DB2 internal performance factors, external subsystem performance (TSO/ISPF, CICS, IMS/DC), and the MVS operating system itself.

DB2 ACCESS PATH SELECTION

The access paths to the DB2 table data are created by the DB2 'optimizer.' This optimizer is a part of DB2 which creates the access path for the application plan at bind time. The information used in the process is a combination of the following processes:

1. Access catalog statistics.
2. Compute numbers of rows for each predicate.
3. Choose set of reasonable paths.
4. Calculate access path cost (includes I/Os and CPU).
5. Add required Sort cost.
6. Select least cost path.

---

* Reprinted from Jeff D. Vowell, Jr., *DB2: The Complete Guide to Implementation and Use* (QED Information Sciences, Inc., 1990).

CATALOG STATISTICS USED

In access path selection, the following DB2 system catalog columns are used:

- SYSINDEXES columns
  - FULLKEYCARD
  - FIRSTKEYCARD
  - NLEAF
  - NLEVELS
  - CLUSTERED
- SYSTABLES columns
  - NPAGES
  - PCTPAGES
  - CARD
- SYSTABLESPACE columns
  - NACTIVE
- SYSCOLUMNS columns
  - COLCARD
  - HIGH2KEY
  - LOW2KEY

RUNSTATS updates these columns when it is run.

USE OF RUNSTATS FOR PERFORMANCE

As is shown above, the catalog statistics are important factors in access path selection. As such, it is necessary to make sure those statistics are as up to date as possible. This is done through the execution of the RUNSTATS utility. For the best results overall, you should run RUNSTATS just after the data has been loaded and before the BIND is done, or if this is a REBIND on plans using data which has a heavy amount of insert, delete, and/or update activity, the RUNSTATS should be run before the REBIND.

USING INDEXES FOR PERFORMANCE

Creating the right index can have very large performance implications because of the changes to the access path selection process that can occur.

ACCESS PATH SELECTION

Three major factors influence the cost of the access path that is selected. These are:

- The Filter Factor
- Sargability
- Matching indexability

If predicate combinations have the above features, they will be much more efficient than other access path selection combinations. Otherwise, access path selection will be based on other considerations.

Filter Factor

Filter Factor represents a fraction based upon an assumption of the rows that will be selected to satisfy a given predicate. This factor is based upon the assumptions that there is a uniform distribution of values over a given range and that you have columns value independence.

This factor is not very accurate or helpful in reality, but if the above mentioned factors do exist, you have a better probability of selecting the best access path.

In any case, a default filter factor is used in the following situations:

- Queries against DB2 system catalogs.
- Column selected is NOT first column of index.
- No RUNSTATS have been executed.
- Host variables are used in range expression.

Sargability

An SQL statement is considered sargable when the Internal Data Manager can process the request. When the Internal Data Manager can process the request, CPU time is reduced considerably. If the Internal Data Manager cannot process the request, then the Relational Data Manager must process the request, thus increasing CPU.

Current sargable predicates are as follows:

- COLUMN1 = value1
- COLUMN1 = COLUMN 2 (different tables)
- COLUMN1 > , > = , —< value1
- COLUMN1 range-operator COLUMN2 (different tables)
- COLUMN1 BETWEEN (value1, value2)
- COLUMN1 IS NULL (if COLUMN1 defined as nullable)
- COLUMN1 IS NOT NULL (if COLUMN1 defined as nullable)
- COLUMN1 = (subquery)
- COLUMN1 > , < (subquery)
- COLUMN1 IN (list)
- COLUMN1 —IN (list)
- COLUMN1 LIKE 'CHAR'
- COLUMN1 —LIKE 'CHAR'
- COLUMN1 LIKE '%CHAR'
- COLUMN1 LIKE '—CHAR'
- COLUMN1 LIKE :host-variable
- COLUMN1 —BETWEEN (value1, value2)
- COLUMN1 > value1 OR COLUMN1 < value1
- COLUMN1 = value1 OR COLUMN2 = value2
- NOT basic predicate
- NOT COLUMN1 > value1
- NOT COLUMN1 like 'CHAR'
- NOT COLUMN1 IN (list)
- NOT COLUMN1 BETWEEN (value1, value2)

Other predicate combinations will be non-sargable and thus have to be processed by the Relational Data Manager.

As is obvious, most times you will get the default because of the column limitation.

Matching Indexability

The first point to be made in this case is that the predicates used must be sargable before they can have matching indexability. However, just because a predicate is sargable does not mean that it also has matching indexability. You have matching indexability when a predicate matches an index access path, is sargable, and when the columns defined in the predicate are an initial substring of the set defined in an index key that will be used.

Current matching indexability predicates are as follows:

- COLUMN1 = value1
- COLUMN1 = COLUMN2 (different tables)
- COLUMN1 > , >= , —< value 1
- COLUMN1 range-operator COLUMN2 (different tables)
- COLUMN1 BETWEEN (value1, value2)
- COLUMN1 = (subquery)
- COLUMN1 > , < (subquery)
- COLUMN1 IN (list)
- COLUMN1 LIKE 'CHAR'
- COLUMN1 —LIKE 'CHAR'
- NOT basic predicate — (sometimes)
- NOT COLUMN1 > value1

Access Path Efficiency

The preferred access paths in order of efficiency are as follows:

1. Match UNIQUE INDEX with Equal (=) Predicate — No data access
2. Non-matching INDEX—No data access
3. Match UNIQUE INDEX with Equal (=) Predicate — With data access
4. Match on clustering index
5. Match on non-clustered index
6. Non-matching clustered index
7. Non-matching index — With data access
8. Tablespace scan

The resultant scan that is chosen performs the following functions:

MATCHING INDEX SCAN — NO DATA ACCESS

- Reads index only
- Returns index values
- Does NOT access data rows

MATCHING INDEX SCAN — DATA ACCESS

- Reads index
- Selects subset data based on Select requirements
- Reads appropriate data rows
- Returns selected data

NON-MATCHING INDEX SCAN—NO DATA ACCESS

- Retrieves all index rows by processing leaf pages
- Returns index values
- Does NOT access data rows

NON-MATCHING INDEX SCAN — DATA ACCESS

- Retrieves all index rows by processing leaf pages
- Selects subset data based on Select requirements
- Reads appropriate data rows
- Returns selected data

TABLESPACE SCAN

- Retrieves data rows
- Selects subset data based on Select requirements
- Returns selected data
- DO NOT DO for large tables

Appendix 3
SQL RESERVED WORDS

The following words are reserved words in SQL. They may not be used as ordinary identifiers in forming names. They may be used as delimited identifiers by enclosing them in double quotation marks.

ACQUIRE
ADD
ALL
ALTER
AND
ANY
AS
ASC
AVG

BETWEEN
BY

COLUMN
COMMENT
COMMIT
CONNECT
COUNT
CREATE
CURRENT

DBA
DBSPACE
DELETE
DESC
DISTINCT
DROP

EXCLUSIVE
EXISTS
EXPLAIN

FOR
FROM

GRANT
GROUP

HAVING

IDENTIFIED
IN
INDEX
INSERT
INTO
IS

LIKE
LOCK

MAX
MIN
MODE

NAMED
NHEADER
NOT
NULL

OF
ON
OPTION
OR
ORDER

PAGE
PAGES
PCTFREE
PCTINDEX
PRIVATE
PRIVILEGES
PROGRAM
PUBLIC

RELEASE
RESOURCE
REVOKE
ROLLBACK
ROW
RUN

SELECT
SET
SHARE
STATISTICS
STORPOOL
SUM
SYNONYM

TABLE
TO

UNION
UNIQUE
UPDATE
USER

VALUES
VIEW

WHERE
WITH
WORK

Appendix 4
GUIDE TO THE DB2 REFERENCE MANUALS

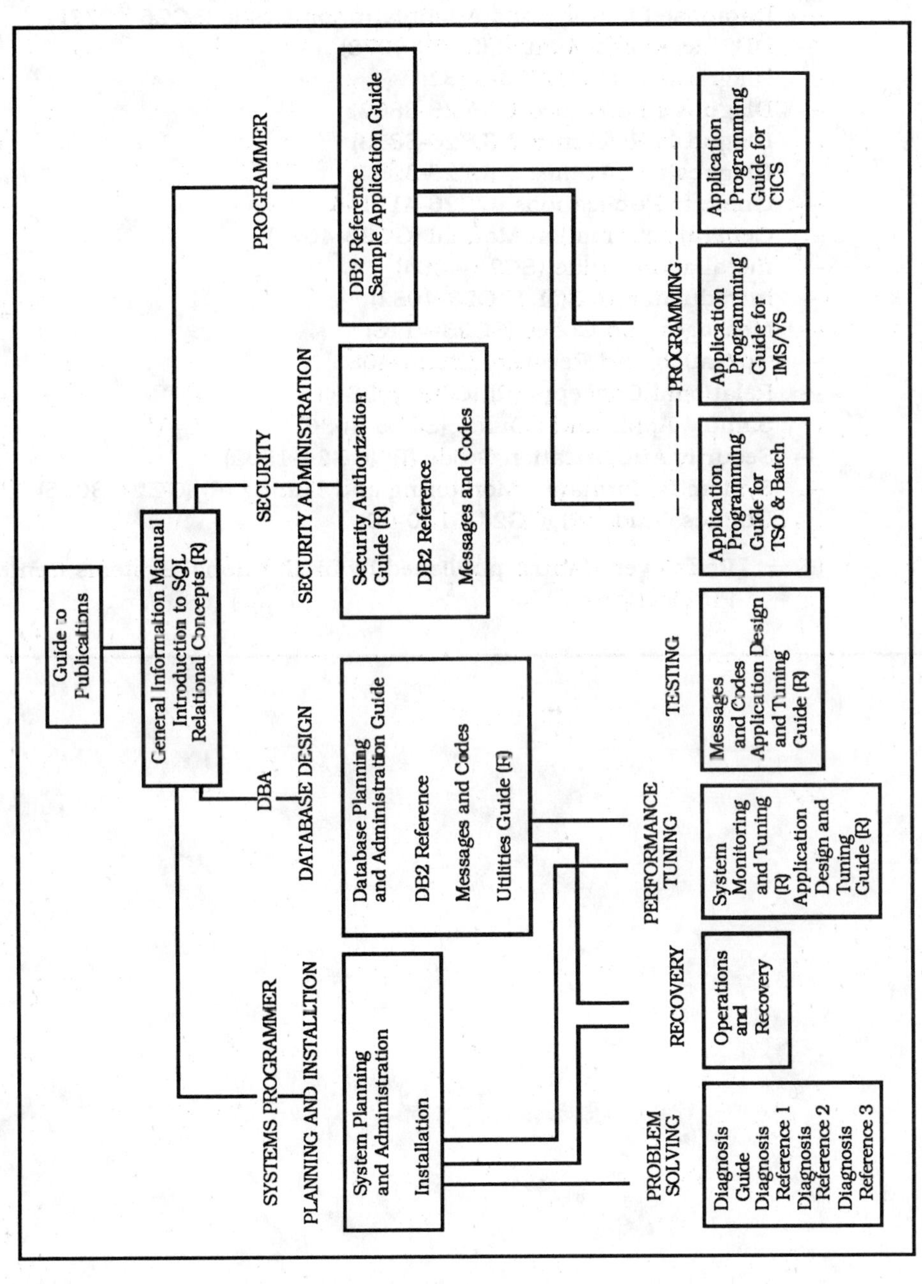

DOCUMENTATION IDENTIFICATION NUMBERS

- Application Design and Tuning Guide (R) (GG24-3004-00)
- Application Programming Guide for CICS (SC26-4080)
- Application Programming Guide for IMS/VS (SC26-4079)
- Application Programming Guide for TSO and Batch (SC26-4081)
- Database Planning and Administration Guide (SC26-4077)
- DB2 Reference Manual (SC26-4078)
- Diagnosis Guide (LY26-3862)
- Diagnosis Reference 1 (LY26-3862)
- Diagnosis Reference 2 (LY26-3863)
- Diagnosis Reference 3 (LY26-3952)
- Guide to Publications (GC26-4111)
- General Information Manual (GC26-4073)
- Installation Guide (SC26-4084)
- Introduction to SQL (GC26-4082)
- Messages and Codes (SC26-4113)
- Operation and Recovery (SC26-4083)
- Relational Concepts (R) (GG24-1581)
- Sample Application Guide (SC26-4086)
- Security Authorization Guide (R) (GG24-1599)
- System Performance Monitoring and Tuning (R) (GG24-3005)
- Utilities Guide (R) (GG24-3130-00)

(R) = Red Cover Manual published by International Systems Center Publications.

Appendix 5
SQL/DS REFERENCE MANUALS

SQL/DS

General Information for VM/SP	GH09-8043
Concepts and Facilities for VM/SP	GH09-8044
Application Programming for VM/SP	SH09-8019
Data Base Services Utility for VM/SP	SH09-8051
Messages and Codes for VM/SP	SH09-8052
System Planning and Administration for VM/SP	SH09-8018
Database Planning and Administration for VM/SP	SH09-8017
Installation for VM/SP	SH09-8046
Operation for VM/SP	SH09-8053
Master Index for VM/SP	SH09-8056
Diagnosis Guide for VM/SP	LH09-8054
Diagnosis Reference for VM/SP	LH09-8055

QMF

General Information	GC26-4229
Reference	SC26-4344
Reference Summary	SX26-3752
Advanced User's Guide	SC26-4343
Application Development Guide for VM	SC26-4238
Learner's Guide	SC26-4231
Guide to Developing QMF Applications using ISPF	GG24-1658
Planning and Administration for VM/SP	SC26-4236
Installation Guide for VM/SP	SC26-4234
Diagnosis Guide for VM/SP	LY26-9500
Diagnosis Reference for VM/SP	LY27-9501

REXX

VM/SP Interpreter Interface to SQL/DS	SH20-7051

Appendix 6

LOAD INSTRUCTIONS FOR SQLBase®

SQLBase® is a relational database management system used in this course with the permission of Gupta Technologies, Inc. The system has been modified specifically to use with this course. *Do not use this for production, as your results may not be accurate.* Also, some of the more advanced commands taught in this course may not work on this scaled down version of SQLBase®. If you want more information about SQLBase® for production use, contact:

Gupta Technologies, Inc.
1040 Marsh Road, Suite 210
Menlo Park, CA 94025

Using the PC Version of this Course

In this appendix you will learn how to start up the PC system ("boot up"), and how to use SQLBase®. You should be seated in front of a PC at this point.

The first step is to load the Disk Operating System (DOS) disk.

Load DOS from a Hard Disk

In a hard drive system DOS is normally stored in the hard drive named drive C. The DOS prompt is C>. This may be different in your installation. If it is, ask your education coordinator for help. If your DOS prompt is already on the screen, you may continue. If not, follow these instructions:

1. Make sure there is no diskette in the diskette drive.
2. Load DOS into the computer by either:
   a. turning the on/off switch on or, if it is already on,
   b. holding down the keys CTRL, ALT, DEL.

3. If the computer stops and requests the date and time, press the **ENTER** key for each request.

   The screen will then display C>

Installing SQLBase® on a Hard Drive

1. You should be seated in front of your PC with it turned on, and the system booted. You should also have with you the QED SQL diskettes (Diskettes 1 and 2).

   Note: **You should have your own set of diskettes that have not been previously used for this course.**

2. Before you begin, get to the root directory by typing CD\ press **ENTER**
3. Insert diskette 1 into the A drive of your PC.
4. Type **A: HD_INSTL** A command procedure will be invoked that will create an SQLBase® directory, a DEMO subdirectory, and copy some files to your hard drive. At a specified time, you will be asked to insert QED SQL Diskette 2 into your drive. Follow the instructions on the screen. After all of the files have been copied, you will be ready to log onto the system and begin SQL training.

System Storage RequirementsThe QED SQL training database will take up a little less than .5 megabytes of disk space on your hard drive. If you are near capacity on your system, we suggest that you clean some files off before running this install program to avoid any possible "disk I/O" problems.

Now that you've installed SQL, it's easy to get going. You will need to change your directory to the SQL directory and start SQLDEMO. You should accept all the SQL defaults.

Follow these directions:

1. Change the directory to SQLBase; type: **CD\SQLBASE** press **ENTER**
2. Start the SQLDEMO program; type: **SQLDEMO** press **ENTER**
3. Accept the system defaults by pressing the enter key at each question prompt:

   Database? — **ENTER**
   Username? — **ENTER**
   Password? — **ENTER**

Your screen should now look like this:

```
C>SQLDEMO

SQLTALK Interactive Data Manager (Version 3.2.2)
Copy (c): Gupta Technologies, Inc. 1985, 1986, 1987, 1988.

Database?

Username?

Password?

Correcting to DEMO in Interactive Line Mode.


Enter a SQL command

SQL>
```

Using the Editor:

To create your SQL statements, type each statement from within SQLDEMO, on the line after the prompt (SQL>). Put on the **CAPS LOCK** key.

Try this example now just to make sure the system is installed correctly:

Type: SQL> **SELECT * FROM CUST;** press **ENTER**

Your screen should look like this.

```
Enter a SQL command

SQL> SELECT * FROM CUST;

ID  LNAME          FNAME      CITY             ST
==  =============  =========  ==============   ===
01  DANGERFIELD    RODNEY     HARTFORD         CT
02  FIELD          SALLY      FRANKLIN         NY
03  NICHOLSON      JACK       HARTFORD         CT
04  MURRAY         FRED       BOZRAH           CT
05  MADDEN         JOHN       BRATTLEBORO      VT
06  WEST           MAE        PARK PLACE       CA
07  WOODWARD       JOANNE     GETTYSBURG       PA
08  ROGERS         ROY        HAPPY TRAILS     TX
09  RINGWALD       MOLLY      PINKSVILLE       CA
10  ROGERS         FRED       PARK PLACE       CA

10 ROWS SELECTED

Enter a SQL command
```

Logging Off of SQLBase®

To log off of the SQL system, simply type: SQL> **EXIT;** press **ENTER**
This accomplishes two things:

1. You are logged off the SQLDEMO system and placed back into the DOS directory.

2. More importantly, any updates you made to your tables are saved. This is known as commit processing, and we will discuss it in much more depth in Chapter 11 of this book. There are points in this course where we do not want to save your work.

Whenever you are asked in this course to rollback your work, enter the command **ROLLBACK;** if you are using SQLBase®. For now, let's exit SQLBase® by using the **EXIT** command.

Glossary

The following terms and abbreviations are defined as they are used in this book. If you do not find the term you are looking for, refer to the index or to the *IBM Dictionary of Computing*, SC20-1699.

access path. The path used to get to data specified in SQL statements. An access path can involve an index or a sequential search.

application-embedded SQL. SQL statements coded within an application program.

authorization ID. A string that designates a set of privileges. It may represent an individual, an organizational group, or a function, but DB2 does not determine this representation.

AUTOCOMMIT. A SPUFI and DBSU option that commits the effects of SQL statements automatically if they are successfully executed.

built-in function. Scalar function or column function.

character string. A sequence of bytes representing bit data, single-byte characters, or a mixture of single and double-byte characters.

clause. In SQL, a distinct part of a statement, such as a SELECT clause or a WHERE clause.

clustering index. An index that determines how rows are physically ordered in a table space.

column. The vertical component of a table. A column has a name and a particular data type (for example, character, decimal, or integer).

column function. An SQL operation that derives its result from a collection of values across one or more rows.

commit. The process that allows data, changed by one application or user, to be referenced by other applications or users. When a COMMIT occurs, locks are freed so that other applications can reference the just-committed data. When data has been committed, a new commit point is established.

comparison operator. A symbol (such as =, >, <) used to specify a relationship between two values.

concurrency. The shared use of resources by multiple application processes at the same time.

correlated subquery. A subquery (part of a WHERE or HAVING clause) applied to a row or group of rows of a table or view named in an outer subSELECT statement.

database. A collection of table spaces and index spaces.

database administrator (DBA). An individual responsible for the design, development, operation, safeguarding, maintenances, and use of a database.

data type. An attribute of columns, literals, and host variables.

date. A three-part value that designates a day, month, and year.

date duration. A decimal integer that represents a number of years, months, and days.

date/time value. A value of the datatype DATE, TIME, or TIMESTAMP.

DBA. Database administrator.

DBCS. Double-byte character set.

DBID. Database identifier.

DBRM. Database request module.

DBSU. The SQL/DS facility that provides for the execution of SQL statements and utility invocations.

DB2 Interactive (DB2I). The DB2 facility that provides for the execution of SQL statements, DB2 (operator) commands, programmer commands, and utility invocation.

DB2I. DATABASE 2 Interactive.

DCLGEN. Declarations generator.

declarations generator (DCLGEN). A subcomponent of DB2 that generates SQL table declarations and COBOL, C, or PL/I data structure declarations that conform to the table. The declarations are generated from DB2 system catalog information. DCLGEN is also a DSN subcommand.

default value. A predetermined value, attribute, or option that is assumed when no other is explicitly specified.

dependent. An object (row, table, or table space) is a dependent if it has at least one parent. The object is also said to be a dependent (row, table, or tablespace) of its parent. See *parent row, parent table, parent table space.*

double-byte character set (DBCS). A set of characters used by national languages such as Japanese and Chinese that have more symbols than can be represented by a single-byte. Each character is two bytes in length, and therefore requires special hardware to be displayed or printed.

duration. A number that represents an interval of time. See *labeled duration, date duration,* and *time duration.*

EBCDIC. Extended binary coded decimal interchange code.

embedded SQL. SQL statements that are embedded within an application program and are prepared during the program preparation process before the program is executed. After it is prepared, the statement itself does not change (although values of host variables specified within the statement might change).

EUR. IBM European Standards.

expression. An operand or a collection of operators and operands that yields a single value.

fixed-length string. A character or graphic string whose length is specified and cannot be changed.

function. A scalar function or column function. Same as *built-in function.*

graphic string. A sequence of DBCS characters.

index. A set of pointers that are logically ordered by the values of a key. Indexes can provide faster access to data and can enforce uniqueness on the rows in a table.

index key. The set of columns in a table used to determine the order of index entries.

indicator variable. A variable used to represent the null value in an application program. If the value for the selected column is null, a negative value is placed in the indicator variable.

ISO. International Standards Organization.

isolation level. The degree to which a unit of work is isolated from the updating operations of other units of work. See also *cursor stability* and *repeatable read.*

JIS. Japanese Industrial Standard.

join. A relational operation that allows retrieval of data from two or more tables based on matching column values.

K. Kilobyte (1024 bytes).

labeled duration. A number that represents a duration of years, months, days, hours, minutes, seconds, or microseconds.

locking. The process by which integrity of data is ensured. Locking prevents concurrent users from accessing inconsistent data.

long string. A string whose actual length, or a varying-length string whose maximum length, is greater than 254 bytes or 127 double-byte characters.

mixed data string. A character string that may contain both single-byte and double-byte characters.

null. A special value that indicates the absence of information.

object. Anything that can be created or manipulated with SQL — that is, tables, views, or indexes.

predicate. An element of a search condition that expresses or implies a comparison operation.

primary key. A unique, non-null key that is part of the definition of a table. A table cannot be defined as a parent unless it has a primary key.

privilege. A capability given to a user by the execution of a GRANT statement.

recovery. The process of rebuilding databases after a system failure.

referential integrity. The enforcement of referential constraints on LOAD, INSERT, UPDATE, and DELETE operations.

result table. The set of rows selected for use by an application program. The application uses a cursor to retrieve the rows one by one into a host structure or a set of host variables.

rollback. Initiates the abort process.

row. The horizontal component of a table. A row consists of a sequence of values, one for each column of the table.

scalar function. An SQL operation that produces a single value from another value and is expressed as a function name followed by a list of arguments enclosed in parentheses.

search condition. A criterion for selecting rows from a table. A search condition consists of one or more predicates.

short string. A string whose actual length, or a varying-length string whose maximum length, is 254 bytes (127 double-byte characters) or less.

SPUFI. SQL Processor Using File Input. A facility of the CMS attachment subcomponent that enables the DBSU user to execute SQL statements without embedding them in an application program.

SQL. Structured Query Language. A language that can be used within host programming languages or interactively to access data and to control access to resources.

SQLCA. SQL communication area.

SQLDA. SQL descriptor area.

SQL/DS. SQL/Data System.

SQL/DS catalog. SQL/DS-maintained tables that contain descriptions of SQL/DS objects such as tables, views, and indexes.

SQL escape character. See *escape character*.

SQL ID. The authorization ID that is used as the implicit qualifier of table, view, and index names in dynamic SQL statements. The SQL ID, along with the other authorization IDs of a process, is used for authorization checking of dynamic SQL statements. It also serves a special role for CREATE, GRANT, and REVOKE statements.

SQL string delimiter. A symbol used to enclose an SQL string constant. The SQL string delimiter is the apostrophe ('), except in COBOL applications, in which case the symbol (either an apostrophe or a quotation mark) may be assigned by the user.

SSI. MVS subsystem interface.

static SQL. See *embedded SQL.*

storage group. A named set of DASD volumes on which DB2 data can be stored.

string. A character or graphic string.

subquery. A SELECT statement within the WHERE or HAVING clause of another SQL statement; a nested SQL statement.

subselect. That form of the SELECT statement that does not include ORDER BY or UNION operators.

sync point. See *commit point.*

synonym. In SQL, an alternative name for a table or view.

system administrator. The person having the second highest level of authority within DB2. System administrators make decisions about how DB2 is to be used and implement those decisions by choosing system parameters. They monitor the system and change its characteristics to meet changing requirements and new data processing goals.

table. A named data object consisting of a specific number of columns and some number of unordered rows.

time. A three-part value that designates a time of day in hours, minutes, and seconds.

time duration. A decimal integer that represents a number of hours, minutes, and seconds.

timestamp. A seven-part value that consists of a date and time expressed in years, months, days, hours, minutes, seconds, and microseconds.

UNION. An SQL operation that combines the results of two select statements. UNION is often used to merge lists of values obtained from several tables.

unique index. An index which ensures that no identical key values are stored in a table.

unit of recovery (UR). A recoverable sequence of operations within an application process. At any time, an application process is a single unit of recovery, but the life of an application process may involve many units of recovery as a result of commit or rollback operations.

value. The smallest unit of data manipulated in SQL.

varying-length string. A character or graphic string whose length varies within set limits.

view. An alternative representation of data from one or more tables. A view can include all or some of the columns contained in the table or tables on which it is defined.

Notice to Consumers

This product cannot be returned for
credit or refund if the perforation
on the vinyl disk holder
is broken or tampered with.

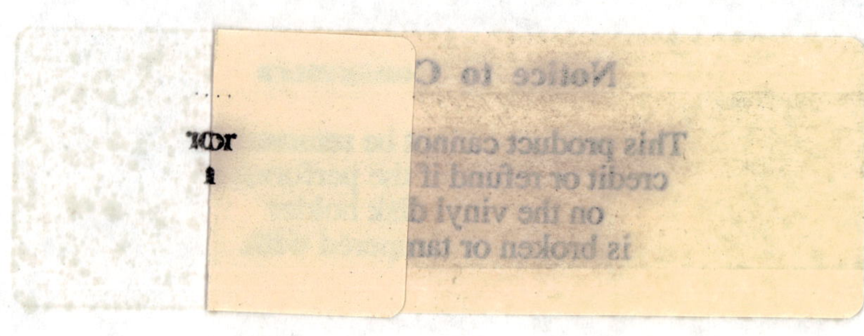
Notice to C
This product cannot
credit or refund if
on the vinyl d
is broken or tam